Ultra-Low-Power ICs for the Internet of Things

Ultra-Low-Power ICs for the Internet of Things

Editor

Orazio Aiello

MDPI • Basel • Beijing • Wuhan • Barcelona • Belgrade • Manchester • Tokyo • Cluj • Tianjin

Editor
Orazio Aiello
University of Genoa
Italy

Editorial Office
MDPI
St. Alban-Anlage 66
4052 Basel, Switzerland

This is a reprint of articles from the Special Issue published online in the open access journal *Journal of Low Power Electronics and Applications* (ISSN 2079-9268) (available at: https://www.mdpi.com/journal/jlpea/special_issues/low_power_iot).

For citation purposes, cite each article independently as indicated on the article page online and as indicated below:

LastName, A.A.; LastName, B.B.; LastName, C.C. Article Title. *Journal Name* **Year**, *Volume Number*, Page Range.

ISBN 978-3-0365-7902-3 (Hbk)
ISBN 978-3-0365-7903-0 (PDF)

Contents

About the Editor

Orazio Aiello

Orazio Aiello received B.Sc. and M.Sc. degrees (cum laude) from the University of Catania, Italy, in 2005 and 2008, respectively, a M.Sc. degree (cum laude) from the Scuola Superiore di Catania, Italy, in 2009, and a Ph.D. degree from the Politecnico di Torino, Italy, in 2013. He has earned his technical background in worldwide universities R&D institutions, consultant activities, and direct work experience in semiconductor companies. He is currently a tenure-track assistant Professor at the University of Genoa, Italy. His main research interests include energy-efficient analog mixed-signal circuits and sensor interfaces.

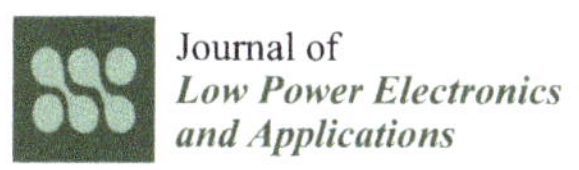

Journal of
Low Power Electronics and Applications

Editorial

Ultra-Low-Power ICs for the Internet of Things

Orazio Aiello

Department of Electrical, Electronics and Telecommunication Engineering and Naval Architecture (DITEN), University of Genoa, 16100 Genova, Italy; orazio.aiello@unige.it

Citation: Aiello, O. Ultra-Low-Power ICs for the Internet of Things. *J. Low Power Electron. Appl.* **2023**, *13*, 38. https://doi.org/10.3390/jlpea13020038

Received: 23 May 2023
Accepted: 24 May 2023
Published: 26 May 2023

The collection of research works in this Special Issue focuses on Ultra-Low-Power (ULP) Integrated Circuits (ICs) operating under a tight budget of power as a criterion to build electronic devices relying less and less on batteries. These enable the Internet of Things (IoT): a view of a world in which we are surrounded by devices that exchange data to enhance our quality of living. Thus, the goals of novel IC design strategies target both reducing the cost and the power consumption of any device. A method to reduce the cost is to minimize the use of a manual design process and maximize the use of a digital (automated) design flow so that the design is transferable across technological nodes. A digital-in-concept design also allows the scale of the supply voltage and offers a performance–power consumption trade-off [1–4]. In particular, a two-stage inverter-based operational transconductance amplifier (OTA) using rail-to-rail output operating with a supply voltage of 0.5 V is presented in [1]. Then, a novel implementation of a digital-based OTA consisting of only digital gates usually available in the standard cell libraries is the focus of [2]. In [3], a novel fully standard-cell-based common-mode feedback (CMFB) loop to improve the CMRR and to stabilize the DC output voltage of pseudo-differential standard-cell-based amplifiers is proposed. To further explore complexity, dynamic performance, and energy efficiency, a fully synthesizable digital–delta (Δ) modulator (ΔM) ADC with noise shaping using passive components (i.e., integrated capacitors and resistors) and standard-cell-based amplifiers is presented in [4].

The other research works exploit other methods, focusing on increasing the energy efficiency for a number of building blocks for general-purpose applications (i.e., amplifiers); more specifically, they target biomedical applications or at the system level. ULP/Ultra-Low-Voltage (ULV) ICs exploring bulk-drive solutions and operating with Sub-1V supply voltage down to 0.3 V were considered [5–8]. In [5], the authors proposed a new technique to improve the DC voltage gain, while keeping the high linearity in symmetrical bulk-driven (BD) OTA topology. A novel tree-based architecture that allows the implementation of a ULV OTA exploiting a body-driven input stage to guarantee a rail-to-rail input common mode range is also described in [6]. A bootstrapped BD Voltage Buffer is used to increase the intrinsic voltage gain of the Second-Order Gm-C Bandpass Filter in [7]. Moreover, a current-controlled CMOS ring oscillator topology, which exploits the bulk voltages of the inverter stages as control terminals to tune the oscillation frequency, is proposed and analyzed in [8]. Then, a fully differential (FD) instrumentation amplifier aimed at electrical impedance measurements in an IoT biomedical scenario is presented in [9].

To assist the ULP IC design flow, a compact and simplified approach that contains only four parameters and is based on the Advanced Compact MOSFET (ACM) model was implemented in Verilog-A and compared with the BSIM model in [10].

Sinusoidal oscillators based on second-generation voltage conveyors are investigated in [11], while a relaxation oscillator with valuable line sensitivity for Low Power Applications is shown in [12].

The last two studies in this Special Issue consider the IC as part of a ULP/ULV sensor system that needs to interact with the surrounding environment.

A wideband cascaded receiver including an inverter-based low-noise transconductance amplifier and a stacked receiver using an improved clock strategy with reduced mixer switches is described in [13]. Hardware solutions for Low-Power Smart Edge Computing are presented in [14].

In summary, the published research works cover a wide area of the ULP/ULV IC field, offering the reader many ideas inspired by these innovative design approaches.

Acknowledgments: The Guest Editor of the Special Issue "Ultra-Low-Power ICs for the Internet of Things" thanks the Multidisciplinary Digital Publishing Institute (MDPI) for the invitation to write this Editorial as a presentation of the 14 papers published in this Special Issue which is freely available at https://www.mdpi.com/journal/jlpea/special_issues/low_power_iot. Moreover, based on the success of this Special Issue, a "Volume 2" has been launched at https://www.mdpi.com/journal/jlpea/special_issues/919Q5756T0.

Conflicts of Interest: The author declares no conflict of interest.

References

1. Ballo, A.; Pennisi, S.; Scotti, G. 0.5 V CMOS Inverter-Based Transconductance Amplifier with Quiescent Current Control. *J. Low Power Electron. Appl.* **2021**, *11*, 37. [CrossRef]
2. Palumbo, G.; Scotti, G. A Novel Standard-Cell-Based Implementation of the Digital OTA Suitable for Automatic Place and Route. *J. Low Power Electron. Appl.* **2021**, *11*, 42. [CrossRef]
3. Centurelli, F.; Della Sala, R.; Scotti, G. A Standard-Cell-Based CMFB for Fully Synthesizable OTAs. *J. Low Power Electron. Appl.* **2022**, *12*, 27. [CrossRef]
4. Correia, A.; Tavares, V.G.; Barquinha, P.; Goes, J. All-Standard-Cell-Based Analog-to-Digital Architectures Well-Suited for Internet of Things Applications. *J. Low Power Electron. Appl.* **2022**, *12*, 64. [CrossRef]
5. Sanchotene Silva, R.; Rodovalho, L.H.; Aiello, O.; Ramos Rodrigues, C. A 1.9 nW, Sub-1 V, 542 pA/V Linear Bulk-Driven OTA with 154 dB CMRR for Bio-Sensing Applications. *J. Low Power Electron. Appl.* **2021**, *11*, 40. [CrossRef]
6. Centurelli, F.; Della Sala, R.; Monsurrò, P.; Scotti, G.; Trifiletti, A. A Tree-Based Architecture for High-Performance Ultra-Low-Voltage Amplifiers. *J. Low Power Electron. Appl.* **2022**, *12*, 12. [CrossRef]
7. Carrillo, J.M.; de la Cruz-Blas, C.A. 0.6-V 1.65-μW Second-Order Gm-C Bandpass Filter for Multi-Frequency Bioimpedance Analysis Based on a Bootstrapped Bulk-Driven Voltage Buffer. *J. Low Power Electron. Appl.* **2022**, *12*, 62. [CrossRef]
8. Ballo, A.; Pennisi, S.; Scotti, G.; Venezia, C. A 0.5 V Sub-Threshold CMOS Current-Controlled Ring Oscillator for IoT and Implantable Devices. *J. Low Power Electron. Appl.* **2022**, *12*, 16. [CrossRef]
9. Corbacho, I.; Carrillo, J.M.; Ausín, J.L.; Domínguez, M.Á.; Pérez-Aloe, R.; Duque-Carrillo, J.F. A Fully-Differential CMOS Instrumentation Amplifier for Bioimpedance-Based IoT Medical Devices. *J. Low Power Electron. Appl.* **2023**, *13*, 3. [CrossRef]
10. Adornes, C.M.; Alves Neto, D.G.; Schneider, M.C.; Galup-Montoro, C. Bridging the Gap between Design and Simulation of Low-Voltage CMOS Circuits. *J. Low Power Electron. Appl.* **2022**, *12*, 34. [CrossRef]
11. Stornelli, V.; Barile, G.; Pantoli, L.; Scarsella, M.; Ferri, G.; Centurelli, F.; Tommasino, P.; Trifiletti, A. A New VCII Application: Sinusoidal Oscillators. *J. Low Power Electron. Appl.* **2021**, *11*, 30. [CrossRef]
12. Liao, Y.; Chan, P.K. A 1.1 V 25 ppm/°C Relaxation Oscillator with 0.045%/V Line Sensitivity for Low Power Applications. *J. Low Power Electron. Appl.* **2023**, *13*, 15. [CrossRef]
13. Abbasi, A.; Nabki, F. Wideband Cascaded and Stacked Receiver Front-Ends Employing an Improved Clock-Strategy Technique. *J. Low Power Electron. Appl.* **2023**, *13*, 14. [CrossRef]
14. Martin Wisniewski, L.; Bec, J.-M.; Boguszewski, G.; Gamatié, A. Hardware Solutions for Low-Power Smart Edge Computing. *J. Low Power Electron. Appl.* **2022**, *12*, 61. [CrossRef]

Journal of
*Low Power Electronics
and Applications*

Article

A New VCII Application: Sinusoidal Oscillators

Vincenzo Stornelli [1], Gianluca Barile [1], Leonardo Pantoli [1], Massimo Scarsella [1], Giuseppe Ferri [1,*],
Francesco Centurelli [2], Pasquale Tommasino [2] and Alessandro Trifiletti [2]

[1] Department of Industrial and Information Engineering and Economics, University of L'Aquila,
67100 L'Aquila, Italy; vincenzo.stornelli@univaq.it (V.S.); gianluca.barile@univaq.it (G.B.);
leonardo.pantoli@univaq.it (L.P.); massimo.scarsella@student.univaq.it (M.S.)

[2] Department of Information Engineering, Electronics and Telecommunications, Sapienza University of Rome,
00185 Rome, Italy; francesco.centurelli@uniroma1.it (F.C.); pasquale.tommasino@uniroma1.it (P.T.);
alessandro.trifiletti@uniroma1.it (A.T.)

* Correspondence: giuseppe.ferri@univaq.it

Abstract: The aim of this paper is to prove that, through a canonic approach, sinusoidal oscillators based on second-generation voltage conveyor (VCII) can be implemented. The investigation demonstrates the feasibility of the design results in a pair of new canonic oscillators based on negative type VCII (VCII$^-$). Interestingly, the same analysis shows that no canonic oscillator configuration can be achieved using positive type VCII (VCII$^+$), since a single VCII$^+$ does not present the correct port conditions to implement such a device. From this analysis, it comes about that, for 5-node networks, the two presented oscillator configurations are the only possible ones and make use of two resistors, two capacitors and a single VCII$^-$. Notably, the produced sinusoidal output signal is easily available through the low output impedance Z port of VCII, removing the need for additional voltage buffer for practical use, which is one of the main limitations of the current mode (CM) approach. The presented theory is substantiated by both LTSpice simulations and measurement results using the commercially available AD844 from Analog Devices, the latter being in a close agreement with the theory. Moreover, low values of THD are given for a wide frequency range.

Keywords: VCII; RC oscillator; sinusoidal oscillator; current mode; voltage conveyor application

Citation: Stornelli, V.; Barile, G.;
Pantoli, L.; Scarsella, M.; Ferri, G.;
Centurelli, F.; Tommasino, P.; Trifiletti,
A. A New VCII Application:
Sinusoidal Oscillators. *J. Low Power
Electron. Appl.* **2021**, *11*, 30. https://
doi.org/10.3390/jlpea11030030

Academic Editor: Orazio Aiello

Received: 15 June 2021
Accepted: 6 July 2021
Published: 8 July 2021

Publisher's Note: MDPI stays neutral
with regard to jurisdictional claims in
published maps and institutional affiliations.

1. Introduction

There has always been an interest in designing sinusoidal oscillators due to several applications in different areas such as communication, instrumentation, biomedical, etc. [1–3]. Compared to LC and RLC sinusoidal oscillators, RC-active type oscillators are advantageous from the integration point of view. In the early implementations of RC-active sinusoidal oscillators, operational amplifiers (Op-Amps) were used as active elements [4–6]. A systematic approach was introduced in [5] to design Op-Amp-based oscillators with a single active element and the minimum number of passive elements. The design method of [5] resulted in Op-Amp-based oscillator configurations composed of one active device, two capacitors and four resistors.

However, the limited frequency performance and slew rate of Op-Amps as well as their high power consumption imposed a restriction in the application of Op-Amp-based sinusoidal oscillators. A literature survey shows that, after revealing the potential capabilities of current-mode (CM) signal processing, efforts have been made to design RC-active sinusoidal oscillators using various CM active building blocks (ABBs) [7–34]. Undoubtedly, second-generation current conveyor (CCII) as the main ABB of CM signal processing is the most widely used one for this purpose. Different approaches were employed to realize CCII-based oscillators. For example, in [8], the Op-Amps were replaced with composite current conveyors, resulting in CM oscillators. Unfortunately, this approach did not reach a simple realization because each amplifier could only be implemented with at least two CCIIs and two resistors. The extension of the approach presented in [5] was employed in [9] to synthesize CCII-based oscillators. Although the resulting sinusoidal oscillators enjoyed a canonic structure with the minimum possible number of elements, they were still not readily cascadable, i.e., they required additional voltage buffers to be actually usable in a real-world application. Most of the other CM oscillator realizations reported in [10–34] using different ABBs instead of CCIIs also suffered from a large number of active and/or passive elements.

Recently, the dual circuit of CCII, called second-generation voltage conveyor (VCII), has attracted the attention of researchers [35–44]. In particular, the recent study reported in [35,36] showed that this device helps to benefit from CM signal processing features and overcome the limitations in CCII-based circuits. Particularly, unlike CCII, there is a low-impedance voltage output port in VCII which allows it to be easily cascaded with other high-impedance processing blocks, without the need for extra voltage buffers in voltage output applications. Compared to CCII, VCII has proven superior performance in many applications [37]; up to now, this device has not been employed in the realization of sinusoidal oscillators.

However, the VCII, combining the advantages of CM processing with a voltage-mode interfacing, could provide sinusoidal oscillators operating up to higher frequency than Op-Amp-based ones. Moreover, breaking the gain-bandwidth tradeoff, it could ease decoupling oscillation frequency and oscillation condition even at the higher end of the spectrum. Among the possible implementations of VCII-based sinusoidal oscillators, those requiring the minimum number of (active and) passive components, so-called canonic, are of particular interest to minimize silicon area and power consumption. The aim of this work is only to present possible VCII-based canonic sinusoidal oscillator realizations, replicating the general approach presented in [5,9] which, as previously anticipated, has been used to synthesize Op-Amp-based and CCII-based sinusoidal oscillators. We will show that it is possible to implement sinusoidal oscillators with a minimum number of elements using a single negative type VCII (VCII$^-$), two resistors and two capacitors, so demonstrating a new practical application of the VCII. The notable advantage of the proposed VCII$^-$-based oscillator is that it is easily cascadable from port Z of VCII$^-$, alleviating the need for any extra voltage buffer. Moreover, THD values are low also for higher frequency oscillators. However, the results of this study show that the applied approach does not reach any canonic configuration using positive type VCII (VCII$^+$). The effect of non-idealities in the VCII has been considered, and the proposed approach has been tested by both simulations and measurement results.

The organization of this paper is the following: in Section 2, an introduction on the VCII as active building block as well as the basics of the general configuration of the VCII-based oscillator is introduced. Section 3 proposes, in detail, the study on the possible realizations of VCII-based oscillators, and the effects of non-idealities in VCII are considered in Section 4. Simulations and measurement results are given in Section 5. Finally, Section 6 concludes the paper.

2. General Configuration of the VCII-Based Oscillator

The symbolic representation and internal structure of VCII are shown in Figure 1. In this block, Y is a low-impedance (ideally zero) current input terminal. The current entering into Y node is transferred to X terminal which is a high-impedance (ideally infinity) current output port. The voltage produced at X terminal is transferred to Z terminal which is a low-impedance (ideally zero) voltage output terminal. The relationship between port currents and voltages are given by: $v_Z = \alpha v_X$, $i_X = \beta i_Y$ and $v_Y = 0$. In the ideal case we have $\alpha = 1$ and $\beta = \pm 1$. If $\beta = 1$ we are considering a VCII$^+$, whereas if $\beta = -1$ we have a VCII$^-$.

Using the approach presented in [5,9], the general configuration of an RC-active oscillator based on a single VCII is shown in Figure 2, where N_{GC} represents 4-terminal network consisting of only capacitors and conductances.

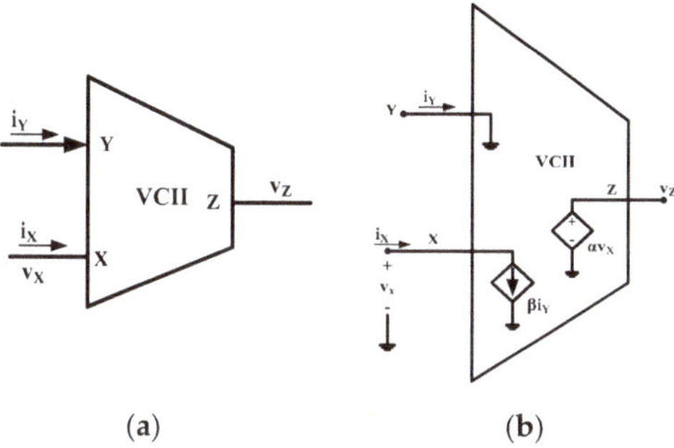

(**a**)   (**b**)

Figure 1. VCII: (**a**) symbol; (**b**) internal structure.

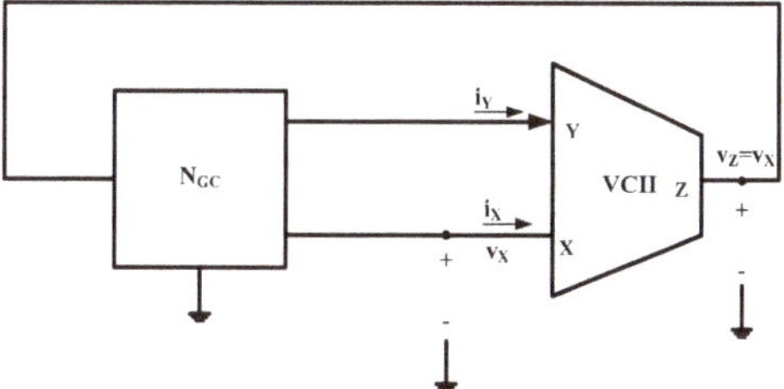

Figure 2. General configuration of RC-VCII oscillator.

The characteristic equation (CE) of the whole system can be calculated replacing, in the circuit of Figure 2, the equivalent model of a VCII of Figure 1b and considering a fictitious input at the Y node (of course, no input signal will be present in an actual oscillator circuit), as shown in Figure 3a at the building block level and Figure 3b in more detail.

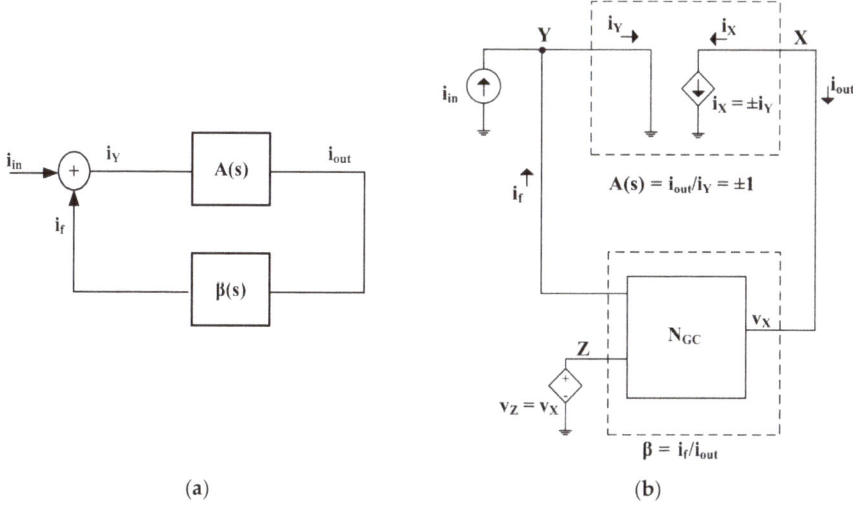

(**a**)   (**b**)

Figure 3. Positive feedback system: (**a**) general schematic; (**b**) positive feedback in the VCII-based oscillator.

The configurations in Figures 2 and 3 can hence be seen as a positive feedback system for which the current transfer function (TF) is given by:

$$T_I(s) = \frac{i_{out}(s)}{i_{in}(s)} = \frac{A(s)}{1 - A(s)\beta(s)}.$$ (1)

Since $A(s) = \pm 1$ and $\beta(s) = i_f(s)/i_{out}(s)$, (1) becomes:

$$T_I(s) = \frac{\pm 1}{1 \mp i_f(s)/i_{out}(s)}.$$ (2)

However, since from Figure 3b $i_{out} = -i_X$ and in an oscillator circuit there is no input $(i_{in} = 0)$, we have $i_f = i_Y$ and the TF is given by:

$$T_I(s) = \frac{\pm 1}{1 \mp i_Y(s)/i_{out}(s)} = \frac{\pm i_X(s)}{i_X(s) \pm i_Y(s)}.$$ (3)

From (3), we can derive the condition of existence (CE) as:

$$i_X(s) \pm i_Y(s) = 0.$$ (4)

By assuming $v_Z = v_X$, $v_Y = 0$, the transconductance functions of the passive network in Figure 2 can be expressed by a rational expression as:

$$\frac{i_X(s)}{v_Z(s)} = \frac{N_X(s)}{D(s)}$$ (5)

$$\frac{i_Y(s)}{v_Z(s)} = \frac{N_Y(s)}{D(s)}$$ (6)

where $N_X(s)$ and $N_Y(s)$ are the numerators at X and Z nodes, respectively, while $D(s)$ is a common denominator. Using (5) and (6) in (4), the CE becomes:

$$N_X(s) \pm N_Y(s) = 0$$ (7)

In (7), the plus and minus signs are for VCII$^-$ and VCII$^+$ respectively. To ensure a pure sinusoidal oscillation, the CE in (7) should be a second-order polynomial with purely imaginary roots. This requires the network N_{GC} to include at least two capacitors. It has to be noted that, in Figure 2, by using a VCII$^+$ rather than a VCII$^-$, at least three capacitors are required to provide a phase shift to generate a positive feedback loop. Therefore, no canonic oscillator is possible using VCII$^+$, and for the following, we will consider the VCII in Figure 2 as a VCII$^-$. By then assuming a network with only two capacitors, Equation (7) will be in the form:

$$as^2 + bs + c = 0.$$ (8)

In order to start the oscillation, the following commonly known criteria must be satisfied:

$$b = 0$$ (9)

$$\frac{c}{a} > 0$$ (10)

with $c \neq 0$, $a \neq 0$, so that, according to the Barkhausen criterion, purely imaginary poles for the closed-loop transfer function are obtained. The oscillation frequency is:

$$\omega_o = \sqrt{\frac{c}{a}}.$$ (11)

3. Oscillator Circuits

In this section we analyze the possible VCII$^-$-based oscillators based on the scheme of Figure 2. The passive N_{GC} is assumed as a general n-node network consisting of b possible branches between two nodes. Each node is a junction where two or more branches are connected, and each branch is an admittance connected between two nodes represented as:

$$Y_i = sC_i + G_i. \tag{12}$$

In the following, we analyze the CE to see if oscillation is possible for the particular case study of a five-node network. From this analysis we see that for a four-node network it is not possible to obtain a second-order polynomial for (7), whereas for a six-node network (or more) only non-canonic oscillators using more than the minimum number of passive components are possible.

N_{GC} as a Five-Node Network

In Figure 4 we assume N_{GC} as a five-node network. We start analyzing this network by performing KCL at node Y as reported in the following:

$$i_Y = i_3 + i_6 + i_7 = i_3 + Y_6 V_Z + Y_7 V_Z. \tag{13}$$

Since no current is flowing into Y_8 and Y_9, these admittances can be assumed as open circuit ($Y_8 = Y_9 = 0$). Routine analysis of Figure 4 results in i_3 as:

$$i_3 = \frac{Y_3(Y_1 + Y_2)}{Y_1 + Y_2 + Y_3 + Y_4} V_Z. \tag{14}$$

Using (13)–(14), we have:

$$\frac{i_Y}{V_Z} = \frac{Y_3(Y_1 + Y_2)}{Y_1 + Y_2 + Y_3 + Y_4} + Y_6 + Y_7. \tag{15}$$

Similar analysis for i_x results:

$$\frac{i_X}{V_Z} = -\frac{Y_2(Y_3 + Y_4)}{Y_1 + Y_2 + Y_3 + Y_4} - Y_5 - Y_7 \tag{16}$$

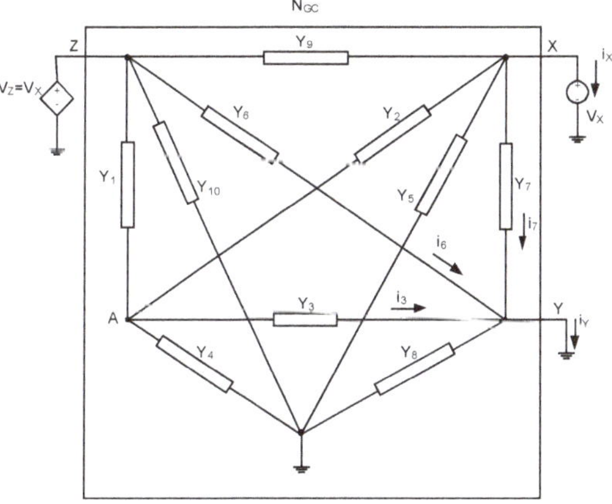

Figure 4. The N_{GC} as a five-node network.

Using (15) and (16) in (7), the CE of the five-node network is found as:

$$- Y_2 Y_4 + Y_3 Y_1 + (Y_6 - Y_5)(Y_1 + Y_2 + Y_3 + Y_4) = 0. \tag{17}$$

It can be noticed that CE does not depend on $Y_7, \ldots, Y_{10}$ which means that these branches can be assumed to be open circuit. For the other branches we can make different choices. If two branches have non-zero admittances, the following CEs are possible:

$$Y_1, Y_3 \neq 0 \rightarrow CE : Y_1 Y_3 = 0 \tag{18a}$$

$$Y_1, Y_5 \neq 0 \rightarrow CE : Y_1 Y_5 = 0 \tag{18b}$$

$$Y_1, Y_6 \neq 0 \rightarrow CE : Y_1 Y_6 = 0 \tag{18c}$$

$$Y_2, Y_4 \neq 0 \rightarrow CE : Y_2 Y_4 = 0 \tag{18d}$$

$$Y_2, Y_5 \neq 0 \rightarrow CE : Y_2 Y_5 = 0 \tag{18e}$$

$$Y_2, Y_6 \neq 0 \rightarrow CE : Y_2 Y_6 = 0 \tag{18f}$$

$$Y_3, Y_5 \neq 0 \rightarrow CE : Y_3 Y_5 = 0 \tag{18g}$$

$$Y_3, Y_6 \neq 0 \rightarrow CE : Y_3 Y_6 = 0 \tag{18h}$$

$$Y_4, Y_5 \neq 0 \rightarrow CE : Y_4 Y_5 = 0 \tag{18i}$$

$$Y_4, Y_6 \neq 0 \rightarrow CE : Y_4 Y_6 = 0 \tag{18j}$$

In the general case, (18) can be expressed as

$$CE : \quad Y_a Y_b = 0 \tag{19}$$

By assuming $Y_a = sC_a + G_a$ and $Y_b = sC_b + G_b$, (19) can be written as:

$$s^2 C_a C_b + s[C_a G_b + C_b G_a] + G_a G_b = 0 \tag{20}$$

From (20) it is not possible to have imaginary roots. Therefore, in case of two non-zero branches, no oscillation is possible.

Finally, we investigate the possibility of achieving oscillations from (17) in the case that three branches of N_{GC} present non-zero admittance.

For $(Y_1 = Y_5 = Y_6 = 0)$ or $(Y_3 = Y_5 = Y_6 = 0)$, we have $Y_2 Y_4 = 0$, while for $(Y_2 = Y_5 = Y_6 = 0)$ or $(Y_4 = Y_5 = Y_6 = 0)$, we have $Y_1 Y_3 = 0$. In both these cases, the CE has the general form of (19).

For $(Y_1 = Y_2 = Y_5 = 0)$, $(Y_1 = Y_2 = Y_6 = 0)$, $(Y_1 = Y_4 = Y_5 = 0)$, $(Y_1 = Y_4 = Y_6 = 0)$, $(Y_2 = Y_3 = Y_5 = 0)$, $(Y_2 = Y_3 = Y_6 = 0)$, $(Y_3 = Y_4 = Y_5 = 0)$, $(Y_3 = Y_4 = Y_6 = 0)$ the CE has the following form:

$$Y_c(Y_a + Y_b) = 0. \tag{21}$$

For $(Y_1 = Y_3 = Y_6 = 0)$ and $(Y_2 = Y_4 = Y_5 = 0)$, the CE is obtained as:

$$Y_a Y_b + Y_c(Y_a + Y_b) = 0. \tag{22}$$

The CEs of (21) and (22) do not result in pure imaginary roots; therefore, these cases cannot give oscillator topologies.

Considering instead the cases $(Y_1 = Y_2 = Y_3 = 0)$, $(Y_1 = Y_2 = Y_4 = 0)$, $(Y_1 = Y_3 = Y_4 = 0)$ and $(Y_2 = Y_3 = Y_4 = 0)$, the CE has the following general form:

$$Y_c(Y_a - Y_b) = 0. \tag{23}$$

It is easy to verify that the CE in (23) cannot be associated with an oscillator topology if only two capacitors are used (we need three of them at least).

Finally, for $(Y_1 = Y_3 = Y_5 = 0)$ and $(Y_2 = Y_4 = Y_6 = 0)$, the CEs will be given by (24a) and (24b), respectively:

$$Y_2 Y_4 - Y_6(Y_2 + Y_4) = 0 \tag{24a}$$

$$Y_1 Y_3 - Y_5(Y_1 + Y_3) = 0 \tag{24b}$$

which are equations with the general form:

$$Y_a Y_b - Y_c(Y_a + Y_b) = 0. \tag{25}$$

In (25), oscillation condition is related to the choice of Y_c and Y_a or Y_b as a capacitance.

In order to design an oscillator with the minimum number of components, we now have to verify the choice of the components in (25). It can be demonstrated that it is possible to have a minimum of two capacitors and at least two resistors in order to have a constant term in the constituting equation. In fact, with this choice we obtain a complete polynomial. In this case, having only three branches of the type $sC + G$ with $C \geq 0, G \geq 0$, it is a matter of choosing an admittance between Y_a, Y_b, Y_c of the type $sC + G$; the two remaining admittances will be a capacitance sC, and a conductance G. Inserting all possible combinations of options into (25), two sets of CEs which show imaginary roots are obtained.

For $Y_c = sC_c + G_c$; $Y_a = sC_a$; $Y_b = G_b$, the CE becomes

$$- s^2 C_a C_c + s[C_a G_b - C_a G_c - C_c G_b] - G_b G_c = 0. \tag{26}$$

For $Y_c = sC_c + G_c$; $Y_b = sC_b$; $Y_a = G_a$, the CE becomes

$$- s^2 C_b C_c + s[C_b G_a - C_c G_a - C_b G_c] - G_a G_c = 0. \tag{27}$$

From (26) and (27), the oscillation condition (C_o) and oscillation frequency (ω_0) for the two cases are obtained respectively as:

$$C_o : \frac{G_c}{G_b} + \frac{C_c}{C_a} = 1, \; \omega_0 = \sqrt{\frac{G_b G_c}{C_a C_c}} \tag{28a}$$

$$C_o : \frac{G_c}{G_a} + \frac{C_c}{C_b} = 1, \; \omega_0 = \sqrt{\frac{G_a G_c}{C_b C_c}} \tag{28b}$$

Thus, the minimum number of elements necessary to obtain an oscillator based on the scheme of Figure 2 is four, being two of these capacitors and two resistors. Considering the two cases $(Y_1 = Y_3 = Y_5 = 0)$ and $(Y_2 = Y_4 = Y_6 = 0)$ and the possible choices for Y_a and Y_b, we obtain a total of four canonic oscillators, corresponding to the following CEs:

$$- s^2 C_1 C_5 + s[C_1 G_3 - C_1 G_5 - C_5 G_3] - G_3 G_5 = 0 \tag{29}$$

$$- s^2 C_3 C_5 + s[C_3 G_1 - C_3 G_5 - C_5 G_1] - G_1 G_5 = 0 \tag{30}$$

$$s^2 C_2 C_6 + s[C_6 G_4 + C_2 G_6 - C_2 G_4] + G_4 G_6 = 0 \tag{31}$$

$$s^2 C_4 C_6 + s[C_6 G_2 + C_4 G_6 - C_4 G_2] + G_2 G_6 = 0 \tag{32}$$

However, this number is reduced again to two if we consider that from each of the cases $(Y_1 = Y_3 = Y_5 = 0)$ and $(Y_2 = Y_4 = Y_6 = 0)$ we obtain two equal oscillators if we exchange the order of the elements which are connected in series. These two configurations are shown in Figure 5, and the corresponding transfer functions, oscillation frequencies ω_0 and oscillation conditions are reported in Table 1. The oscillation frequencies and oscillation conditions in (28) show a strong interdependence since they are functions of the same parameters. Since the oscillation condition requires that the sum of the ratios of the capacitances and of the conductances is constant and equal to 1, a possible strategy for frequency tuning requires varying both resistors or both capacitors, maintaining their ratio constant. For example, a ratio of 2 between C_a and C_c can be obtained by using two parallel capacitors equal to C_a to obtain C_c; all three capacitors can be varied together; thus their ratio remains constant unless there are mismatches and the effect of parasitics.

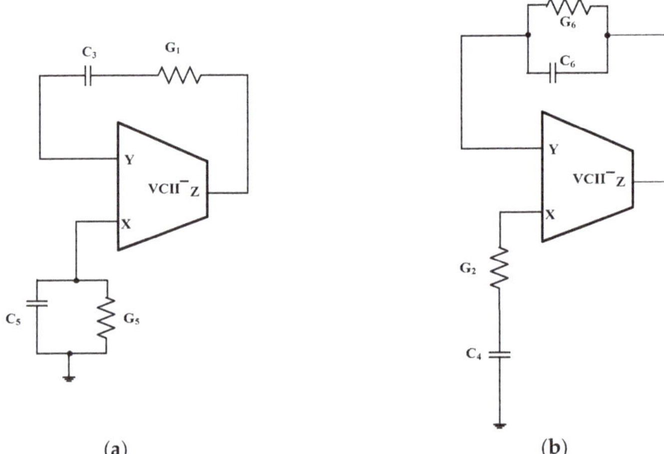

(a)   (b)

Figure 5. VCII-based oscillators obtained in the case NGC is a five-node network. (**a**)Series-parallel impedances configuration; (**b**) parallel-series impedances configuration.

Table 1. Main equations of the canonic VCII-based oscillators.

Figure 5a	Figure 5b
$T_I(s) = -\dfrac{s^2 C_5 C_3 + s[C_5 G_1 + C_3 G_5] + G_5 G_1}{s^2 C_5 C_3 + G_5 G_1}$	$T_I(s) = \dfrac{s C_2 G_2}{s^2 C_4 C_6 + G_2 G_6}$
$\omega_0 = \sqrt{\dfrac{G_5 G_1}{C_5 C_3}}$	$\omega_0 = \sqrt{\dfrac{G_2 G_6}{C_4 C_6}}$
$C_o: \dfrac{G_5}{G_1} + \dfrac{C_5}{C_3} = 1$	$C_o: \dfrac{G_6}{G_2} + \dfrac{C_6}{C_4} = 1$

4. Analysis of Parasitic Effects: A Case Study

The only two possible canonic topologies for the VCII-based oscillator are synthesized in Figure 6, where Z_A and Z_B are a series-connected RC network and a parallel-connected RC network; we define $t_A = R_A C_A$ and $t_B = R_B C_B$ as the time constants associated with these networks. The two oscillator topologies shown in Figure 6 correspond to the cases:

$$\text{Type I}: \quad Z_A = \frac{R_5}{1 + s R_5 C_5}, \quad Z_B = R_1 + \frac{1}{s C_3} \tag{33a}$$

$$\text{Type II}: \quad Z_A = R_2 + \frac{1}{s C_4}, \quad Z_B = \frac{R_6}{1 + s R_6 C_6}. \tag{33b}$$

where $R_i = 1/G_i$. From Figure 6, the oscillation condition can be obtained as:

$$\alpha \, |\beta| \frac{Z_A}{Z_B} = 1 \tag{34}$$

where β and α are the VCII current and voltage gains (ideally both equal to 1), and the oscillation frequency is given by:

$$\omega_0 = \frac{1}{\sqrt{\tau_A \tau_B}}. \tag{35}$$

The oscillation condition and the oscillation frequency are affected by the non-idealities of the VCII, i.e., finite port impedances, gain errors ($a < 1$, $|b| < 1$) and poles of the voltage and current buffers. In order to analyze the effects of these non-idealities on the oscillator behavior, a model of a real VCII has been developed and implemented (see Figure 7), able to take into account the non-idealities.

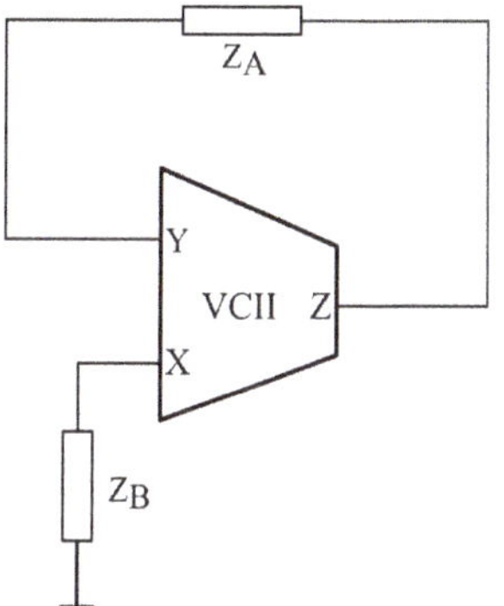

Figure 6. General topology of the canonic VCII-based oscillators.

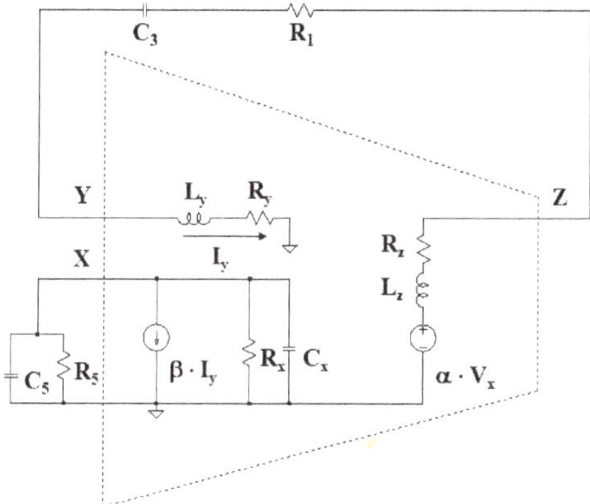

Figure 7. Model for the Type I canonic oscillator with non-ideal VCII.

In the general case, we can model the VCII with the first-order transfer functions

$$\alpha(s) = \alpha_0 / (1 + s\tau_z) \tag{36}$$

$$\beta(s) = -\beta_0 / (1 + s\tau_x) \tag{37}$$

and complex port impedances

$$Y_x(s) = G_x + sC_x \tag{38}$$

$$Z_y(s) = R_y + sL_y \tag{39}$$

$$Z_z(s) = R_z + sL_z. \tag{40}$$

In order to better understand the effects of non-idealities and to compare the performance of the two topologies in Figure 5, different cases have been considered under the hypothesis that the ideal design has been carried out starting from the oscillation condition (34). When the non-idealities of the VCII are taken into account, Equation (34) becomes

$$\alpha(s) |\beta(s)| \frac{1}{Z_B + Z_y + Z_z} \frac{1}{1/Z_A + Y_x} = 1. \tag{41}$$

By a simple inspection of the impedances Z_A and Z_B given by (33), and of the port impedances (38)–(40), it is evident that the Type I canonic oscillator should be less affected by non-idealities. In fact, in this case Y_x can be absorbed in Z_A (G_x and C_x are summed to $1/R_5$ and C_5, respectively), and Z_y and Z_z in Z_B: a parallel RC network is used in parallel to a port impedance modeled as an RC parallel network, and a series impedance is connected in series to port impedances modeled as RL series networks. In contrast, for the Type II canonic oscillator, a series network is connected in parallel to the parallel RC port impedance, and a parallel RC network is connected in series to LR series port impedances, thus non-ideal port impedances alter Z_A and Z_B more significantly. The Type I canonic VCII-based oscillator seems therefore more suited to a practical realization, and it has been selected for further analysis.

4.1. Resistive Port Impedances

If only the resistive parasitics G_x, R_y and R_z in (39)–(41) are considered, the oscillation condition for the Type I canonic oscillator becomes:

$$\frac{\alpha \, |\beta| \, s(R_5||R_x)C_3}{[1 + s(R_5||R_x)C_5]\,[1 + s(R_1 + R_y + R_z)C_3]} = 1. \tag{42}$$

It is evident from (42) that the effect of the port impedances is limited, since they are simply summed to the ones from the NGC network (that have to be chosen as much larger than the corresponding parasitics to make them negligible). The oscillation frequency in Table 1 is modified as follows:

$$\omega_0{}' = \sqrt{\frac{G_5{}'}{C_5 C_3 R_1{}'}} = \omega_0 \cdot \sqrt{\frac{1 + G_x/G_5}{1 + G_1(R_y + R_z)}} \tag{43}$$

where $R_1{}' = R_1 + R_y + R_z$ and $G_5{}' = 1/R_5{}' = G_5 + G_x$, and the oscillation condition becomes

$$\frac{\alpha \, |\beta|}{\frac{R_1}{R_5}\left(1 + \frac{R_5}{R_x}\right)\left(1 + \frac{R_y + R_z}{R_1}\right) + \frac{C_5}{C_3}} = 1. \tag{44}$$

If the parasitic capacitance C_x at the X terminal is also considered, Equations (43) and (44) have to be slightly modified by considering $G_5{}' = C_5 + C_x$ instead of C_5. Inductances L_y and L_z can be neglected in several applications and have not been considered in the following. However, for the sake of completeness, we report below the expression for the oscillation frequency when inductive parasitics are also considered:

$$\omega_0{}' = \omega_0 \cdot \sqrt{\frac{1 + G_x/G_5}{(L_y + L_z)(1 + G_x/G_5)\frac{G_5 G_1}{C_5} + (1 + C_x/C_5)\,[1 + G_1(R_y + R_z)]}} \tag{45}$$

4.2. Single-Pole Transfer Functions

If the non-ideal transfer functions in (36) and (37) are also considered in addition to the terminal resistive parasitics in (38)–(40), the denominator of the oscillation condition in (34) becomes of fourth degree:

$$\frac{\alpha \, |\beta| \, sG_5{}'R_5{}'}{as^4 + bs^3 + cs^2 + ds + e} = 1 \tag{46}$$

Prime variables are considered for $R_5{}'$, $G_5{}'$ and $R_1{}'$ to account for parasitic resistances R_y and R_z and admittance Y_x, as in the previous subsection, and we have

$$a = R_5{}'C_5{}'R_1{}'C_3\tau_x\tau_z \tag{47a}$$

$$b = R_5{}'C_5{}'R_1{}'C_3 \cdot (\tau_x + \tau_z) + \tau_x\tau_z \cdot (R_5{}'C_5{}' + R_1{}'C_3) \tag{47b}$$

$$c = R_5{}'C_5{}'R_1{}'C_3 + \tau_x\tau_z + (R_5{}'C_5{}' + R_1{}'C_3)\cdot(\tau_x + \tau_z) \tag{47c}$$

$$d = R_5{}'C_5{}' + R_1{}'C_3 + \tau_x + \tau_z \tag{47d}$$

$$e = 1 \tag{47e}$$

A real value is obtained for the left-hand side, under the hypothesis of a purely imaginary denominator. By equating to zero the real part of the denominator at $\omega = \omega_0$, we get:

$$\omega_0'^2 = \frac{c}{2a}\left(1 - \sqrt{1 - \frac{4a}{c^2}}\right) \cong \frac{c}{2a}\left(\frac{1}{2}\frac{4a}{c^2}\right) = \frac{1}{c} \tag{48}$$

where c is given by (47c). The approximation $\frac{4a}{c^2} \ll 1$ is justified under the hypothesis that the parasitic time constants τ_x and τ_z are significantly lower than the time constants $\tau_A = R_5{}'C_5{}'$ and $\tau_B = R_1{}'C_3$. Finally, the oscillation frequency ω_0' can be expressed in terms of the ideal value ω_0, by using the expression of coefficient c:

$$\omega_0' \cong \frac{1}{\sqrt{c}} = \omega_0 \frac{1}{\sqrt{1 + \frac{\tau_x\tau_z+(\tau_A+\tau_B)(\tau_x+\tau_z)}{\tau_A\tau_B}}} \tag{49}$$

Under the simplifying assumptions $\tau_x = \tau_y = \tau_{par}$ and $\tau_A = \tau_B = \tau$, the relative error on the oscillation frequency $(1 - \omega_0'/\omega_0)$ can be readily expressed as a function of the ratio τ_{par}/τ, thus providing a design guideline for the bandwidth of the VCII transfer functions. The graph in Figure 8 shows that errors lower than 10% can be obtained if the time constant ratio is lower than 0.06.

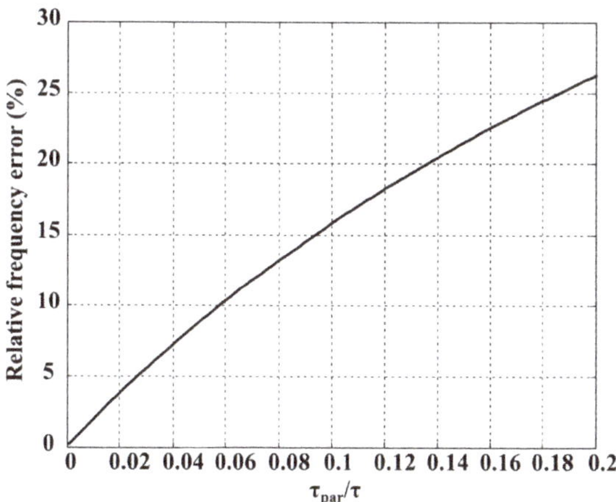

Figure 8. Relative error on the oscillation frequency vs. the time constant ratio τ_{par}/τ.

5. Experimental Results

The performance of the Type I canonic oscillator of Figure 5a has been verified by both LTSpice simulations and experimental results. In particular, the approximated expression for ω_0 in (49) has been checked for different values of τ and $\tau_x = \tau_z$, and errors lower than 1% have been found.

Then, we have used the commercially available AD844 to configure a VCII$^-$ as shown in Figure 9. A single VCII is realizable using two AD844 ICs, whose Spice model can be found in [45]. The situation is quite different in the case of an integrated design, where a single VCII block can be exploited to design the oscillator, as shown in the previous sections.

The circuit was supplied with a dual ± 5 V voltage, achieving a total power consumption of 14 mA.

Firstly, simulation of the topology in Figure 5a has been carried out to evaluate performance in terms of robustness to parasitics, and to estimate the achievable THD. In particular, the circuit has been designed with $C_3 = 2C_5 = 2$ nF and $R_5 = 2R_1 = 15$ kΩ, and an oscillation frequency $f_0 = 10.6$ kHz was expected.

However, AD844 parasitics can slightly change the oscillation frequency and/or cause failing of oscillation condition: in this case, starting from the nominal design, the resistance R_1 can be changed (to 7.3 kΩ in the present case, see the schematic in Figure 10) to allow fulfillment of oscillation condition in (41): the obtained oscillation frequency is $f_0 = 10.8$ kHz, as shown in Figure 11.

A model for the VCII composed of AD844 components, shown in Figure 9, has been extracted from Spice simulations according to the equations presented in Section 4. At terminal X, we have found $C_x = 5.5$ pF in parallel with a resistor $R_x = 3$ MΩ. Purely resistive input impedances have been extracted at node Y ($R_y = 50$ Ω) and Z ($R_z = 15$ Ω). Finally, a dominant pole has been found for both the transfer function $\alpha(s)$ at $f = 49$ MHz (corresponding to $\tau_x = 3.25$ ns), and for $\beta(s)$ at $f = 764$ MHz ($\tau_z = 208$ ps).

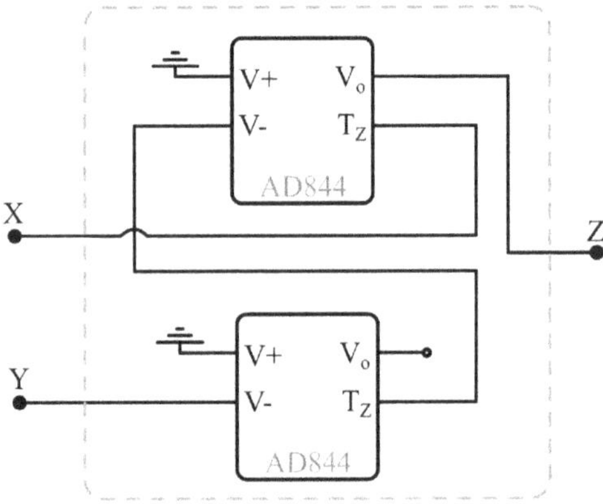

Figure 9. Realization of a VCII$^-$ using the AD844.

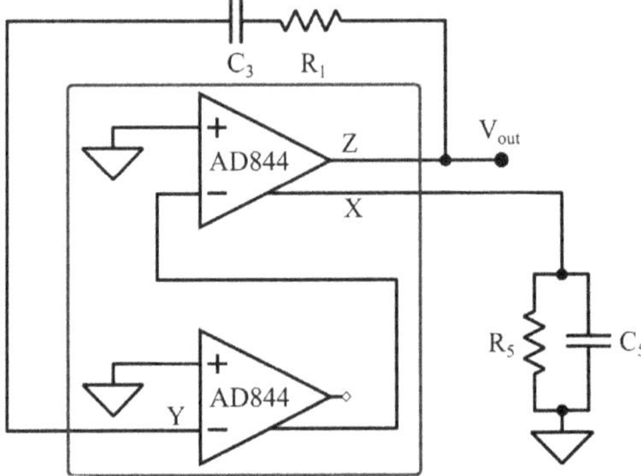

Figure 10. VCII oscillator based on the AD844.

The element values used for the different design case studies, the simulated THD and the oscillation frequency evaluated with both the LTSpice AD844 non-linear model and with the VCII linear model, including parasitics, are summarized in Table 2. The linear model is accurate enough to be used for circuit design, and excellent simulated performance has been achieved in terms of THD with the proposed VCII topology.

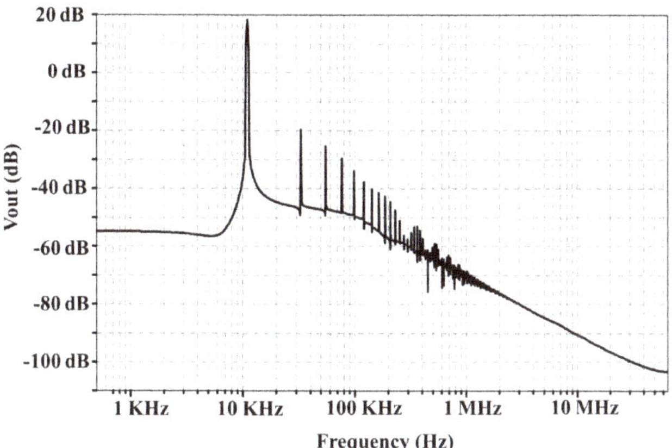

Figure 11. Simulated output spectrum of the oscillator shown in Figure 10.

Table 2. Simulation results at different frequencies.

$C_3 = 2C_5$ (nF)	R_5 (kΩ)	R_1 (kΩ)	f_0 (Spice) (Hz)	f_0 (Model) (Hz)	f_0 Error (%)	THD (%)
2000	15	7.3	10.8	10.74	0.6	0.7
200	15	7.35	107.3	107.0	0.3	0.4
20	15	7.35	1.073 K	1.070 K	0.3	0.4
2	15	7.3	10.80 K	10.70 K	0.9	0.5
0.2	15	6.9	108.1 K	107.2 K	0.9	0.7
0.2	1.5	0.65	1.080 M	1.040 M	3.7	1.7

Finally, experimental verification of performance has been carried out, exploiting the test bench shown in Figure 12: for data acquisition, the Digilent Analog Discovery 2™ board was used [46]. The design of Figure 5a was implemented as the reference topology for the oscillator. Measurements were carried out in the range $(10–10^6)$ Hz and are reported in Table 3. In agreement with simulation results, the oscillator shows a very low THD value even at 1 MHz (considering 10 harmonics). The average relative frequency error between measured and ideal values is −5.2% and is comparable with tolerances of the passive components.

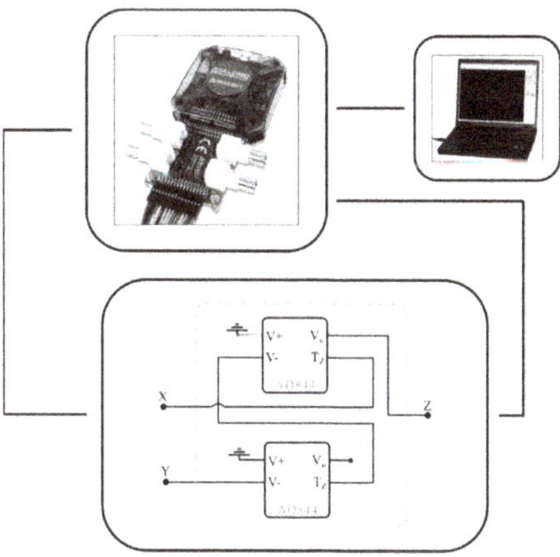

Figure 12. Test bench for the experimental verification of the VCII oscillator based on the AD844.

Table 3. Measured results.

R1 (kΩ)	R5 (kΩ)	C3 (F)	C5 (F)	Ideal Frequency (Hz)	Measured Frequency (Hz)	Error (%)	THD (%)
15	6.8	2 μ	1 μ	11.1	10.8	−3	1.12
15	6.8	200 n	100 n	111	109	−1.9	0.94
15	6.8	20 n	10 n	1.11 k	1.08 k	−2.7	0.92
15	6	2 n	1 n	11.9 k	11.5 k	−3.3	0.47
15	6	200 p	100 p	119 k	109 k	−7.8	0.56
15	0.64	200 p	100 p	1.15 M	1.0 M	−12.9	2.24

An example of the output signal, both in the time and frequency domains, is reported in Figure 13a,b for a frequency of 1 MHz.

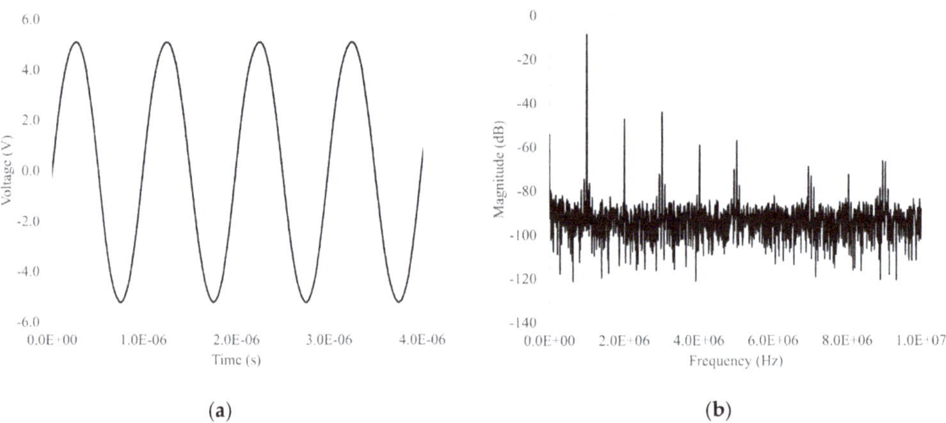

(a)   (b)

Figure 13. Output waveform of the canonic VCII-based oscillator of Figure 5a. (**a**) Time domain, (**b**) frequency domain for an output frequency of 1 MHz.

Figure 14 shows the THD and frequency error trends vs. frequency.

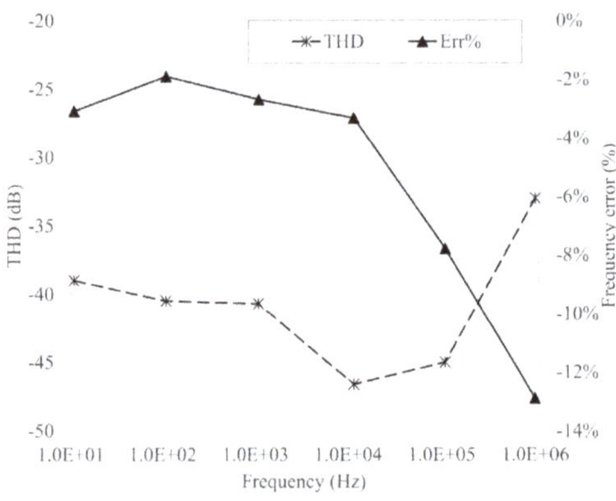

Figure 14. THD and frequency error vs. frequency.

6. Conclusions

By means of a systematic analysis, the possibility of realizing VCII-based oscillators is studied and demonstrated. The investigation results in a pair of new canonic oscillators based on VCII$^-$. However, it is shown that, using the systematic approach, no oscillator configuration is possible using VCII+. The two found oscillator configurations are the only possible ones which use only two resistors, two capacitors and a single VCII$^-$. Compared to Op-Amp-based oscillators, designed using the same systematic approach which employs two capacitors and four resistors, the proposed VCII-based oscillator is preferred in terms of low number of capacitors and resistances. Another interesting feature of the found VCII-based oscillator is that the produced sinusoidal output signal is easily available through the low output impedance Z port, while the CCII-based oscillators designed using the same systematic approach requires an additional voltage buffer for practical use. Simulations and experimental results using AD844 as VCII are reported to validate the theory.

A comparison with oscillator topologies based on different ABBs, with particular attention to canonic topologies, is reported in Table 4. The table reports the type of active building block (ABB) the oscillator is based on, the number of active and passive components, specifying how many of them are grounded, the availability of a quadrature output and the independence of oscillation condition from oscillation frequency that allows tuning the oscillator acting on a single component. It has to be noted that the independence from the oscillation condition on oscillation frequency often requires additional passive (and sometimes also active) components, thus resulting in non-canonic topologies. Notable exceptions are the oscillators of [21,26] that use complex ABBs with gain, whose value contributes to satisfying the oscillation condition.

Table 4. Comparative table of sinusoidal oscillator topologies.

Ref.	ABB Type	ABB Number	C (Grounded)	R (Grounded)	Output Phases	Indep. w_0/C_0
[5]	Op-Amp	1	2 (2)	4 (2)	1	NO
[7]	OTA	3	2 (2)	–	2	YES
[9]	CCII	1	2 (2)	2 (1)	1	NO
[9]	CCII	1	2 (1)	3 (3)	1	YES
[12]	FTFN	1	2 (1)	5 (1)	1	YES
[13]	CCII	2	2 (2)	2 (1)	1	YES
[16]	CDTA	2	2 (1)	–	2	NO
[19]	OTRA	1	2 (0)	3 (1)	1	YES
[21]	CCCCTA	1	2 (2)	1 (1)	1	YES
[23]	CCIII	2	2 (2)	3 (3)	1	YES
[25]	UVC	1	2 (1)	3 (1)	1	YES
[26]	VDTA	1	2 (2)	–	2	YES
[29]	CFOA	1	3 (2)	4 (2)	1	YES
[47]	CFOA	1	2 (2)	2 (1)	1	NO
[47]	CFOA	1	2 (1)	3 (1)	1	YES
[48]	OTRA	1	2 (0)	2 (0)	1	NO
[48]	OTRA	1	2 (0)	3 (1)	1	YES
[49]	CFOA	1	3 (2)	3 (3)	1	YES
[50]	CDBA	2	2 (2)	3 (0)	2	YES
[51]	OTRA	1	3 (1)	3 (0)	1	YES
This Work	VCII	1	2 (1)	2 (1)	1	NO

Op-Amp: operational amplifier; OTA: operational transconductance amplifier; CCII: second generation current conveyor; FTFN: four terminal floating nullor; CDTA: current differencing transconductance amplifier; OTRA: operational transresistance amplifier; CCCCTA: current controlled current conveyor transconductance amplifier; CCIII: third generation current conveyor; UVC: universal voltage conveyor; VDTA: voltage dependent transconductance amplifier; CFOA: current feedback operational amplifier; CDBA: current differencing buffered amplifier; VCII: second generation voltage conveyor.

Author Contributions: Conceptualization, M.S. and G.B.; methodology, G.B. and F.C.; software, G.B.; validation, G.F., P.T., V.S. and L.P.; formal analysis, M.S. and F.C.; investigation, P.T. and V.S.; resources, V.S. and G.F.; data curation, V.S. and L.P.; writing—original draft preparation, L.P. and G.B.; writing—review and editing, V.S., G.B., L.P., M.S., G.F., F.C. and A.T.; visualization, F.C. and L.P.; supervision, L.P., G.F., V.S. and P.T.; project administration, G.F., V.S. and A.T.; funding acquisition, V.S. All authors have read and agreed to the published version of the manuscript.

Funding: The work did not receive any external funding.

Data Availability Statement: No new data were created or analyzed in this study. Data sharing is not applicable to this article.

Conflicts of Interest: The authors declare no conflict of interest.

References

1. Jaikla, W.; Adhan, S.; Suwanjan, P.; Kumngern, M. Current/voltage controlled quadrature sinusoidal oscillators for phase sensitive detection using commercially available IC. *Sensors* **2020**, *20*, 1319. [CrossRef] [PubMed]
2. Mohsen, M.; Said, L.; Elwakil, A.; Madian, A.; Radwan, A. Extracting optimized bio-impedance model parameters using different topologies of oscillators. *IEEE Sensors J.* **2020**, *20*, 9947–9954. [CrossRef]
3. Li, G.; Cui, J.; Yang, H. A new detecting method for underwater acoustic weak signal based on differential double coupling oscillator. *IEEE Access* **2021**, *9*, 18842–18854. [CrossRef]
4. Senani, R. New canonic sinusoidal oscillator with independent frequency control through a single grounded resistor. *Proc. IEEE* **1979**, *67*, 691–692. [CrossRef]
5. Bhattacharyya, B.; Tavakoli Darkani, M. A unified approach to the realization of canonic RC-active, single as well as variable, frequency oscillators using operational amplifiers. *J. Frankl. Inst.* **1984**, *317*, 413–439. [CrossRef]
6. Senani, R.; Bhaskar, D.R.; Gupta, M.; Singh, A.K. Canonic OTA-C sinusoidal oscillators: Generation of new grounded-capacitor versions. *Am.J. Electr. Electron. Eng.* **2015**, *3*, 137–146.
7. Gupta, S.; Senani, R. State variable synthesis of single-resistance-controlled grounded capacitor oscillators using only two CFOAs: Additional new realisations. *IEE Proc. Circuits Devices Syst.* **1998**, *145*, 135–138. [CrossRef]
8. Soliman, A.M. Current-mode oscillators using single output current conveyors. *Microelectron. J.* **1998**, *29*, 907–912. [CrossRef]
9. Celma, S.; Martinez, P.A.; Carlosena, A. Approach to the synthesis of canonic RC-active oscillators using CCll. *IEE Proc. Circuits Devices Syst.* **1994**, *141*, 493–497. [CrossRef]
10. Khan, A.A.; Bimal, S.; Dey, K.K.; Roy, S.S. Novel RC sinusoidal oscillator using second-generation current conveyor. *IEEE Trans. Instrum. Meas.* **2005**, *54*, 2402–2406. [CrossRef]
11. Bajer, J.; Lahiri, A.; Biolek, D. Current-mode CCII+ based oscillator circuits using a conventional and a nodified Wien-bridge with all capacitors grounded. *Radio Eng.* **2011**, *20*, 245–251.
12. Çam, U.G.; Toker, A.; Çicekoglu, O.G.; Kuntman, H. Current-mode high output impedance sinusoidal oscillator configuration employing single FTFN. *Analog Integr. Circuits Signal Process.* **2000**, *24*, 231–238. [CrossRef]
13. Lahiri, A. New canonic active RC sinusoidal oscillator circuits using second-generation current conveyors with application as a wide frequency digitally controlled sinusoidal generator. *Act. Passiv. Electron. Compon.* **2011**, *2011*, 274394. [CrossRef]
14. Horng, J.-W. Current-mode third-order quadrature oscillator using CDTAs. *Act. Passiv. Electron. Compon.* **2009**, *2009*, 789171. [CrossRef]
15. Uttaphut, P.; Mekhum, W.; Jaikla, W. Current-mode multiphase sinusoidal oscillator using CCCCTAs and grounded elements. In Proceedings of the NEWCAS 11 IEEE 9th Int. New Circuits and Systems Conference, Bordeaux, France, 26–29 June 2011; pp. 345–349.
16. Jin, J.; Wang, C. Current-mode universal filter and quadrature oscillator using CDTAs. *Turk. J. Electr. Eng. Comput. Sci.* **2014**, *22*, 276–286. [CrossRef]
17. Agrawal, D.; Maheshwari, S. An active-C current-mode universal first order filter and oscillator. *J. Circuits Systems Comput.* **2019**, *28*, 1950219. [CrossRef]
18. Soliman, A.M. Current-mode oscillators using inverting CCII. *J. Act. Passive Electron. Devices* **2011**, *6*, 305–320.
19. Cam, U. A novel single-resistance-controlled sinusoidal oscillator employing single operational transresistance amplifier. *Analog Integr. Circ. Sig. Proc.* **2002**, *32*, 183–186. [CrossRef]
20. Gupta, S.S.; Sharma, R.K.; Bhaskar, D.R.; Senani, R. Sinusoidal oscillators with explicit current output employing current-feedback op-amps. *Int. J. Circuit Theory Appl.* **2010**, *38*, 131–147. [CrossRef]
21. Li, Y. Derivation for current-mode Wien oscillators using CCCCTAs. *Analog Integr. Circ. Sig. Proc.* **2015**, *84*, 479–490. [CrossRef]
22. Abduelma'atti, M.T.; Alsuhaibani, E.S. New current-feedback operational-amplifier based sinusoidal oscillators with explicit current output. *Analog Integr. Circ. Sig. Proc.* **2015**, *85*, 513–523. [CrossRef]
23. Sharma, R.K.; Arora, T.S.; Senani, R. On the realization of canonic single-resistance-controlled oscillators using third generation current conveyors. *IET Circuits Devices Syst.* **2017**, *11*, 10–20. [CrossRef]
24. Chen, H.-P.; Hwang, Y.-S.; Ku, Y.-T. Voltage-mode and current-mode resistorless third-order quadrature oscillator. *Appl. Sci.* **2017**, *7*, 179. [CrossRef]
25. Pushkar, K.L. Single-resistance controlled sinusoidal oscillator employing single universal voltage conveyor. *SCIRP Circuits Syst.* **2018**, *9*, 1–7. [CrossRef]
26. Banerjee, K.; Singh, D.; Paul, S.K. Single VDTA based resistorless quadrature oscillator. *Analog Integr. Circ. Sig. Proc.* **2019**, *100*, 495–501. [CrossRef]
27. Komanapalli, G.; Pandey, R.; Pandey, N. New sinusoidal oscillator configurations using operational transresistance amplifier. *Int. J. Circuit Theory Appl.* **2019**, *47*, 666–685. [CrossRef]

28. Yuce, F.; Yuce, E. Supplementary CCII based second-order universal filter and quadrature oscillators. *AEU Int. J. Electron. Commun.* **2020**, *118*, 153138. [CrossRef]

29. Srivastava, D.K.; Singh, V.K.; Senani, R. A class of single-CFOA-based sinusoidal oscillators. *Am.J. Electr. Electron. Eng.* **2020**, *8*, 1–10.

30. Kumari, S.; Gupta, M. A new CMOS design of high transconductance current follower transconductance amplifier and its applications. *Analog Integr. Circ. Sig. Process.* **2018**, *95*, 325–349. [CrossRef]

31. Pushkar, K.L.; Bhaskar, D.R. Voltage-mode third-order quadrature sinusoidal oscillator using VDIBAs. *Analog Integr. Circ. Sig. Process.* **2019**, *98*, 201–207. [CrossRef]

32. Bhagat, R.; Bhaskar, D.R.; Kumar, P. Quadrature sinusoidal oscillators using CDBAs: New realizations. *Circuits Systems Sig. Process.* **2021**, *40*, 2634–2658. [CrossRef]

33. Gupta, S.; Arora, T.S. Design and experimentation of VDTA based oscillators using commercially available integrated circuits. *Analog Integr. Circ. Sig. Process.* **2021**, *106*, 713–728. [CrossRef]

34. Roy, S.; Pal, R.R. Electronically tunable third-order dual-mode quadrature sinusoidal oscillators employing VDCCs and all grounded components. *Integr. VLSI J.* **2021**, *76*, 99–112. [CrossRef]

35. Safari, L.; Barile, G.; Stornelli, V.; Ferri, G. An overview on the second generation voltage conveyor: Features, design and applications. *IEEE Trans. Circuits Syst. Part II Express Briefs* **2019**, *66*, 547–551. [CrossRef]

36. Safari, L.; Barile, G.; Ferri, G.; Stornelli, V. A new low-voltage low-power dual-mode VCII-based SIMO universal filter. *Electronics* **2019**, *8*, 765. [CrossRef]

37. Safari, L.; Barile, G.; Ferri, G.; Stornelli, V. Traditional Op-Amp and new VCII: A comparison on analog circuits applications. *AEU Int. J. Electron. Commun.* **2019**, *110*, 152845. [CrossRef]

38. Pantoli, L.; Barile, G.; Leoni, A.; Muttillo, M.; Stornelli, V. A novel electronic interface for micromachined Si-based photomultipliers. *Micromachines* **2018**, *9*, 507. [CrossRef] [PubMed]

39. Barile, G.; Ferri, G.; Safari, L.; Stornelli, V. A new high drive class-AB FVF-based second generation voltage conveyor. *IEEE Trans. Circuits Syst. Part II Express Briefs* **2020**, *67*, 405–409. [CrossRef]

40. Centurelli, F.; Monsurrò, P.; Tommasino, P.; Trifiletti, A. On the use of voltage conveyors for the synthesis of biquad filters and arbitrary networks. In Proceedings of the ECCTD 17 23rd Eur. Conf. Circuit Theory and Design, Catania, Italy, 4–6 September 2017.

41. Yesil, A.; Minaei, S. New simple transistor realizations of second-generation voltage conveyor. *Int. J. Circuit Theory Appl.* **2020**, *48*, 2023–2038. [CrossRef]

42. Al-Shahrani, S.M.; Al-Absi, M.A. Efficient inverse filter based on second-generation voltage conveyor (VCII). *Arab J. Sci. Eng.* **2021**. [CrossRef]

43. Al-Absi, M.A. Realization of inverse filters using second generation voltage conveyor (VCII). *Analog Integr. Circ. Sig. Process.* **2021**. [CrossRef]

44. Kumngern, M.; Torteanchai, U.; Khateb, F. CMOS class AB second generation voltage conveyor. In Proceedings of the 2019 IEEE International Circuits and Systems Symposium (ICSyS), Kuantan, Malaysia, 18–19 September 2019.

45. Analog Devices. AD844 Datasheet and Product Info | Analog Devices. Analog.com. 2020. Available online: https://www.analog.com/en/products/ad844.html (accessed on 6 June 2021).

46. Analog Discovery 2 [Digilent Documentation]. Available online: https://reference.digilentinc.com/reference/instrumentation/analog-discovery-2/start (accessed on 6 June 2021).

47. Soliman, A.M. Current feedback operational amplifier based oscillators. *Analog Integr. Circ. Sig. Process.* **2000**, *23*, 45–55. [CrossRef]

48. Chien, N.-C. New realizations of single OTRA-based sinusoidal oscillators. *Act. Passiv. Electron. Compon.* **2014**, *2014*, 938987. [CrossRef]

49. Srivastava, D.K.; Singh, V.K.; Senani, R. Novel single-CFOA-based sinusoidal oscillator capable of absorbing all parasitic impedances. *Am.J. Electr. Electron. Eng.* **2015**, *3*, 71–74.

50. Arora, T.S.; Gupta, S. A new voltage mode quadrature oscillator using grounded capacitors: An application of CDBA. *Eng. Sci. Technol.* **2018**, *21*, 43–49. [CrossRef]

51. Komanapalli, G.; Pandey, N.; Pandey, R. New realization of third order sinusoidal oscillator using single OTRA. *AEU Int. J. Electron. Commun.* **2020**, *93*, 182–190. [CrossRef]

Journal of
*Low Power Electronics
and Applications*

Article

0.5 V CMOS Inverter-Based Transconductance Amplifier with Quiescent Current Control

Andrea Ballo [1], Salvatore Pennisi [1,*] and Giuseppe Scotti [2]

[1] DIEEI (Dipartimento di Ingegneria Elettrica Elettronica e Informatica), University of Catania, 95125 Catania, Italy; andrea.ballo@unict.it
[2] DIET (Dipartimento di Ingegneria dell'Informazione Elettronica e Telecomunicazioni), Sapienza University of Rome, 00184 Rome, Italy; giuseppe.scotti@uniroma1.it
* Correspondence: salvatore.pennisi@unict.it; Tel.: +39-0957382318

Abstract: A two-stage CMOS transconductance amplifier based on the inverter topology, suitable for very low supply voltages and exhibiting rail-to-rail output capability is presented. The solution consists of the cascade of a noninverting and an inverting stage, both characterized by having only two complementary transistors between the supply rails. The amplifier provides class-AB operation with quiescent current control obtained through an auxiliary loop that utilizes the MOSFETs body terminals. Simulation results, referring to a commercial 28 nm bulk technology, show that the quiescent current of the amplifier can be controlled quite effectively, even adopting a supply voltage as low as 0.5 V. The designed solution consumes around 500 nA of quiescent current in typical conditions and provides a DC gain of around 51 dB, with a unity gain frequency of 1 MHz and phase margin of 70 degrees, for a parallel load of 1 pF and 1.5 MΩ. Settling time at 1% is 6.6 μs, and white noise is $125\,\mathrm{nV}/\sqrt{Hz}$.

Keywords: feedback amplifier; analog; CMOS; bulk; class AB; low voltage

Citation: Ballo, A.; Pennisi, S.; Scotti, G. 0.5 V CMOS Inverter-Based Transconductance Amplifier with Quiescent Current Control. *J. Low Power Electron. Appl.* **2021**, *11*, 37. https://doi.org/10.3390/jlpea11040037

Academic Editor: Orazio Aiello

Received: 26 August 2021
Accepted: 27 September 2021
Published: 28 September 2021

Publisher's Note: MDPI stays neutral with regard to jurisdictional claims in published maps and institutional affiliations.

1. Introduction

It is known that CMOS technology scaling, together with supply voltage reduction, is principally aimed at improving the performance of digital circuits and that, in this framework, the design of analog and mixed-signal blocks becomes increasingly demanding. It is indeed very difficult to obtain high linearity and high precision under near- and sub-threshold supply.

For this reason, operational transconductance amplifiers (OTAs) remain indispensable blocks for the implementation of high-accuracy closed-loop analog circuits, and several techniques have been proposed for the implementation of (ultra) low-voltage solutions. These include subthreshold-operated MOS transistors [1,2], bulk (body) driven [3,4], floating gate and quasi-floating gate MOS transistors [5,6], threshold lowering [7,8], level shifting [9], complementary pairs with body-driven gain boosting, and non-tailed pairs [10]. Additional approaches have also been proposed to replace OTAs, though not for general purpose usage, including dynamic amplifiers [11], ring amplifiers [12], and zero-crossing based circuits [13]. In addition, one interesting trend is the use of inverter-based topologies [14–28]. (A good review of the principal techniques for low-voltage OTAs can be found in the last reference.) At the basis of this approach is the single inverter (CMOS NOT gate), which is topologically simple, as it requires only two transistors between the supply rails, it provides a quite good voltage gain (though multi-stage topologies are usually required for 40 dB or more), and it exhibits class-AB and full swing operation. Therefore, it is rather effective under low supply voltages. However, the main drawback of the inverter-based solutions is related to the difficult control of the quiescent current feature that is especially required in low-power applications with a restricted current budget.

In this paper, a body-biasing technique, originally developed in [29] and utilized in [30], is applied to set the quiescent current of the generic inverter stage. Starting from this generic stage, a gate-driven, two-stage, inverter-based transconductance amplifier, suitable for switched-capacitor applications, is designed. Simulations results are also provided taking into account process and temperature variations. The proposed amplifier is designed in a 28-nm bulk process and is powered by a 0.5 V supply voltage. Typical quiescent current is 488 nA and, with a 1-pF//1.5-MΩ load, it provides 51-dB DC gain with a unity gain frequency of 1 MHz and phase margin of 70 degrees. Settling time at 1% is 6.6 μs and white noise is 125 nV/$\sqrt{Hz}$.

2. The Proposed Solution

Figure 1 shows the circuit schematic of the proposed amplifier. It consists of a first noninverting stage, made up of transistors M_1-M_6, and a second inverting stage, made up of transistors M_7-M_8. As it is seen, the second stage is a straight CMOS NOT gate while the first one is based also onto the NOT topology, but rearranged to invert the gain trough two complementary p-channel and n-channel current mirrors M_3, M_5 and M_4, M_6. In quiescent conditions, the input terminal is set to $V_{DD}/2$ and thanks to the overall negative feedback (not shown) also the output and intermediate node, *out1*, are all biased at $V_{DD}/2$.

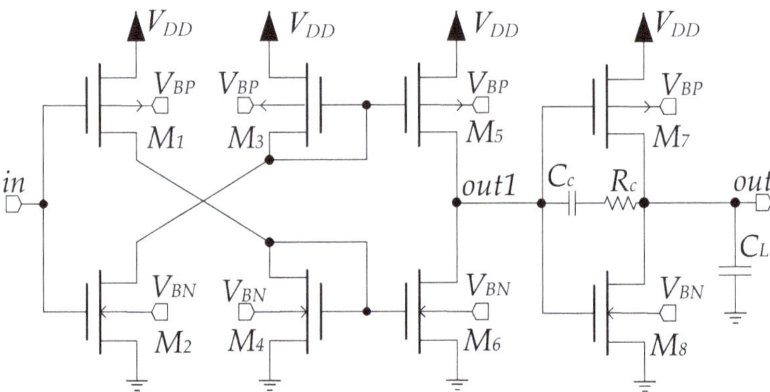

Figure 1. Simplified schematic of the proposed solution.

As far as the quiescent current control of the two stages is concerned, it is implemented through the bulk terminals via voltage V_{BP}, for p-channel transistors, and V_{BN}, for the n-channel ones. These voltages are generated by exploiting a technique proposed in [29] and utilized also in [10,30]. The basic working principle can be inferred with the aid of Figure 2, showing the simplified schematic of the amplifier's biasing section.

M_{R1} and M_{R2} are two reference transistors both with their $|V_{GS}|$ equal to $V_{DD}/2$. Their quiescent drain current is equal to I_{BIAS} thanks to the local feedback loop operated by the auxiliary amplifiers A_1 and A_2, which generate the required bulk voltages, V_{BP} and V_{BN}, under the following summarized constraints:

(a)   assigned aspect ratios $(W/L)_{R1}$ and $(W/L)_{R2}$;
(b) $I_{D1,2} = kI_{BIAS}$, where k is the ratio of the transistors aspect ratio as in (1);
(c) $V_{SGR1} = V_{GSR2} = V_{DD}/2$;
(d) $V_{SDR1} = V_{DSR2} = V_{DD}/2$, assuming ideal input virtual short in A_1 and A_2.

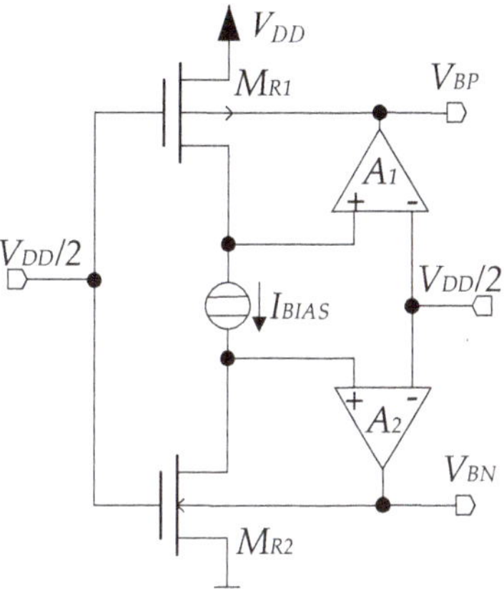

Figure 2. Simplified schematic of the biasing section generating V_{BN} and V_{BP} for the main amplifier in Figure 1.

Of course, aspect ratios of M_{R1} and M_{R2} must be set so that the required bulk voltages are within V_{DD} and ground. Moreover, the auxiliary amplifiers A_1 and A_2 should provide a maximum (rail-to-rail) output voltage range, whereas input common mode range is not a concern as input voltage is kept constant to $V_{DD}/2$. Therefore, simple two-stage OTAs biased in subthreshold can be profitably used. An example of implementation of this type of amplifier is found in [10], albeit operating with MOSFETs in saturation.

Consider now transistor M_1 of the main amplifier in Figure 1 and remember that in quiescent conditions *Vin* is equal to $V_{DD}/2$. As a consequence, M_{R1} and M_1 have respectively the same source, gate, and bulk voltage and hence the drain current of M_1 is related to that of M_{R1} in a mirror-like condition

$$I_{D1} = \frac{(W/L)_1}{(W/L)_{R1}} I_{BIAS} \tag{1}$$

where equality is accurately verified because the source-drain voltage of M_1 is also equal to $V_{DD}/2$, thanks to the diode-connected transistor M_4 in Figure 1 which absorbs I_{D1} and is designed so that

$$\frac{(W/L)_2}{(W/L)_{R2}} = \frac{(W/L)_1}{(W/L)_{R1}} \tag{2}$$

and consequently $V_{GS4} = V_{DD}/2$.

Similar considerations hold for all the transistors in the main amplifier, in practice, all p-channel and n-channel devices have their current linked to I_{BIAS} via the current-mirror-like relations

$$I_{Di\_P} = \frac{(W/L)_{i\_P}}{(W/L)_{R1}} I_{BIAS} \tag{3a}$$

$$I_{Dj\_N} = \frac{(W/L)_{j\_N}}{(W/L)_{R2}} I_{BIAS} \tag{3b}$$

where $(W/L)_{i\_P}$ ($i = 1,3,5,7$) and $(W/L)_{j\_N}$ ($j = 2,4,6,8$) are respectively the aspect ratios of the generic p-channel and n-channel MOSFET in the main amplifier.

As a concluding remark, closed loop stability is ensured thanks to the conventional frequency compensation network made up of the Miller capacitor, C_C, and nulling resistor, R_C, around the last inverting stage.

3. Validation Results

The proposed solution was designed in a 28-nm triple-well CMOS technology provided by STMicroelectronics and simulated at the schematic level. Threshold voltages of the n- and p-channel devices were 445 mV and -462 mV, respectively. Single power supply was set to 0.5 V, I_{BIAS} was 60 nA, and transistor dimensions, together with other component values, were set as summarized in Table 1. All p-channel (n-channel) MOSFETS are equal to the reference device 990/90 (210/90) nm/nm, except for the last stage transistors that have four times greater aspect ratios. This is important to increase the output current drive capability and the output transconductance to reduce the required value of the nulling resistor (to avoid introducing a positive zero), whose value is in the range of $1/g_{m2}$. Observe that the DC gain of the auxiliary amplifiers, A_1 and A_2, is around 40 dB. As a consequence of the transistor's dimension, the nominal quiescent current in each branch of the first stage is 60 nA, while it is 240 nA in the last stage, resulting in a total nominal quiescent current of 420 nA. The small-signal parameters of the amplifier stages are summarized in Table 2. Load capacitor C_L was 1 pF in parallel to a load resistor of 1.5 MΩ, and the compensation capacitor and the nulling resistor were set to 1.5 pF and 50 kΩ, respectively.

Table 1. Design parameters used in simulations.

Parameter	Value
V_{DD}	0.5 V
I_{BIAS}	60 nA
$(W/L)_{R1}, (W/L)_1, (W/L)_3, (W/L)_5$	990/90 nm/nm
$(W/L)_{R2}, (W/L)_2, (W/L)_4, (W/L)_6$	210/90 nm/nm
$(W/L)_7$	$4 \times (990/90)$ nm/nm
$(W/L)_8$	$4 \times (210/90)$ nm/nm
R_C, C_C	50 kΩ, 1.5 pF
A1, A2	40 dB
$C_L//R_L$	1 pF//1.5 MΩ
V_{DD}	0.5 V

Table 2. Small signal parameters of the amplifier.

Parameter	Value
g_{m1}	3.55 μA/V
r_{O1}	7.7 MΩ
g_{m2}	18.12 μA/V
r_{O2}	1.47 MΩ

The robustness of the quiescent conditions were validated at first. The nominal bulk voltages, V_{BP} and V_{BN}, generated by a circuit in Figure 2 were 256.4 mV and 231.9 mV, respectively. The simulated quiescent current in the main amplifier in Figure 1 was 488 nA, on average, with a standard deviation of 93.7 nA, after running 1000 Monte Carlo iterations. The difference with respect to the expected value of 420 nA is due to the low DC gains of the auxiliary amplifiers, which cause a closed-loop gain error.

Figure 3 shows the Bode plots (magnitude and phase) of the amplifier open-loop gain at the standard temperature (27 $^\circ$C) and nominal component models with a 1-pF and 1.5-MΩ parallel load. DC gain is 51 dB, unity gain frequency (UGF) is 1 MHz and phase margin (PM) is 70 degrees. Note that the load resistance is almost equal to r_{o2} in Table 2, hence causing a 6-dB reduction in the maximum achievable gain.

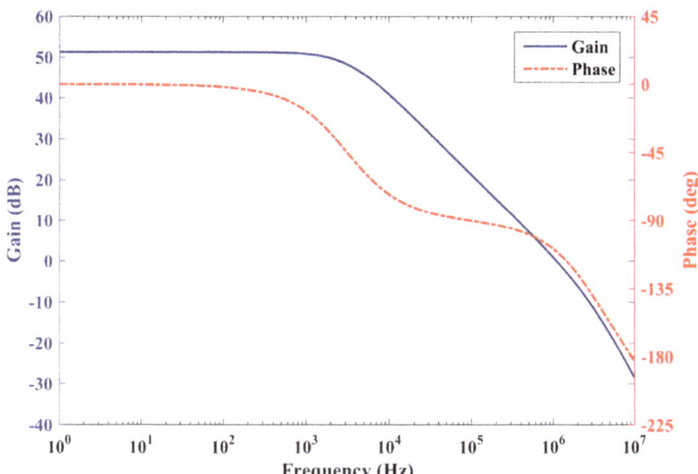

Figure 3. Bode plots (magnitude and phase versus frequency) of the amplifier open-loop gain with 1-pF and 15-MΩ parallel load.

Figure 4 shows the time transient response of the amplifier with the closed-loop gain set to −2. These plots are achieved with two feedback resistors, as in an inverting closed-loop amplifier topology, one of 1 MΩ (connected between the input and output) and the other of 0.5 MΩ (connected between the signal source and the input). The almost rail-to-rail output behavior is apparent. Positive/negative settling time at 1% of the final value is symmetrical and equal to 6.6 μs.

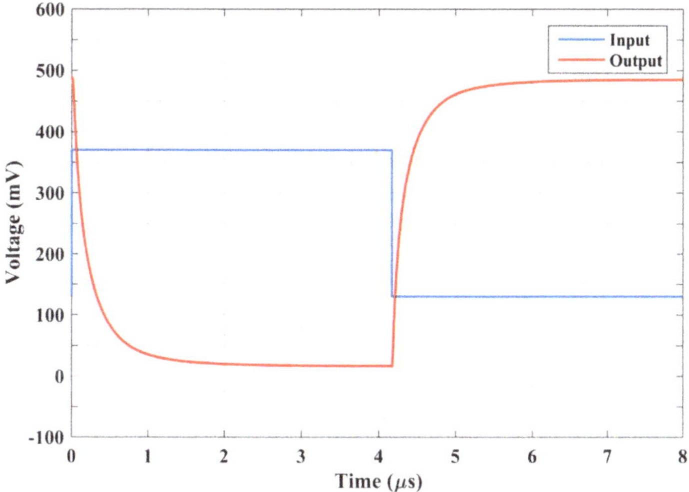

Figure 4. Time response to a 240-mV$_{\text{p-p}}$ input step (closed-loop gain is set to −2).

Power Supply Rejection Ratio was also evaluated from both supply rails. Magnitude versus frequency of PSRR is shown in Figure 5. PSRR$^+$ was 56 dB at DC, while PSRR$^-$ was 58 dB. Equivalent input noise is also simulated and depicted in Figure 6. The white component is 125 nV/$\sqrt{Hz}$ and is dominated by the voltage noise of transistors M1–M6 forming the input stage.

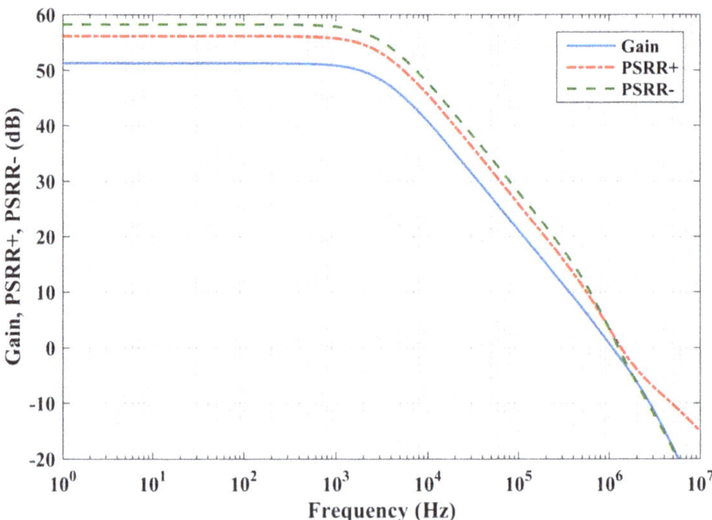

Figure 5. Magnitude versus frequency of the Power Supply Rejection Ratio (PSRR) from positive (PSRR$^+$) and negative (PSRR$^-$) supply rail. Open loop gain is also shown.

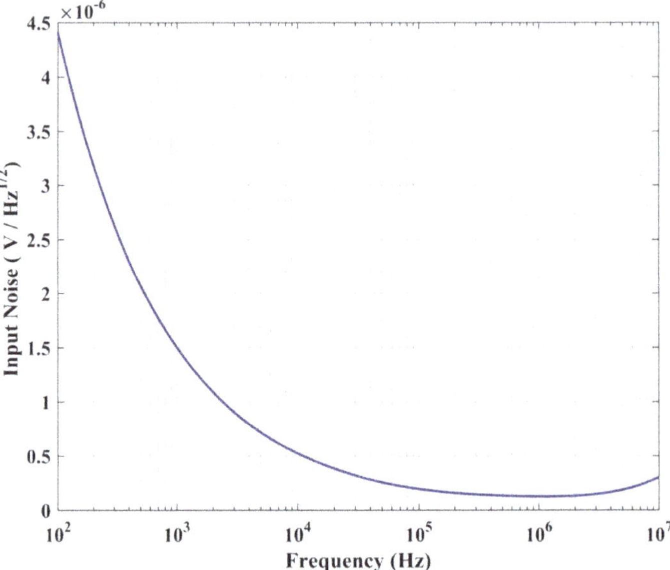

Figure 6. Equivalent input noise voltage spectral density.

The effect of mismatches was also simulated through 1000 Monte Carlo iterations. Table 3 summarizes the results. The largest variation is experienced by the unity gain frequency and settling times (more than 30%).

Table 3. Statistical analysis of main performance parameters due to mismatches (1000 Monte Carlo iterations).

Parameter	μ	σ	σ/μ
V_{out} (mV)	250.1	11.8	4.7%
I_{DD} (nA)	488.1	93.7	19.2%
DC Gain (dB)	51.3	0.56	1.1%
UGF (MHz)	1.13	0.34	30.1%
PM (degrees)	68.9	5.2	7.5%
PSRR$^+$ (dB)	56.1	0.56	1%
PSRR$^-$ (dB)	58.2	0.56	0.9%
1% Ts$^+$/Ts$^-$ (ns) [1]	522/348	206/135	39.5/38.8%

[1] with 100-mV$_{p-p}$ input and in inverting unity gain configuration.

Temperature and process variations were also evaluated via corner simulations under three different temperatures (-20 °C, $+27$ °C and $+85$ °C). Results are summarized in Table 4. It is seen that the quiescent current is sensitive to temperature and to FF and SS corners. In particular, the total amplifier nominal current (which was approximately 488 nA) decreases to 249 nA at -20 °C, SS corner, and increases to 2.4 μA at $+80$ °C, FF corner. DC gain, PM and PSRR exhibit only quite negligible changes, whereas UGF and settling time are affected by these standby current variations. This problem is mainly related to the large threshold voltage excursion induced by temperature variation that cannot be counteracted by the restricted range of the bulk control voltages limited to V_{DD}.

Table 4. Corner simulations (Typical, Fast-Fast, Fast-Slow, Slow-Fast, and Slow-Slow) under three different operating temperatures.

Corner T = -20 °C	TT	FF	FS	SF	SS
V_{out} (mV)	244.4	248.6	229.7	264.7	249.3
I_{DD} (nA)	256	475	243	227	104
DC Gain (dB)	49.8	52	50	50.4	47.4
UGF (MHz)	0.67	1.58	0.63	0.59	0.22
PM (degrees)	69.2	64.6	69	69.9	76.9
PSRR+ (dB)	54.4	56.7	54.7	55.2	52
PSRR- (dB)	56.9	58.9	57.1	57.4	54.7
1% Ts+/Ts- (ns)	685/438	272/182	566/337	854/336	2632/880
Corner T = 27 °C	**TT**	**FF**	**FS**	**SF**	**SS**
V_{out} (mV)	249.9	244.5	249.3	249.9	250
I_{DD} (nA)	488	579	505	479	485
DC Gain (dB)	51.3	51.7	52.1	50.4	51
UGF (MHz)	1.09	1.68	1.13	1.08	0.88
PM (degrees)	69.2	64.3	69.5	69	73.1
PSRR+ (dB)	56.1	56.7	56.9	55.2	55.7
PSRR- (dB)	58.2	58.5	59.8	58.9	58.3
1% Ts+/Ts- (ns)	520/319	240/207	506/322	519/321	719/490
Corner T = 80 °C	**TT**	**FF**	**FS**	**SF**	**SS**
V_{out} (mV)	255.8	249.1	233.9	277.1	259.8
I_{DD} (nA)	1177	2417	1338	1061	621
DC Gain (dB)	52.6	53	53	52.1	51.5
UGF (MHz)	2.2	5.68	2.5	1.98	0.97
PM (degrees)	74.3	77.3	75.8	73.3	73.4
PSRR+ (dB)	57.4	57.7	57.9	56.7	56.3
PSRR- (dB)	59.4	59.7	59.8	58.9	58.4
1% Ts+/Ts- (ns)	356/235	130/123	230/239	233/210	838/436

4. Conclusions

A novel inverter-based two-stage CMOS transconductance amplifier, with quiescent current control and suitable for very low supply voltages was presented. The solution consists of the cascade of a noninverting and an inverting stage both characterized by having only two complementary transistors between the supply rails, thus providing rail-to-rail and class-AB output capability. The designed solution is supplied from 0.5 V and in quiescent conditions consumes (typically) approximately 488 nA, while providing a DC gain of approximately 51 dB, with a unity gain frequency of 1 MHz and phase margin of 70 degrees, for a 1-pF//1.5-MΩ load.

The quiescent current control loop proved to be effective against mismatches and process variations. Further investigation is currently being carried out to reduce the quiescent current sensitivity to temperature. This drawback is caused by the limited variation allowed to the body biasing control voltage, which is of course restricted to V_{DD} and ground. Once V_{BP} and V_{BN} reach these limits and saturate, the control loop becomes ineffective. For this reason, making I_{BIAS} with a coefficient negative to absolute temperature (NTAT) could be a favorable solution and subject for further study.

Author Contributions: Conceptualization: S.P. and G.S.; data curation: A.B.; original draft preparation: S.P. and A.B.; writing—review and editing: all authors; supervision: S.P. and G.S. All authors have read and agreed to the published version of the manuscript.

Funding: This research received no external funding.

Data Availability Statement: The data presented in this study are available in article.

Conflicts of Interest: The authors declare no conflict of interest.

References

1. Vittoz, E.; Fellrath, J. CMOS analog integrated circuits based on weak inversion operations. *IEEE J. Solid-State Circuits* **1977**, *12*, 224–231. [CrossRef]
2. Kumar, A.R.A.; Sahoo, B.D.; Dutta, A. A Wideband 2–5 GHz Noise Canceling Subthreshold Low Noise Amplifier. *IEEE Trans. Circuits Syst. II Express Briefs* **2018**, *65*, 834–838. [CrossRef]
3. Zuo, L.; Islam, S.K. Low-Voltage Bulk-Driven Operational Amplifier With Improved Transconductance. *IEEE Trans. Circuits Syst. I Regul. Pap.* **2013**, *60*, 2084–2091. [CrossRef]
4. Akbari, M.; Hussein, S.M.; Hashim, Y.; Tang, K.-T. An Enhanced Input Differential Pair for Low-Voltage Bulk-Driven Amplifiers. *IEEE Trans. Very Large Scale Integr. (VLSI) Syst.* **2021**, *29*, 1601–1611. [CrossRef]
5. Ramirez-Angulo, J.; Lopez-Martin, A.; Carvajal, R.; Chavero, F. Very low-voltage analog signal processing based on quasi-floating gate transistors. *IEEE J. Solid-State Circuits* **2004**, *39*, 434–442. [CrossRef]
6. Miguel, J.M.A.; Lopez-Martin, A.J.; Acosta, L.; Ramirez-Angulo, J.; Carvajal, R.G. Using Floating Gate and Quasi-Floating Gate Techniques for Rail-to-Rail Tunable CMOS Transconductor Design. *IEEE Trans. Circuits Syst. I Regul. Pap.* **2011**, *58*, 1604–1614. [CrossRef]
7. Lehmann, T.; Cassia, M. 1-V power supply CMOS cascode amplifier. *IEEE J. Solid-State Circuits* **2001**, *36*, 1082–1086. [CrossRef]
8. Chatterjee, S.; Tsividis, Y.; Kinget, P. 0.5-V analog circuit techniques and their application in OTA and filter design. *IEEE J. Solid-State Circuits* **2005**, *40*, 2373–2387. [CrossRef]
9. Carrillo, J.; Duque-Carrillo, J.; Torelli, G.; Ausin, J. Constant-gm constant-slew-rate high-bandwidth low-voltage rail-to-rail CMOS input stage for VLSI cell libraries. *IEEE J. Solid-State Circuits* **2003**, *38*, 1364–1372. [CrossRef]
10. Grasso, A.D.; Pennisi, S.; Scotti, G.; Trifiletti, A. 0.9-V Class-AB Miller OTA in 0.35-µm CMOS With Threshold-Lowered Non-Tailed Differential Pair. *IEEE Trans. Circuits Syst. I Regul. Pap.* **2017**, *64*, 1740–1747. [CrossRef]
11. Lin, J.; Paik, D.; Lee, S.; Miyahara, M.; Matsuzawa, A. An Ultra-Low-Voltage 160 MS/s 7 Bit Interpolated Pipeline ADC Using Dynamic Amplifiers. *IEEE J. Solid-State Circuits* **2015**, *50*, 1399–1411. [CrossRef]
12. Lim, Y.; Flynn, M.P. A 100 MS/s, 10.5 Bit, 2.46 mW Comparator-Less Pipeline ADC Using Self-Biased Ring Amplifiers. *IEEE J. Solid-State Circuits* **2015**, *50*, 2331–2341. [CrossRef]
13. Lee, S.; Chandrakasan, A.P.; Lee, H.S. A 12 b 5-to-50 MS/s 0.5-to-1 V Voltage Scalable Zero-Crossing Based Pipelined ADC. *IEEE J. Solid-State Circuits* **2012**, *47*, 1603–1614. [CrossRef]
14. Michel, F.; Steyaert, M.S.J. A 250 mV 7.5 W 61 dB SNDR SC $\Delta\Sigma$ Modulator Using Near-Threshold-Voltage-Biased Inverter Amplifiers in 130 nm CMOS. *IEEE J. Solid-State Circuits* **2012**, *47*, 709–721. [CrossRef]
15. Yaul, F.M.; Chandrakasan, A.P. A Noise-Efficient 36 nV/Hz Chopper Amplifier Using an Inverter-Based 0.2-V Supply Input Stage. *IEEE J. Solid-State Circuits* **2017**, *52*, 3032–3042. [CrossRef]

16. Wang, P.; Ytterdal, T. A 54-μW Inverter-Based Low-Noise Single-Ended to Differential VGA for Second Harmonic Ultrasound Probes in 65-nm CMOS. *IEEE Trans. Circuits Syst. II Express Briefs* **2016**, *63*, 623–627. [CrossRef]
17. Ng, K.A.; Xu, Y.P. A Low-Power, High CMRR Neural Amplifier System Employing CMOS Inverter-Based OTAs With CMFB Through Supply Rails. *IEEE J. Solid-State Circuits* **2016**, *51*, 724–737.
18. Crovetti, P.S. A Digital-Based Virtual Voltage Reference. *IEEE Trans. Circuits Syst. I Regul. Pap.* **2015**, *62*, 1315–1324. [CrossRef]
19. Aiello, O.; Crovetti, P.S.; Alioto, M. Fully Synthesizable Low-Area Digital-to-Analog Converter With Graceful Degradation and Dynamic Power-Resolution Scaling. *IEEE Trans. Circuits Syst. I Regul. Pap.* **2019**, *66*, 2865–2875. [CrossRef]
20. Ismail, A.; Mostafa, I. A Process-Tolerant, Low-Voltage, Inverter-Based OTA for Continuous-Time $\Sigma\Delta$ ADC. *IEEE Trans. Very Large Scale Integr. (VLSI) Syst.* **2016**, *24*, 2911–2917. [CrossRef]
21. Guo, Y.; Jin, J.; Liu, X.; Zhou, J. An Inverter-Based Continuous Time Sigma Delta ADC With Latency-Free DAC Calibration. *IEEE Trans. Circuits Syst. I Regul. Pap.* **2020**, *67*, 3630–3642. [CrossRef]
22. Toledo, P.; Crovetti, P.; Klimach, H.; Bampi, S.; Aiello, O.; Alioto, M. 300 mV-Supply, sub-nW-Power Digital-Based Operational Transconductance Amplifier. *IEEE Trans. Circuits Syst. II Express Briefs* **2021**, *5*, 1.
23. Rodovalho, L.H.; Aiello, O.; Rodrigues, C.R. Ultra-Low-Voltage Inverter-Based Operational Transconductance Amplifiers with Voltage Gain Enhancement by Improved Composite Transistors. *Electronics* **2020**, *9*, 1410. [CrossRef]
24. Rodovalho, L.H.; Ramos Rodrigues, C.; Aiello, O. Self-Biased and Supply-Voltage Scalable Inverter-Based Operational Transconductance Amplifier with Improved Composite Transistors. *Electronics* **2021**, *10*, 935. [CrossRef]
25. Bae, W. CMOS inverter as analog circuit: An overview. *J. Low Power Electron. Appl.* **2019**, *9*, 26. [CrossRef]
26. Palani, R.K.; Harjani, R. *Inverter-Based Circuit Design Techniques for Low Supply Voltages*; Springer: Berlin/Heidelberg, Germany, 2018.
27. Zheng, K. An Inverter-Based Analog Front-End for a 56-Gb/s PAM-4 Wireline Transceiver in 16-nm CMOS. *IEEE Solid-State Circuits Lett.* **2018**, *12*, 249–252. [CrossRef]
28. Zhou, X.; Qiao, Z.; Li, Q. Inverter-Based Subthreshold Amplifier Techniques and Their Application in 0.3-V $\Delta\Sigma$-Modulators. *IEEE J. Solid-State Circuits* **2019**, *54*, 1436–1445.
29. Monsurró, P.; Scotti, G.; Trifiletti, A.; Pennisi, S. Biasing technique via bulk terminal for minimum supply CMOS amplifiers. *Electron. Lett.* **2005**, *41*, 779–780. [CrossRef]
30. Monsurró, P.; Pennisi, S.; Scotti, G.; Trifiletti, A. Exploiting the Body of MOS Devices for High Performance Analog Design. *IEEE Circuits Syst. Mag.* **2011**, *11*, 8–23. [CrossRef]

Journal of
*Low Power Electronics
and Applications*

Article

A 1.9 nW, Sub-1 V, 542 pA/V linear Bulk-Driven OTA with 154 dB CMRR for Bio-Sensing Applications

Rafael Sanchotene Silva [1,*], Luis Henrique Rodovalho [1], Orazio Aiello [2], Cesar Ramos Rodrigues [1]

[1]   Biomedical Engineering Institute, Federal University of Santa Catarina (IEB-UFSC),
      Florianópolis 88040-900, Brazil; luis.henrique.rodovalho@posgrad.ufsc.br (L.H.R.); cesar@ieee.org (C.R.R.)
[2]   Department of Electrical and Computer Engineering, National University of Singapore (NUS),
      Singapore 117583, Singapore; orazio.aiello@ieee.org
*   Correspondence: rafael.sanchotene@posgrad.ufsc.br; Tel.: +55-48-3721-8686

Abstract: In this paper, a new technique for improvement on the DC voltage gain, while keeping the high-linearity in symmetrical operational transconductance amplifier (OTA) bulk-driven (BD) topology is proposed. These features are achieved by allying two topological solutions: enhanced forward-body-biasing self-cascode current mirror, and source degeneration. The proposed concept is demonstrated through simulations with typical process parameters and Monte Carlo analysis on nominal transistors of the CMOS TSMC 180 nm node. Results indicate that the proposed OTA can achieve a very small transconductance, only 542 pA/V while keeping a voltage gain higher than 60 dB, 150 dB CMRR, and high linearity of 475 mVpp (1% THD), consuming only 1.9 nW for a supply voltage of 0.6 V. This set of features allows the proposed OTA to be an attractive solution for implementing OTA-C filters for the analog front-ends in wearable devices and bio-sensing.

Keywords: bulk-driven OTA; transconductor; self-cascode mirror

Citation: Sanchotene Silva, R.; Rodovalho, L.H.; Aiello, O.; Ramos Rodrigues, C. A 1.9 nW, Sub-1 V, 542 pA/V linear Bulk-Driven OTA with 154 dB CMRR for Bio-Sensing Applications. *J. Low Power Electron. Appl.* **2021**, *11*, 40. https://doi.org/10.3390/jlpea11040040

Academic Editor: Stylianos D. Assimonis

Received: 1 August 2021
Accepted: 17 October 2021
Published: 20 October 2021

Publisher's Note: MDPI stays neutral with regard to jurisdictional claims in published maps and institutional affiliations.

1. Introduction

The effort to develop implantable or bio-sensing battery-less biomedical instrumentation systems has been continuously challenging analog designers because of the intensified constraints arising from CMOS scaling [1–3]. Topological solutions for endowing operational transconductance amplifiers (OTAs) to process μV signals with common-mode swings in the range of tens of volts, allied to features like ultra-low power consumption, low-noise, enhanced linearity, high common-mode rejection ratio (CMRR), tiny silicon footprint, and large common-mode range (CMR) are frequently pursued by the analog circuit designers [4–18].

As a basic block in analog front-ends (AFEs) for biosensing, the OTA-C filter with large time constants is among the most important applications for OTAs with reduced transconductance [19]. Such circuits when used in implantable/wearable biomedical applications have their design challenged by the restricted-sized on-chip integrated capacitors. In order to decrease the size of such filters, OTAs must output a very small transconductance in the order of a few nA/V, which is achieved with very low biasing currents [20] at the cost of the OTA linearity.

Among the typical OTA design techniques to increase linearity is the use of non-unity gain current mirrors [21–23] to allow higher biasing currents and maintain a low transconductance. Another well-known technique that is used to improve both OTA linearity and input signal voltage swing is the bulk-driven differential pair [1,24–29]. Unlike the gate-driven OTA topologies, the bulk-driven OTAs outputs are an alternative for a relatively lower transconductance [20,30]. In this case, the main drawback of this approach is a poor DC voltage gain, which can be improved by using several techniques [25]. An interesting and widely employed technique relies on a self-cascode topology known as trapezoidal or composite transistor [31–34]. Additionally, an improvement for the self-cascode transistor association was proposed in [13,35–37] allowing to increase voltage gain and decrease area usage. Therefore, in this paper, we propose a new symmetrical bulk-driven OTA topology that takes the advantages of previously described techniques, i.e., the combination of the topology presented by [23], with a bulk-driven differential pair [24], and the bulk-driven active source degeneration linearization technique adapted from [1,38]. Besides the employed combination of techniques in the OTA topology, we propose an innovative improved self-cascode current mirror (ISCCM) which is based on [35,37].

This paper is organized as follows: Section 2 describes current mirror topologies made of rectangular transistor arrays (composite transistor). The improved self-cascode current mirror that sources the proposed OTA is introduced. Section 3 presents the bulk-driven symmetrical OTAs topologies. Simulations and comparisons among the proposed BD topology, the conventional bulk-driven, and state-of-art transconductors are shown in Section 4. Finally, Section 5 presents the conclusions.

2. Current Mirrors

Current mirrors are the essential component of CMOS OTAs, and their output impedance improvement leads to OTAs with superior voltage gain and common-mode rejection. Implementing current mirrors with series-parallel associations of transistors is a design solution that allows for high current gain, reduced area usage, and less process variability compared to parallel-only current mirrors [22]. This technique was employed by [23] to achieve very low transconductance OTAs without sacrificing linearity and process variability tolerance [13,14]. Since the output transistor array has a large equivalent channel length (L_{eq}), the output current I_o is less dependent on the output voltage V_{out} variation.

Rectangular transistor arrays as illustrated in Figure 1 can be considered understood as a single transistor [31] with a higher output impedance [2,23,39]. The rectangular array, shown in Figure 1, is an m by n matrix of single transistors composed of m parallel columns of n series single transistors. The rectangular equivalent transistor aspect ratio S_{eq-R} is a function of the single transistor aspect ratio S_u, as shown in Equation (1). The total gate area of the rectangular array is $A_T = (mn)A_u$, where A_u is the gate area of the single transistor.

$$S_{eq-R} = \frac{W_{eq}}{L_{eq}} = \frac{mW_u}{nL_u} = \frac{m}{n}S_u.$$ (1)

Figure 2 represents an N-type improved composite transistor. It consists of a series connection of two independently forward-body-biased N-type MOS transistors M_{N1} and M_{N2}, as first proposed in [35], and described in detail in [13,14,37], by using the ACM (advanced compact model) all-region transistor model [40].

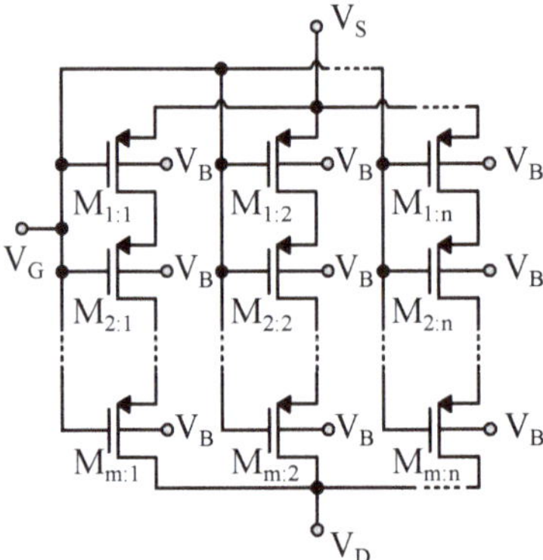

Figure 1. Rectangular $1 \times m : n$ transistor array.

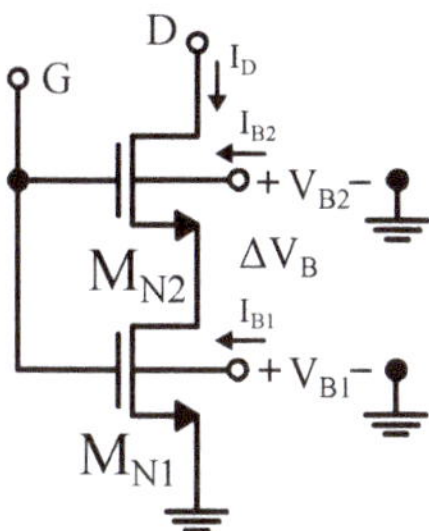

Figure 2. Improved composite transistor.

The improved composite transistor equivalent aspect ratio S_{eq} is defined as:

$$S_{eq} = \frac{S_{N1} \cdot \beta S_{N2}}{S_{N1} + \beta S_{N2}} = \frac{\beta k}{\beta k + 1} \cdot S_{N1} \tag{2}$$

where

$$\beta \approx e^{\frac{(n-1)\Delta V_B}{n\phi_t}} \tag{3}$$

represents a correction factor for the current drain I_D definition due to the difference between the body-bias of the series transistors M_{N2} and M_{N1} $\Delta V_B = V_{B2} - V_{B1}$, assuming the transistors are operating in weak inversion, and

$$k = \frac{S_{N2}}{S_{N1}} \tag{4}$$

is the ratio between transistors M_{N1} and M_{N2} and physical aspect ratios S_{N1} and S_{N2}. Figure 3a shows the conventional current mirror (CM). The ratio between transistors M_{1B}, and M_{1A} aspect ratios S_{1A} and S_{1B} define the current mirror gain $A_I = S_{1B}/S_{1A}$ and attenuation $1/A_I$. In order to have a better matching for non-unity current gain, the current mirror transistors should be replaced with rectangular transistor arrays [22].

A higher current attenuation is achieved by combining parallel transistor arrays at the current mirror input, and series transistor arrays at the output. This scheme is a desirable feature for ultra-low transconductance OTAs [21,23], as it provides transconductance attenuation without decreasing linearity.

The typical cascode current mirror is a variation of the Wilson current mirror first proposed by [41]. The topology increases the output impedance in order to decrease the output current gain error. On the other hand, its drawback is a lower output voltage swing, which will be solved by the proposed current mirror as follows.

An alternative topology to a typical cascode, is the self-biased self-cascode current mirror (SCCM), first proposed by [42], which uses composite transistor arrays in a trapezoidal shape, which are equivalent to single transistors with increased output impedance. The trapezoidal geometry means that the top composite transistors, i.e., those related to drain portion must have a greater aspect ratio than the bottom transistors, i.e., corresponding to source portion, so this kind of composite transistor can be made by arranging their drain transistors in an array connected to a series array corresponding to source transistors (the smaller base of the trapezoid) [43]. This topology is recommended for low input currents and unity current mirror gain, but it is not appropriate for higher currents or very large current gains, since it would require a very large area. Nevertheless, the trapezoidal current mirror can still use the parallel-series technique for current attenuation [21,22] by replacing the output series transistor array with a trapezoidal transistor array, as shown in Figure 3b. This is possible because there is no need for trapezoidal arrays at the mirror input for non-unity gains.

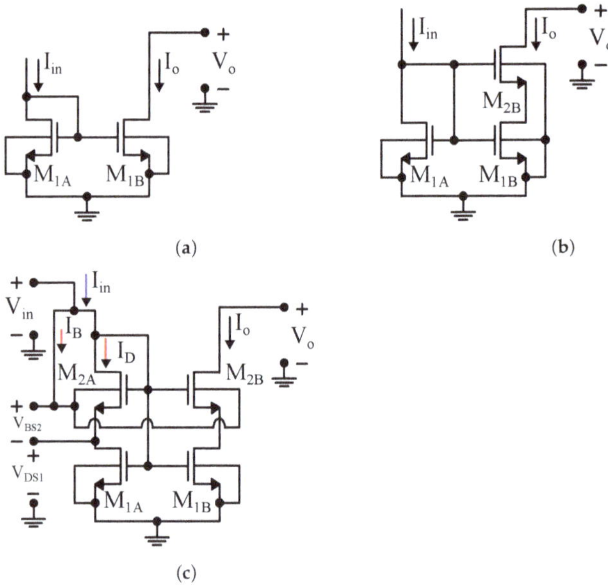

Figure 3. Self-biased current mirrors: (**a**) conventional current mirror with rectangular transistor arrays (CM) [22], (**b**) trapezoidal output current mirror (SCCM) [42], and (**c**) improved self-cascode current mirror (ISCCM).

By taking (2), $\beta = 1$, and since M_{1B} and M_{2B} bulk terminals are connected to each other, the current gain A_I can be expressed as

$$A_I = \left(\frac{S_{2B}}{S_{1B} + S_{2B}} \right) \cdot \frac{S_{1B}}{S_{1A}} \tag{5}$$

For $S_{2B} \gg S_{1B}$, the SCCM current gain is approximately S_{1B}/S_{1A}, as in the conventional parallel-series current mirror. However, this current mirror has a relatively larger output resistance, consequently, it is more tolerant to output voltage variation.

The SCCM output resistance can be further increased by independently forward-body-biasing transistors M_{1B} and M_{2B} by connecting their shared gate terminals to their shared bulk-terminals [37], as shown in Figure 3c. In its turn, the k factor is increased by a β factor function of the bulk-to-source voltage V_{BS2}, accordingly to (3), and hence the gain of the current mirror, A_I is defined as

$$A_I = \frac{S_{1B} \cdot \beta S_{2B}}{S_{1B} + \beta S_{2B}} \cdot \frac{S_{1A} + \beta S_{2A}}{S_{1A} \cdot \beta S_{2A}} = \left(\frac{S_{1A} + \beta S_{2A}}{S_{1B} + \beta S_{2B}} \cdot \frac{S_{2B}}{S_{2A}} \right) \cdot \frac{S_{1B}}{S_{1A}} \tag{6}$$

Again, considering a high value of β, the current gain A_I is approximately S_{1B}/S_{1A}.

For proof of concept, the above current mirrors were designed for the TSMC 180 nm technology and simulated for typical process parameters and room temperature. Table 1 summarizes the transistor arrays dimensions for each circuit.

First, by considering a fixed 1.6 nA input current I_{in} and an output voltage V_o sweeping from 0 to 600 mV, Figure 4a shows the output current mirrors. According to the transistor arrays dimensions, the conventional rectangular parallel-series current mirror (CM) should attenuate the input current by a 16× factor, and provide a 100 pA current. However, due to non-ideal behavior, it outputs about 125 nA, which is close to 13× attenuation. The self-cascode current mirror (SCCM) behaves similarly to CM, as $S_{2B} = 16 \times S_{1B}$. The improved self-cascode current mirror (ISCCM) has a slightly smaller attenuation, close to 12×. The main difference between these current mirrors is the output resistance $R_o = 1/(dI_o/dV_o)$, shown in Figure 4b. At the saturation region, the SCCM R_o is much higher than CM, while the ISCCM is more than one order of magnitude higher.

Table 1. Transistor sizes (Figure 3).

Mirror		Size (m × (M:N) × W/L)		Size (m × (M:N) × W/L)
RCM	M_{1A}	4 × (4:1) × 1.0 μm/8.0 μm	M_{1B}	4 × (1:4) × 1.0 μm/8.0 μm
TCM	M_{1A}	4 × (4:1) × 1.0 μm/8.0 μm	M_{1B}	4 × (1:4) × 1.0 μm/8.0 μm
	M_{2B}	4 × (4:1) × 1.0 μm/8.0 μm		
ISCCM	M_{1A}	4 × (4:1) × 1.0 μm/8.0 μm	M_{1B}	4 × (1:4) × 1.0 μm/8.0 μm
	M_{2A}	4 × (4:1) × 1.0 μm/8.0 μm	M_{2B}	4 × (4:1) × 1.0 μm/8.0 μm

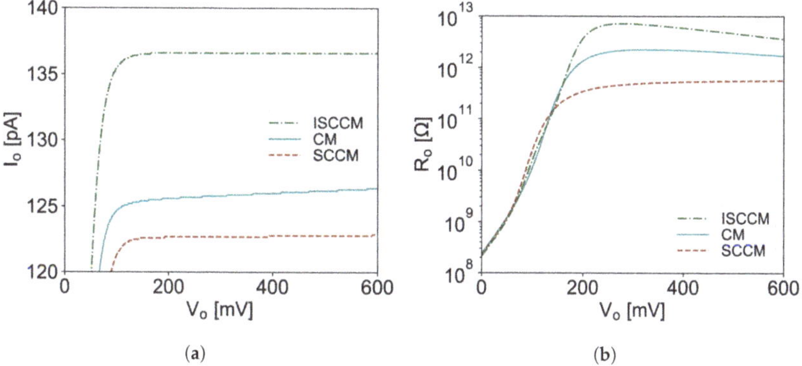

(a)   (b)

Figure 4. Current mirrors comparison: (**a**) output current × output voltage, and (**b**) output resistance × output voltage for I_{in} = 1.6 nA.

Nonetheless, the ISCCM is not perfect. Figure 5 shows the current attenuation $1/A_I$ as a function of the input current I_{in}. As can be seen, the current attenuation is practically constant for the CM and SCCM, but it varies for the ISCCM, as the β is indirectly a function of the input current.

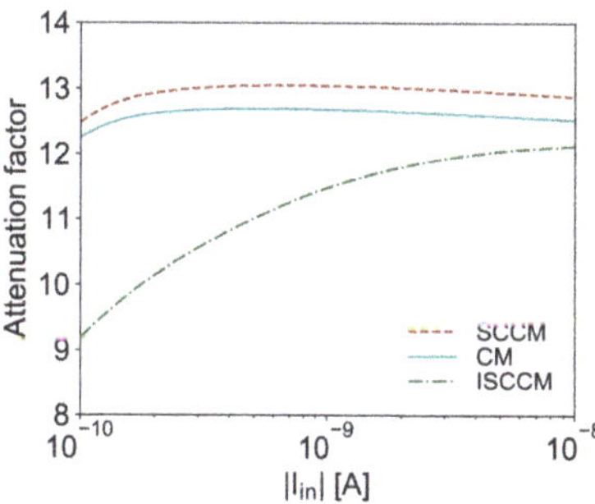

Figure 5. Current mirrors attenuation as a function of input current for V_{out} = 0.3 V.

The ISCCM differential bulk voltage is defined as $\Delta V_B = V_{BS2} = V_{in} - V_{DS1}$. As the input current I_{in} increases exponentially, V_{in} increases linearly, as shown in Figure 6a. For 1 nA input, ΔV_B is approximately 100 mV. As V_{BS2} is always positive, the transistors M_{2A} and M_{2B} are forward-body-biased. In spite of that, the drain current I_D is orders of magnitude higher than I_B (see Figure 6b), so $I_{in} \approx I_D$.

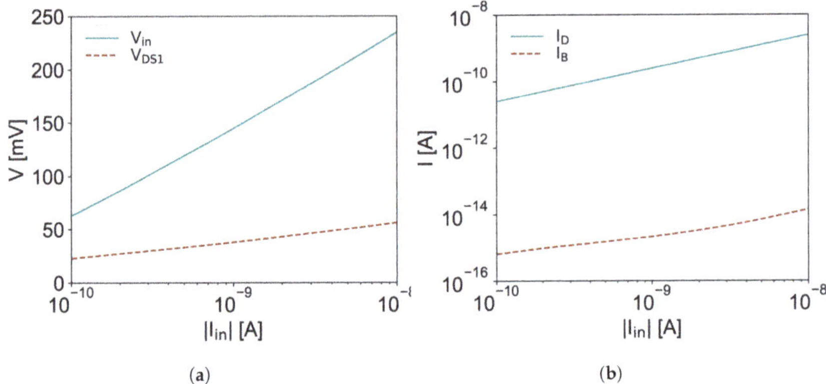

Figure 6. Improved self-cascode current mirror: (**a**) V_{in} and V_{DS1} voltages × input current, and (**b**) drain (I_D) and bulk currents (I_B) × input current.

3. Bulk-Driven Symmetrical Operational Transconductance Amplifiers

Bulk-driven OTA topologies as illustrated in Figure 7 take advantage of the transistor bulk terminal of the differential pair to achieve higher transconductance linearity and input range rather than conventional gate-driven topologies [25,28].

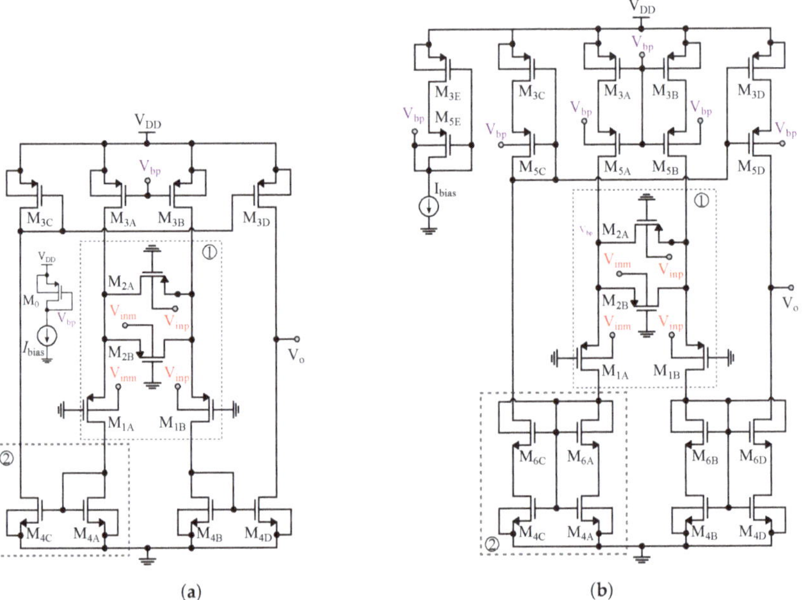

Figure 7. Symmetrical OTA topologies: (**a**) with parallel-series current mirrors [21], bulk-driven inputs and active source degeneration [1,38], and (**b**) proposed topology with the addition of improved current mirrors.

The intrinsic drawback of this scheme is the reduced transconductance due to its equivalent gate-driven OTA, hence, a lower DC voltage gain. Nonetheless, biomedical applications frequently involve slow varying quantities and the supposed disadvantage, i.e., the very-low transconductance turns beneficial as analog signal filters with very low cut-off frequencies using relatively small-sized integrated capacitors are essential. Moreover, the lower voltage gain can be addressed with techniques such as positive feedback [25], cascode gain stages [44], and transistor arrays [31,45].

Figure 7b shows the proposed topology which relies on the conventional symmetrical OTA shown in Figure 7a with a key aspect. The current mirrors are built by improved self-cascode configuration [36], according to Figure 3c. This scheme allows increasing the OTA DC voltage gain as also the CMRR. Further, in this work, the conventional BD-OTA (see Figure 7b) makes use of the same active source degeneration technique [1,38,46] employed in the input differential pair to keep fair comparisons between the topologies.

To describe the topology behavior, we use the ACM transistor model (more details in [47]), hence the BD-OTA topology can be explained as follows: the transconductance G_{mB} is a function of the differential pair transconductance g_{mb1}, the source degeneration factor a [46], and the current mirror factor N, as defined by (7). The differential pair transconductance g_{mb1}, defined by (8) and is attenuated relative to the gate-driven OTA by a factor of $(n-1)$.

$$G_{mB} = \frac{g_{mb1}}{aN},\tag{7}$$

$$g_{mb1} \approx \frac{n-1}{n}g_{ms1} \approx \frac{n-1}{n}\frac{2I_S}{\phi_t}\left(\sqrt{1+i_f}-1\right).\tag{8}$$

Another advantage of the bulk-driven topology over the gate-driven approach is the reduced minimum supply voltage needed for operation,since the differential pair transistors M_{1A-B} and source degeneration transistors M_{2A-B} gate terminals are connected to the ground instead of to the input signal voltages, which has a typical common-mode voltage of half the supply voltage. It is worth noting that, in order to M_{1A-B} operate in the saturation region, V_{GS1} should be greater than the sum of V_{GS4} and V_{DSAT1}, which is achieved by assuring that i_{f1} is sufficiently greater than i_{f4} [38,46].

The differential pair is composed of the transistors M_{1A-B}, and the active source degeneration transistors M_{2A-B}. The ratio between the differential pair and the source degeneration transistor aspect ratios S_1/S_2 is 4 for the earlier explained reasons and is achieved by using rectangular transistor arrays with the same area. The ratio between the differential pair and the tail current source transistors aspect ratios S_1/S_3 is sixteen, consequently, since the drain currents are the same, the ratio of their forward inversion level i_{f1}/i_{f3} is also 16. The current mirrors use the series-parallel technique [23] to achieve a $16\times$ current attenuation.

As sketched in Figure 7, the conventional BD-OTA, and the proposed one differs because of their body biasing and their current mirror schemes according to Figure 7a,b, respectively. Both the designed conventional and proposed OTAs are composed of the same transistors with the same dimensions. The transistors' sizes of both topologies are summarized in Table 2.

Table 2. Transistor sizes (Figure 7).

Trans.	Size (m $\times$ (M:N) $\times$ W/L)	Trans.	Size (m $\times$ (M:N) $\times$ W/L)
M_{1A-B}	$4 \times (1{:}4) \times 3.0\ \mu m/8.0\ \mu m$	M_{2A-B}	$2 \times (1{:}8) \times 3.0\ \mu m/8.0\ \mu m$
M_{3A-B}	$4 \times (4{:}1) \times 3.0\ \mu m/8.0\ \mu m$	M_{3C-E}	$4 \times (1{:}4) \times 3.0\ \mu m/8.0\ \mu m$
M_{4A-B}	$4 \times (4{:}1) \times 3.0\ \mu m/8.0\ \mu m$	M_{4C-D}	$4 \times (1{:}4) \times 3.0\ \mu m/8.0\ \mu m$
M_{5A-B}	$4 \times (4{:}1) \times 3.0\ \mu m/8.0\ \mu m$	M_{5C-E}	$4 \times (1{:}4) \times 3.0\ \mu m/8.0\ \mu m$
M_{6A-B}	$4 \times (4{:}1) \times 3.0\ \mu m/8.0\ \mu m$	M_{6C-D}	$4 \times (1{:}4) \times 3.0\ \mu m/8.0\ \mu m$

Figure 8 illustrates the layout of the conventional and the proposed BD-OTA. It is possible to observe a very small difference between both topologies, with the tiny occupied area of only $0.00867\ mm^2$ and $0.0143\ mm^2$, for conventional BD-OTA and the proposed BD-OTA, respectively.

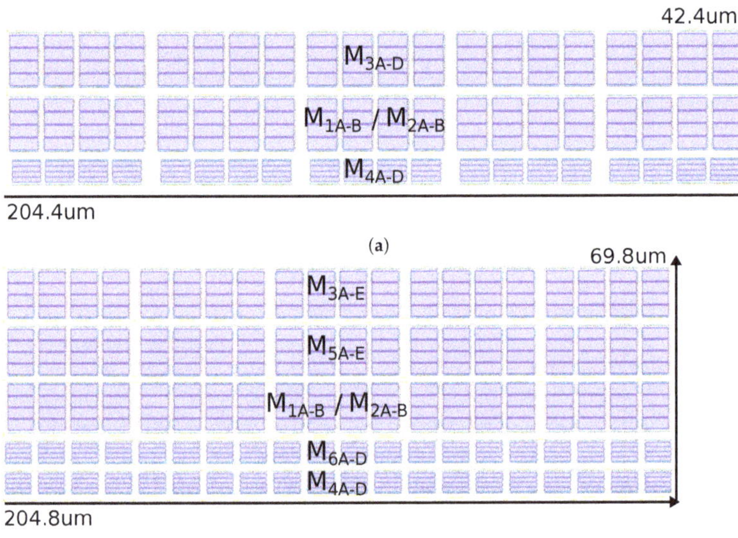

Figure 8. Layout designs of the BD-OTAs. (**a**) Conventional BD-OTA layout. (**b**) Proposed BD-OTA layout.

4. Simulation Results

In this section, the post-layout simulation results referring to a TSMC 180 nm CMOS process for the conventional BD-OTA, and the proposed one, are reported. The circuits are considered to operate under the same conditions, i.e., 27 °C temperature, V_{DD} equal to 0.6 V, I_{bias} equal to 100 pA, besides the typical process parameters. Characteristics from both OTAs were obtained by simulating the four testbenches shown in Figure 9. Figure 9a shows the integrator test bench used in the AC and DC simulations. This scheme allows the evaluation and comparison of DC open-loop gain, as also the gain-bandwidth product (GBW) of each OTA version. Then, Figure 10a shows the open-loop gain AC simulation results, and Figure 10b shows the DC simulation results.

It can be noted that the use of improved mirrors increases DC gain without changing considerably the gain-bandwidth product of the OTA versions using the same differential pair, as they are biased with the same current. As expected, the proposed BD OTA with the enhanced mirror has lower transconductance, while keeping higher gain and the same linearity than the typical BD topology.

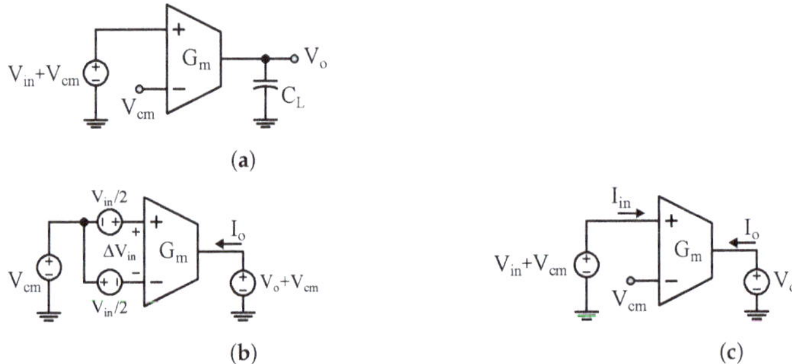

Figure 9. *Cont.*

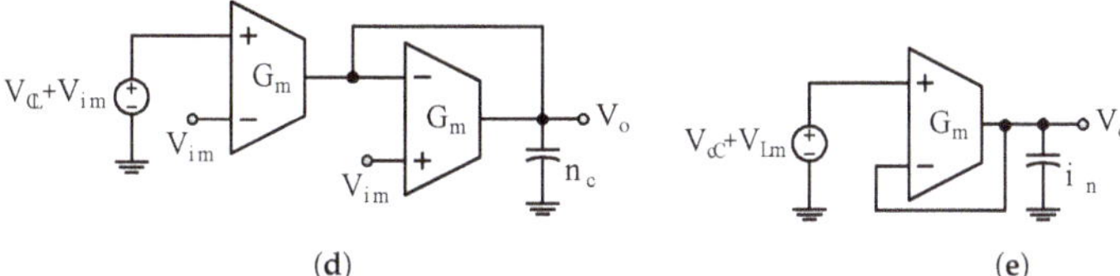

Figure 9. OTA testbenches. (**a**) OTA-C integrator. (**b**) Transconductor (Symmetrical). (**c**) Transconductor (asymmetrical). (**d**) OTA-C low-pass filter. (**e**) Unity gain buffer.

As the power supply rejection ratio (PSRR) is equal to the OTAs DC gain, there is a unity gain voltage between supply and output voltages. The common-mode rejection ratio (CMRR) is inherently increased by the use of improved mirrors, as the current source transistors also use improved self-biased cascode configuration. The CMRR and PSRR can be noted in Figure 10c,d, respectively. Table 3 summarizes the AC simulation results.

Table 3. Integrator simulation results summary.

	Conventional BD-OTA	**Proposed BD-OTA**
DC gain (dB)	44.5	64.2
CMRR (dB)	114	154
PSRR (dB)	88.7	124
GBW (Hz)	78.47	83.14

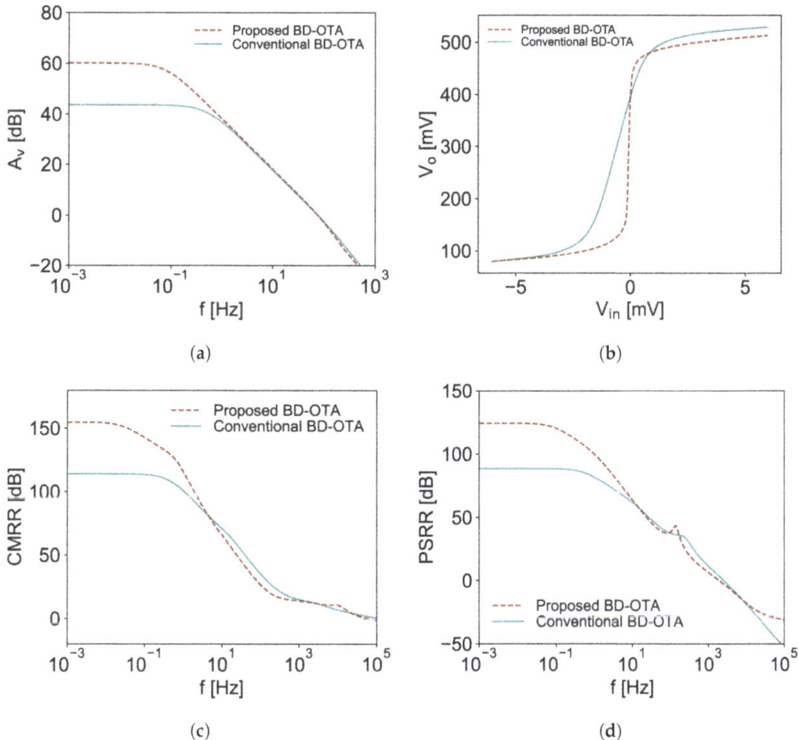

Figure 10. Integrator test bench simulation results. (**a**) AC voltage gain transfer function. (**b**) DC voltage gain transfer function. (**c**) AC CMRR transfer function. (**d**) AC PSRR transfer function.

Figure 9c shows the testbench used in the DC simulations to compare the transconductance linearity of each OTA version. Figure 11a–c show, respectively, the output current, transconductance, and transconductance error for the conventional BD-OTA and for the proposed one.

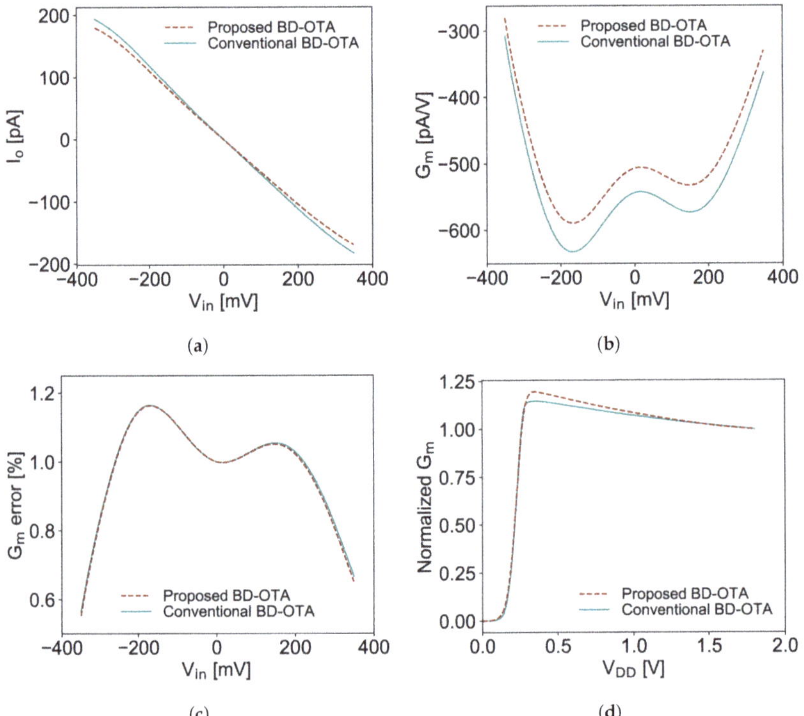

Figure 11. Transconductor testbench simulation results. (**a**) Output current. (**b**) Tranconductance. (**c**) Transconductance error. (**d**) Normalized transconductance.

Table 4 summarizes the transconductance and impedance simulation results for I_{bias} equal to 100 pA. Notice that BD OTAs have finite DC input impedances ($1/G_i$) as large as their output impedances ($1/G_o$), which reduces considerably the effectiveness of the gain improving technique in practical use, where the OTAs are cascaded in OTA-C filters.

Table 4. BD-OTAs DC results.

	Conventional BD-OTA	**Proposed BD-OTA**
G_i (fA V^{-1})	77.3	78.1
G_m (pA V^{-1})	506	542
G_o (fA V^{-1})	3024.5	311.6
A_v (G_m/G_o)	167.3	1739.4

In Figure 11d, the transconductance normalized with respect to the supply voltage V_{DD} is shown. It is possible to note that both OTA versions work properly from a minimum V_{DD} of about 300 mV, which is feasible for implants and wearable biomedical trends. Unlike conventional gate-driven OTA topologies, which are limited by the minimum common-mode input voltage V_{cmi}, and in which frequently are set to half V_{DD} to allow the current source transistors to operate in saturation, according to mentioned this limitation is mitigated in BD topologies. Besides the mentioned aspects, it is worth noticing that the minimum operational voltage, V_{DD}, is directly influenced by the current source, and the differential pair transistors channel inversion, hence which are themselves a function of the bias current, i.e., I_{bias}. In this way, a higher biasing current would result in a larger linear input range and greater transconductance, on the other hand, also a higher minimum V_{DD}.

Figures 12a,b and 13a,b show the nominal output current and its resulting transconductance for symmetrical and asymmetrical input voltage, according to the testbenches shown in Figure 9b,c, respectively. For the asymmetrical test, the inverting input is kept constant at $V_{cm} = V_{DD}/2$, so $-300 < \Delta V_{in} < 300$ mV, while, for the symmetrical input, both OTA inputs are at V_{cm} for $V_{in} = 0$ V, and the differential input voltage excursion is doubled to $-600 < \Delta V_{in} < 600$ mV. Moreover, for the asymmetrical testbench, the common mode input voltage V_{cmi} varies with the input voltage V_{in}, so $V_{cmi} = V_{in}/2 + V_{cm}$. For the symmetrical testbench, V_{cmi} is constant, as the average of the inverting and non-inverting input voltages are the same. It can be noticed, for both cases, that as the biasing current I_{bias} increases, the transconductance G_m increases almost proportionally.

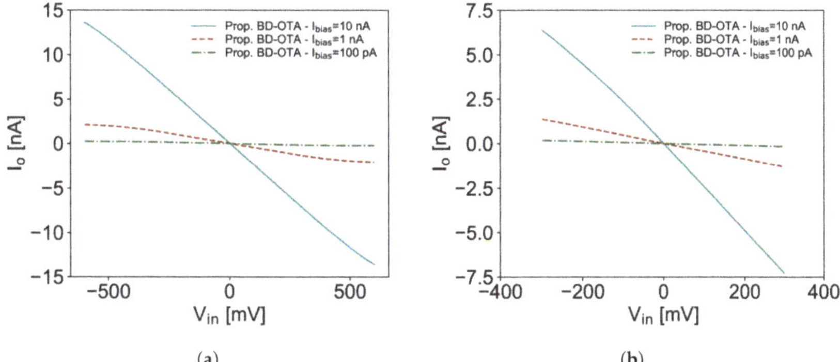

(a)　　　　　　　　　　　　　　(b)

Figure 12. Output current I_0 for (**a**) symmetrical, and (**b**) asymmetrical input voltage V_{in}, as a function of I_{bias}.

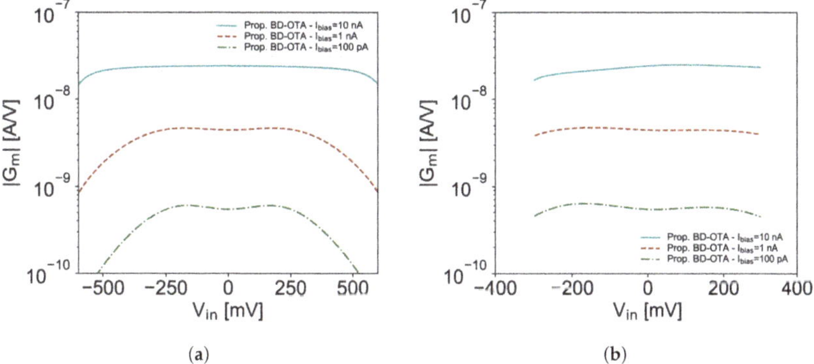

(a)　　　　　　　　　　　　　　(b)

Figure 13. Nominal transconductance G_m for (**a**) symmetrical, and (**b**) asymmetrical input voltage V_{in}, as a function of I_{bias}.

For a better comparison, for different biasing currents, the transconductances were normalized for $\Delta V_{in} = 0$, as shown in Figure 14a,b. It is clear for the asymmetrical input that the error is larger for $\Delta V_{in} < 0$. This happens for two reasons: the parasitic substrate current at the differential pair is extremely non-linear and the common-mode input voltage goes above the limit for $I_{bias} = 10$ nA. For symmetrical input, the resulting G_m is also symmetrical and the range is twice as high. It can also be noted that the shape of the curve changes as the current increases, which is expected, as the differential pair inversion increases.

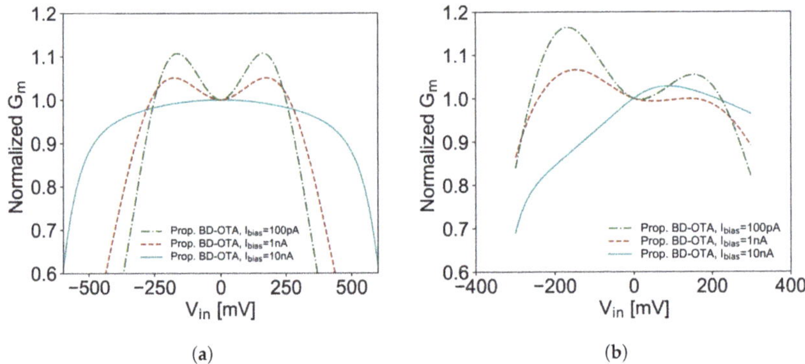

Figure 14. Normalized transconductance G_m for (**a**) symmetrical, and (**b**) asymmetrical input voltage V_{in}, as a function of I_{bias}.

It is also important to notice that for single-ended OTA applications, normally, the input is not symmetrical. This is the case with most OTA-C filters, such as those based on integrators and active loads, as depicted in the testbenches shown in Figure 9a,d. For wider range and linearity, the single-ended OTA should be converted to its fully differential version, which needs extra biasing circuits for its output common-mode definition.

As previously explained, the parasitic input current is one of the causes of transconductance asymmetry. This parasitic current is shown in Figure 15a, and is a function of the input voltage and biasing current. There is a single point where the input current is zero, which happens when the differential pair PMOS transistor bulk-terminal voltage is equal to its source-terminal voltage. For input voltages below this point, the transistor is forward-body-biased and the parasitic current grows exponentially. For voltages above this point, the parasitic current is almost constant, consequently, the input conductance is very small. Figure 15b shows the OTA output current for both inputs at $V_{DD}/2$ and the output sweeping from 0 to 600 mV. As can be seen, the output current, even considering that the current mirrors attenuate the differential pair output current, is considerably larger than the parasitic current for a large range.

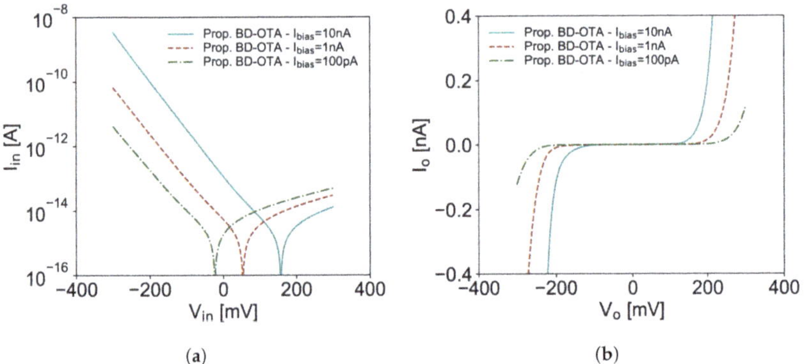

Figure 15. Nominal (**a**) input current I_{in} × input voltage V_{in}, and (**b**) output current I_o × output voltage V_o, as a function of I_{bias}.

The input and output conductances can be derived from the input and output currents, as shown in Figure 16a,b, respectively. It is worth noting that for OTA-C filter applications, the OTA outputs terminals will be connected to other OTAs input terminals. The main advantage of the proposed improved self-cascode current mirror is to decrease the output conductance as it increases the output resistance. If the input conductance of the subsequent stage is greater than the output conductance, the technique effectiveness is reduced.

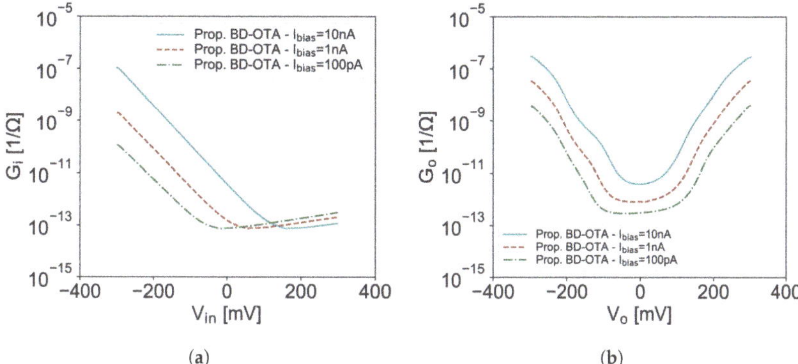

Figure 16. Normalized (**a**) input conductance G_i, and (**b**) output conductance G_o as a function of I_{bias}.

In order to compare the linearity OTAs, the unity gain low-pass OTA-C filter testbench shown in Figure 9d was used in DC and transient simulations. Figure 17a,b show, respectively, the DC transfer functions, transient, and the total harmonic distortion (THD) for both OTAs. It is possible to observe that the BD OTAs have almost the same full input range. Figure 17b shows the total harmonic distortion versus input plotted as a function of input signal amplitude. For both OTAs, one can observe that THD is lower for smaller signal amplitudes. They exhibit approximately the same amount of distortion of 0.07% as a result of a 300 mV amplitude input signal at 100 mHz. As the input voltage amplitude increases, the proposed OTA reaches $\approx$ 1% THD (-39.8 dB), SNR equal to 56.6 dB for a V_{in-pp} = 405 mV at 100 mHz signal.

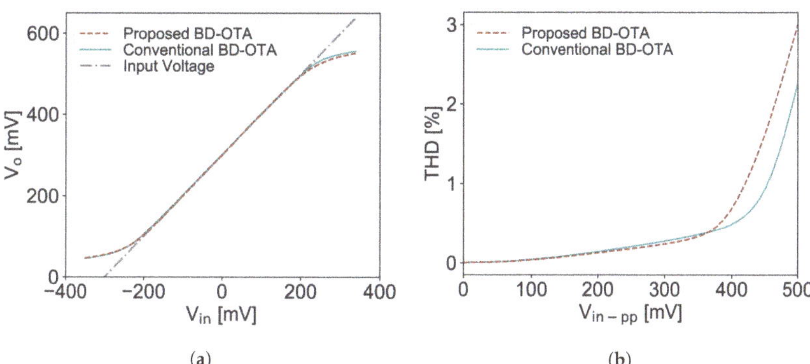

Figure 17. Low-pass testbench simulation results. (**a**) Transfer functions. (**b**) Total harmonic distortion.

Figure 18 shows the input-referred noise (IRN) for both OTA versions configured as a unit-gain buffer. Since both OTA versions have the same transistor dimensions, differing only by the adopted current mirror topology, there is a slight difference in IRN of conventional BD-OTA and the proposed one. The IRN in the proposed topology is equal to 246 μV_{RMS}, and 237 μV_{RMS} in the other, both obtained by integrating noise from 10 mHz–1 kHz.

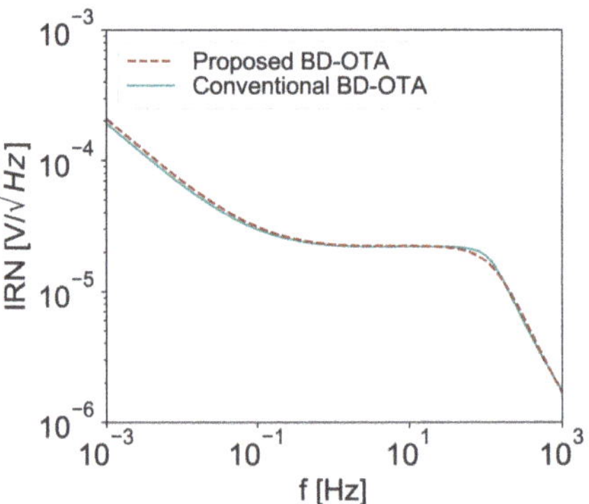

Figure 18. Unity gain buffer testbench simulation results—input-referred noise.

By using the transconductor (Figure 9c) and low-pass filter (Figure 9d) testbenches, 500 runs of Monte Carlo have been carried out for evaluation of transconductance and offset voltage, respectively. Figure 19a,b show the results for the Monte Carlo process and mismatch analysis of the proposed BD-OTA. These results are summarized in Table 5. On this basis, it is possible to conclude that the proposed BD-OTA besides a lower transconductance feature, has a considerably less input voltage offset than the conventional BD-OTA.

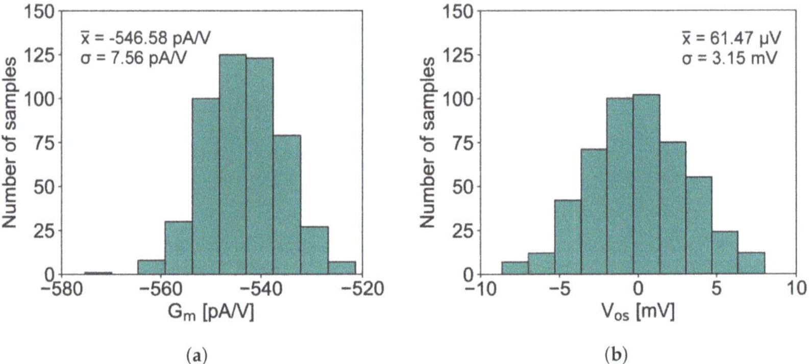

(a) (b)

Figure 19. Five hundred runs of Monte Carlo simulations analyzing process and mismatch with the proposed BD-OTA. (**a**) Transconductance. (**b**) Offset.

Table 5. Monte Carlo simulation results.

		Conventional BD-OTA	Proposed BD-OTA
G_m	$\bar{x}$ (pA/V)	506.6	546.58
G_m	σ (pA/V)	7.61	7.56
G_m	σ/μ (%)	1.5	1.38
V_{os}	$\bar{x}$ (µV)	429.8	61.47
V_{os}	σ (mV)	3.16	3.15

Table 6 compares the performance of the proposed OTA with the state-of-art low-transconductance OTAs. The previous work proposed by us [48] presented a 450 pA/V OTA with small power consumption but lower gain, CMRR, and PSRR, despite being based on non-unitary current gain through the splitting current technique, it achieved a poorer performance with respect to the present work. In [49], a low-transconductance amplifier has been proposed based on the channel-length-modulation effect (Early effect). This solution shows a high IRN. Such an IRN is 3x smaller in the proposed topology while keeping lower transconductance, power consumption, and higher CMRR as also PSRR features. The OTA proposed by [28] is similar to the conventional OTA presented in this work. The difference is in the rectangular arrays used to increase the gain and in no linearization technique employed. Another low-G_m topology presented by [50] uses a linearization technique that relies on a combination of source degeneration with an active attenuator. Despite the valuable linearity and gain achieved, the power consumption, and transconductance may not be suitable to the constraints of biomedical implants or bio-sensing operations. The architecture proposed by [51] is another channel length modulated OTA which contains the same V-I conversion scheme as presented in [49] but requires a higher supply voltage.

Table 6. Comparison of Low-G_m OTA topologies.

Feature	This Work (S)	[48] (S)	[49] (M)	[28] (M)	[50] (M)	[51] (M)
Year	2021	2021	2020	2014	2014	2009
Tech. (nm)	180	180	180	130	350	350
Supply (V)	0.6	1	1	0.25	5	5
G_m (nA/V)	0.542	0.45	0.62–6.28	22	39.5–367.2	0.03–25,000
A_v (dB)	64	37	-	-	52.3–64.7	-
Power (μW)	0.0019	0.032	0.028–0.270	0.01	160	<300
CMRR (dB)	154	56	56	-	>44.8	>80
PSRR (dB)	124	36	47	-	-	>80
GBW (Hz)	83.14	6.9	-	-	-	-
V_{os} (mV)	0.06 ± 3.15	±20	25–50	±10.82	-	
IRN (μV_{RMS})	246	-	760	100	-	635
Linear range sym. input (V_{pp})	0.475	0.3	2	-	2	2.6
THD (%) @ input (V_{pp})	1@0.475	1@0.3	0.18@2	0.53@0.1	0.13@2	<1@2.6
SNR (dB) @ THD (%)	54.8@1	-	59.3@0.47		66.5@0.13	62@1
Layout area (mm^2)	0.0143	-	0.027	0.053	0.006	0.046

THD @ V_{in} = 250 mV$_{pp}$, (M): Measured, (S): Simulated.

5. Conclusions

The paper presented a bulk-driven symmetrical OTA based on a self-cascode current mirror with source degeneration to provide a high-gain, high linearity, and low-G_m topology. The proposed low-G_m topology shown a valuable performance that is suitable for new biomedical IC applications. In particular, the new topology achieved the lowest power consumption compared with the state-of-art topologies, in addition to high gain, linearity, low transconductance, and IRN. Moreover, the circuit obtained the highest CMRR and high PSRR, which turns the proposed OTA into an interesting topology as a basic block for OTA-C filters.

Author Contributions: Conceptualization, R.S.S. and L.H.R.; methodology, R.S.S. and L.H.R.; validation, R.S.S. and L.H.R.; formal analysis, R.S.S. and L.H.R.; investigation, R.S.S. and L.H.R.; resources, C.R.R.; data curation, L.H.R.; writing—original draft preparation, L.H.R.; writing—review and editing, O.A. and C.R.R.; visualization, L.H.R.; supervision, O.A. and C.R.R.; project administration, C.R.R.; funding acquisition, C.R.R. All authors have read and agreed to the published version of the manuscript.

Funding: Brazilian National Council for Scientific and Technological Development (CNPq), and to the Coordination of Improvement of Higher Education Personnel—Brazil (CAPES)—Finance Code 001 for partial financial support. This work also has been partially supported in Brazil by PrInt CAPES-UFSC "Automation 4.0".

Data Availability Statement: Data are contained within the article.

Acknowledgments: The authors are grateful to the Brazilian National Council for Scientific and Technological Development (CNPq), the Coordination of Improvement of Higher Education Personnel—Brazil (CAPES), and CAPES-UFSC PrInt for the financial support.

Conflicts of Interest: The authors declare no conflict of interest.

Abbreviations

The following abbreviations are used in this manuscript:

ACM	Advanced Compact Model
BD	Bulk-driven
CMOS	Complementary Metal-Oxide Semiconductor
CM	Current Mirror
CMR	Common-Mode Range
CMRR	Common-Mode Rejection Ratio
GBW	Gain–Bandwidth Product
OTA	Operational Transconductance Amplifier
SCCM	Self-Cascode Current Mirror
IRF	Equivalent Input Referred Noise
ISCCM	Improved Self-Cascode Current Mirror
PSRR	Power Supply Rejection Ratio
SNR	Signal-to-Noise Ratio
THD	Total Harmonic Distortion

References

1. Khateb, F.; Kulej, T.; Akbari, M.; Kumngern, M. 0.5-V High linear and wide tunable OTA for biomedical applications. *IEEE Access* **2021**, *9*, 103784–103794. [CrossRef]
2. Sanchotene Silva, R.; Pereira Luiz, L.; Cherem Schneider, M.; Galup-Montoro, C. A Test Chip for Characterization of the Series Association of MOSFETs. *IEEE Trans. Very Large Scale Integr. (VLSI) Syst.* **2019**, *27*, 1967–1971. [CrossRef]
3. Bano, S.; Narejo, G.; Shan, S. Power Efficient Fully Differential Bulk Driven OTA for Portable Biomedical Application. *Electronics* **2018**, *7*, 41. [CrossRef]
4. Ballo, A.; Pennisi, S.; Scotti, G. 0.5 V CMOS Inverter-Based Transconductance Amplifier with Quiescent Current Control. *J. Low Power Electron. Appl.* **2021**, *11*, 37. [CrossRef]
5. Giustolisi, G.; Palumbo, G. A gm/ID-Based Design Strategy for IoT and Ultra-Low-Power OTAs with Fast-Settling and Large Capacitive Loads. *J. Low Power Electron. Appl.* **2021**, *11*, 21. [CrossRef]
6. Centurelli, F.; Della Sala, R.; Monsurrò, P.; Scotti, G.; Trifiletti, A. A 0.3 V Rail-to-Rail Ultra-Low-Power OTA with Improved Bandwidth and Slew Rate. *J. Low Power Electron. Appl.* **2021**, *11*, 19. [CrossRef]
7. Centurelli, F.; Della Sala, R.; Monsurrò, P.; Scotti, G.; Trifiletti, A. A Novel OTA Architecture Exploiting Current Gain Stages to Boost Bandwidth and Slew-Rate. *Electronics* **2021**, *10*, 1638. [CrossRef]
8. Centurelli, F.; Della Sala, R.; Scotti, G.; Trifiletti, A. A 0.3 V, Rail-to-Rail, Ultralow-Power, Non-Tailed, Body-Driven, Sub-Threshold Amplifier. *Appl. Sci.* **2021**, *11*, 2528. [CrossRef]
9. Jayasimha, T.; Vijayalakshmi, A. Low Pass Filter Using ECG Detection for OTA-C. *Ir. Interdiscip. J. Sci. Res.* **2018**, *2*, 94–100.
10. Yehoshuva, C.; Rakhi, R.; Anto, D.; Kaurati, S. 0.5 V, Ultra Low Power Multi Standard Gm-C Filter for Biomedical Applications. In Proceedings of the 2016 IEEE International Conference on Recent Trends in Electronics, Information and Communication Technology (RTEICT), Bangalore, India, 20–21 May 2016.
11. Kulej, T.; Khateb, F. A Compact 0.3-V Class AB Bulk-Driven OTA. *IEEE Trans. Very Large Scale Integr. (VLSI) Syst.* **2020**, *28*, 224–232. [CrossRef]
12. Kulej, T.; Khateb, F. A 0.3-V 98-dB Rail-to-Rail OTA in 0.18 μm CMOS. *IEEE Access* **2020**, *8*, 27459–27467. [CrossRef]
13. Rodovalho, L.H.; Aiello, O.; Rodrigues, C.R. Ultra-low-voltage inverter-based operational transconductance amplifiers with voltage gain enhancement by improved composite transistors. *Electronics* **2020**, *9*, 1410. [CrossRef]
14. Rodovalho, L.H.; Ramos Rodrigues, C.; Aiello, O. Self-Biased and Supply-Voltage Scalable Inverter-Based Operational Transconductance Amplifier with Improved Composite Transistors. *Electronics* **2021**, *10*, 935. [CrossRef]
15. Toledo, P.; Crovetti, P.; Aiello, O.; Alioto, M. Fully Digital Rail-to-Rail OTA With Sub-1000-μm² Area, 250-mV Minimum Supply, and nW Power at 150-pF Load in 180 nm. *IEEE Solid-State Circuits Lett.* **2020**, *3*, 474–477. [CrossRef]
16. Fassio, L.; Lin, L.; De Rose, R.; Lanuzza, M.; Crupi, F.; Alioto, M. Trimming-Less Voltage Reference for Highly Uncertain Harvesting Down to 0.25 V, 5.4 pW. *IEEE J.-Solid-State Circuits* **2021**, 3134–3144. [CrossRef]

17. Aiello, O. Design of an Ultra-Low Voltage Bias Current Generator Highly Immune to Electromagnetic Interference. *J. Low Power Electron. Appl.* **2021**, *11*, 6. [CrossRef]

18. Aiello, O.; Crovetti, P.; Alioto, M. A Sub-Leakage PW-Power HZ-Range Relaxation Oscillator Operating with 0.3 V–1.8 V Unregulated Supply. In Proceedings of the 2018 IEEE Symposium on VLSI Circuits, Honolulu, HI, USA, 18–22 June 2018; pp. 119–120. [CrossRef]

19. Lee, S.Y.; Cheng, C.J. Systematic Design and Modeling of a OTA-C Filter for Portable ECG Detection. *IEEE Trans. Biomed. Circuits Syst.* **2009**, *3*, 53–64. [CrossRef]

20. Veeravalli, A.; Sánchez-Sinencio, E.; Silva-Martínez, J. Transconductance amplifier structures with very small transconductances: A comparative design approach. *IEEE J.-Solid-State Circuits* **2002**, *37*, 770–775. [CrossRef]

21. Kinget, P.; Steyaert, M.; Van der Spiegel, J. Full analog CMOS integration of very large time constants for synaptic transfer in neural networks. *Analog Integr. Circuits Signal Process.* **1992**, *2*, 281–295. [CrossRef]

22. Fiorelli, R.; Arnaud, A.; Galup-Montoro, C. Series-parallel association of transistors for the reduction of random offset in non-unity gain current mirrors. In Proceedings of the 2004 IEEE International Symposium on Circuits and Systems (IEEE Cat. No. 04CH37512), Vancouver, BC, Canada, 23–26 May 2004; Volume 1, pp. 1–881.

23. Arnaud, A.; Fiorelli, R.; Galup-Montoro, C. Nanowatt, sub-nS OTAs, with sub-10-mV input offset, using series-parallel current mirrors. *IEEE J.-Solid-State Circuits* **2006**, *41*, 2009–2018. [CrossRef]

24. Blalock, B.J.; Allen, P.E.; Rincon-Mora, G.A. Designing 1-V op amps using standard digital CMOS technology. *IEEE Trans. Circuits Syst. II Analog. Digit. Signal Process.* **1998**, *45*, 769–780. [CrossRef]

25. Chatterjee, S.; Tsividis, Y.; Kinget, P. 0.5-V analog circuit techniques and their application in OTA and filter design. *IEEE J.-Solid-State Circuits* **2005**, *40*, 2373–2387. [CrossRef]

26. Carrillo, J.M.; Torelli, G.; Valverde, R.P.A.; Duque-Carrillo, J.F. 1-V rail-to-rail CMOS opamp with improved bulk-driven input stage. *IEEE J.-Solid-State Circuits* **2007**, *42*, 508–517. [CrossRef]

27. Cotrim, E.D.C.; de Carvalho Ferreira, L.H. An ultra-low-power CMOS symmetrical OTA for low-frequency G_m-C applications. *Analog Integr. Circuits Signal Process.* **2012**, *71*, 275–282. [CrossRef]

28. Colletta, G.D.; Ferreira, L.H.; Pimenta, T.C. A 0.25-V 22-nS symmetrical bulk-driven OTA for low-frequency G_m-C applications in 130-nm digital CMOS process. *Analog Integr. Circuits Signal Process.* **2014**, *81*, 377–383. [CrossRef]

29. Sharan, T.; Bhadauria, V. Sub-threshold, cascode compensated, bulk-driven OTAs with enhanced gain and phase-margin. *Microelectron. J.* **2016**, *54*, 150–165. [CrossRef]

30. del Risco Sánchez, A.; Moreno, R.L.; Ferreira, L.H.; Crepaldi, P.C. Biasing technique to improve total harmonic distortion in an ultra-low-power operational transconductance amplifier. *IET Circuits Devices Syst.* **2019**, *13*, 920–927. [CrossRef]

31. Galup-Montoro, C.; Schneider, M.C.; Loss, I.J. Series-parallel association of FET's for high gain and high frequency applications. *IEEE J.-Solid-State Circuits* **1994**, *29*, 1094–1101. [CrossRef]

32. Comer, D.T.; Comer, D.J.; Li, L. A high-gain complementary metal-oxide semiconductor op amp using composite cascode stages. *Int. J. Electron.* **2010**, *97*, 637–646. [CrossRef]

33. Akbari, M.; Hashemipour, O. A 0.6-V, 0.4-μW bulk-driven operational amplifier with rail-to-rail input/output swing. *Analog Integr. Circuits Signal Process.* **2016**, *86*, 341–351. [CrossRef]

34. Sharan, T.; Chetri, P.; Bhadauria, V. Ultra-low-power bulk-driven fully differential subthreshold OTAs with partial positive feedback for Gm-C filters. *Analog Integr. Circuits Signal Process.* **2018**, *94*, 427–447. [CrossRef]

35. Baek, K.J.; Gim, J.M.; Kim, H.S.; Na, K.Y.; Kim, N.S.; Kim, Y.S. Analogue circuit design methodology using self-cascode structures. *Electron. Lett.* **2013**, *49*, 591–592. [CrossRef]

36. Xu, D.; Liu, L.; Xu, S. High DC gain self-cascode structure of OTA design with bandwidth enhancement. *Electron. Lett.* **2016**, *52*, 740–742. [CrossRef]

37. Niranjan, V.; Kumar, A.; Jain, S.B. Composite transistor cell using dynamic body bias for high gain and low-voltage applications. *J. Circuits Syst. Comput.* **2014**, *23*, 1450108. [CrossRef]

38. Krummenacher, F.; Joehl, N. A 4-MHz CMOS continuous-time filter with on-chip automatic tuning. *IEEE J.-Solid-State Circuits* **1988**, *23*, 750–758. [CrossRef]

39. Braga, R.A.; Ferreira, L.H.; Coletta, G.D.; Dutra, O.O. A 0.25-V calibration-less inverter-based OTA for low-frequency Gm-C applications. *Microelectron. J.* **2019**, *83*, 62–72. [CrossRef]

40. Schneider, M.C.; Galup-Montoro, C. *CMOS Analog Design Using All-Region MOSFET Modeling*; Cambridge University Press: Cambridge UK, 2010.

41. Wilson, G.R. A monolithic junction FET-NPN operational amplifier. *IEEE J.-Solid-State Circuits* **1968**, *3*, 341–348. [CrossRef]

42. De Ceuster, D.; Flandre, D.; Colinge, J.P.; Cristoloveanu, S. Improvement of SOI MOS current-mirror performances using serial-parallel association of transistors. *Electron. Lett.* **1996**, *32*, 278–279. [CrossRef]

43. Rodovalho, L.H. Push–pull based operational transconductor amplifier topologies for ultra low voltage supplies. *Analog Integr. Circuits Signal Process.* **2020**, 1–14. [CrossRef]

44. Haga, Y.; Zare-Hoseini, H.; Berkovi, L.; Kale, I. Design of a 0.8 Volt fully differential CMOS OTA using the bulk-driven technique. In Proceedings of the 2005 IEEE International Symposium on Circuits and Systems, Kobe, Japan, 23–26 May 2005; pp. 220–223.

45. Ferreira, L.H.; Sonkusale, S.R. A 60-dB gain OTA operating at 0.25-V power supply in 130-nm digital CMOS process. *IEEE Trans. Circuits Syst. Regul. Pap.* **2014**, *61*, 1609–1617. [CrossRef]

46. Kuo, K.C.; Leuciuc, A. A linear MOS transconductor using source degeneration and adaptive biasing. *IEEE Trans. Circuits Syst. II Analog Digit. Signal Process.* **2001**, *48*, 937–943.

47. Galup-Montoro, C.; Schneider, M.C.; Cunha, A.I.A.; de Sousa, F.R.; Klimach, H.; Siebel, O.F. The Advanced Compact MOSFET (ACM) Model for Circuit Analysis and Design. In Proceedings of the 2007 IEEE Custom Integrated Circuits Conference, San Jose, CA, USA, 16–19 September 2007; pp. 519–526. [CrossRef]

48. Rodovalho, L.H.; Silva, R.S.; Rodrigues, C.R. A 1V, 450pS OTA Based on Current-Splitting and Modified Series-Parallel Mirrors. In Proceedings of the 2021 IEEE 12th Latin America Symposium on Circuits and System (LASCAS), Arequipa, Peru, 21–24 February 2021; pp. 1–4. [CrossRef]

49. Jakusz, J.; Jendernalik, W.; Blakiewicz, G.; Kłosowski, M.; Szczepański, S. A 1-nS 1-V Sub-1-µW Linear CMOS OTA with Rail-to-Rail Input for Hz-Band Sensory Interfaces. *Sensors* **2020**, *20*, 3303. [CrossRef]

50. Soares, C.F.; de Moraes, G.S.; Petraglia, A. A low-transconductance OTA with improved linearity suitable for low-frequency Gm-C filters. *Microelectron. J.* **2014**, *45*, 1499–1507. [CrossRef]

51. Huang, Y.; Drakakis, E.M.; Toumazou, C. A 30pA/V–25µA/V linear CMOS channel-length-modulation OTA. *Microelectron. J.* **2009**, *40*, 1458–1465. [CrossRef]

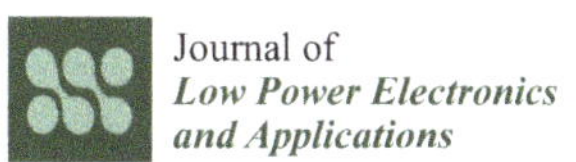

Journal of
*Low Power Electronics
and Applications*

Article

A Novel Standard-Cell-Based Implementation of the Digital OTA Suitable for Automatic Place and Route

Gaetano Palumbo [1] and Giuseppe Scotti [2],*

[1] Dipartimento di Ingegneria Elettrica Elettronica e Informatica (DIEEI), University of Catania, 95125 Catania, Italy; gaetano.palumbo@dieei.unict.it

[2] Dipartimento di Ingegneria dell'Informazione Elettronica e Telecomunicazioni (DIET), Sapienza University of Rome, 00184 Rome, Italy

* Correspondence: giuseppe.scotti@uniroma1.it; Tel.: +39-0644585690

Abstract: This paper presents a novel implementation of a digital-based Operational Transconductance Amplifier (OTA) which has been recently introduced in the technical literature as a fully digital alternative to the conventional differential pair to implement low voltage analog amplifiers and comparators. The proposed implementation does not make use of resistors, floating gate resistors nor C-Muller elements and is made up of only digital gates usually available in the standard cell libraries. The resulting analog circuit schematic can be described using structural VHDL or Verilog languages and is suitable to be integrated in an automatic synthesis and place and route flow for digital circuits. The proposed digital-based amplifier has been implemented in a commercial 130 nm CMOS process by using an automatic place and route flow for layout generation starting from the Verilog netlist. Post layout simulations are presented to show the performance of the proposed circuit and compare it against the state of the art.

Keywords: OTA; low voltage; low power; automatic place and route; standard cell; fully digital

Citation: Palumbo, G.; Scotti, G. A Novel Standard-Cell-Based Implementation of the Digital OTA Suitable for Automatic Place and Route. *J. Low Power Electron. Appl.* **2021**, *11*, 42. https://doi.org/10.3390/jlpea11040042

Academic Editor: Orazio Aiello

Received: 12 October 2021
Accepted: 26 October 2021
Published: 28 October 2021

Publisher's Note: MDPI stays neutral with regard to jurisdictional claims in published maps and institutional affiliations.

1. Introduction

Battery-operated or energy harvested systems such as biomedical implantable devices or sensor nodes for Internet of Things (IoT) applications require the development of low voltage, low power CMOS Systems on Chip (SoCs) in which analog interface circuits are integrated together with the digital processing and communication cores [1].

In the conventional design flow of mixed-signal integrated circuits, the design and implementation of the analog building blocks is usually carried out manually by the analog designer who iterate several times each step of the design flow in order to optimize performance, power and area figures of merit.

Nowadays, due to the continuous scaling of MOS feature size in the nanometer regime, the analog designer has to cope with new challenges in the simulation and implementation steps of the design flow. In fact, the performance of nanometer MOS transistors from an analog designer perspective is worsening with technology scaling, and accurate simulation models are becoming more and more difficult to develop. These challenges often result in analog building blocks which require some form of calibration or programmability after production in order to achieve the required performance [2,3].

If we focus on the design flow of digital circuits, we see that the synthesis and place and route steps are carried out automatically by using CAD tools for the physical synthesis. The netlist of digital circuits is built by the synthesis tool and is made up only of digital gates taken from a standard cell library, which is usually provided by the IC manufacturer.

From a time to market perspective, since the standard cells commonly adopted for the digital design flow exhibits a DRC clean layout, their usage for the implementation of analog building blocks can drastically reduce the layout effort of the analog part and thus the overall time to market of mixed signal SoCs for IoT applications. In addition, for these reasons, recently, several architectures of mixed signal integrated circuits, suited for battery-operated or energy harvested systems that are mostly or completely based on digital standard cells, have been introduced in the technical literature [4–8].

The netlists of analog blocks, which are built using only digital standard cells, can be described using structural VHDL or Verilog languages and are suitable to be integrated in an automatic synthesis and place and route flow for digital circuits. This approach strongly reduces the design effort and brings the advantages of digital circuits, such as design and technology portability, low-voltage operation and effective area shrinkage at more advanced technology generations.

Since the standard cell libraries adopted in semi-custom digital flows allow the usage of a wide set of logic gates with different size ratios (and therefore driving capability), the analog designer can have significant design freedom for different application environments. In addition, the use of digital standard cells can heavily relax the design complexity of analog components, such as amplifiers and voltage comparators requiring ultra-low supply voltage, thus avoiding complex circuit topologies typically developed for ultra-low voltage conditions.

In this paper, we focus on the digital-in-concept approach for the design of analog differential circuits originally presented in [9] and recently exploited in [10–12], but only standard cell libraries are used. Indeed, the pioneering work in [9], which presents the first fully digital alternative to the conventional differential pair to implement low voltage analog amplifiers and comparators, still requires some passive components (resistors or floating-gate resistors). Meanwhile, the evolution of circuits in [9] presented in [10–12], despite not requiring any passive component, exploit the C-Muller element as a fundamental building block, which typically cannot be found among the digital standard cells.

The digital OTA implementation proposed in this paper does not make use of resistors, floating gate resistors nor C-Muller elements and is made up of only digital gates usually available in the standard cell libraries. Being fully standard cell-based, the proposed digital OTA implementation can be integrated in a semi-custom design flow of a mixed signal SoC, and its layout can be automatically generated as is usually done for digital blocks.

In the following, Section 2 describes the proposed standard cell implementation of the digital OTA, Section 3 reports the results of the simulations, whereas the comparison against the state of the art is discussed in Section 4. Finally, some conclusions are drawn in Section 5.

2. Proposed Standard Cell Implementation of Digital OTA

The implementation of the analog amplifier presented in [9] is reported in Figure 1. As explained in [9], the common mode (CM) extractor part in Figure 1 generates a common mode compensation analog signal to be added to the external inputs resulting in a common mode compensation method which is very similar to the common mode rejection mechanism of the conventional analog CMOS differential pair.

The CM compensation signal (V_{CMP}) is added to the external input signals by a summing network so that the actual input signals of the digital buffers can be expressed as:

$$INp = \frac{Vip + V_{CMP}}{2}; \ INn = \frac{Vin + V_{CMP}}{2} \tag{1}$$

and their differential mode (DM) and CM components are related to external DM ($v_D = Vip - Vin$) and CM ($v_{CM} = (Vip + Vin)/2$) components as:

$$v'_D = \frac{v_D}{2}; \ v'_{CM} = \frac{v_{CM} + V_{CMP}}{2} \tag{2}$$

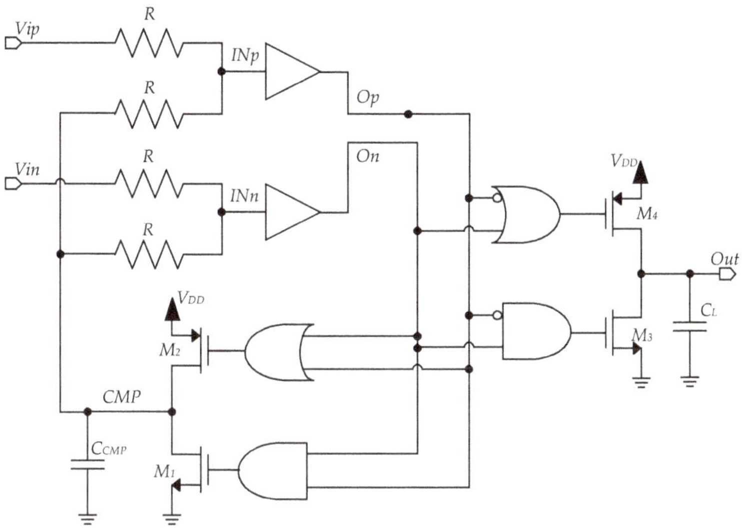

Figure 1. Implementation of the digital-based analog amplifier presented in [9].

In Figure 1, a resistive summing network is included for the sake of simplicity, nonetheless, such a function can be conveniently implemented in CMOS technology by quasi-floating gate (QFG) techniques [9].

The evolution of the circuit in Figure 1, which avoids the passive components, but uses Muller C-elements reported in Figure 2. This idea was proposed in [10] and analyzed in detail in [12].

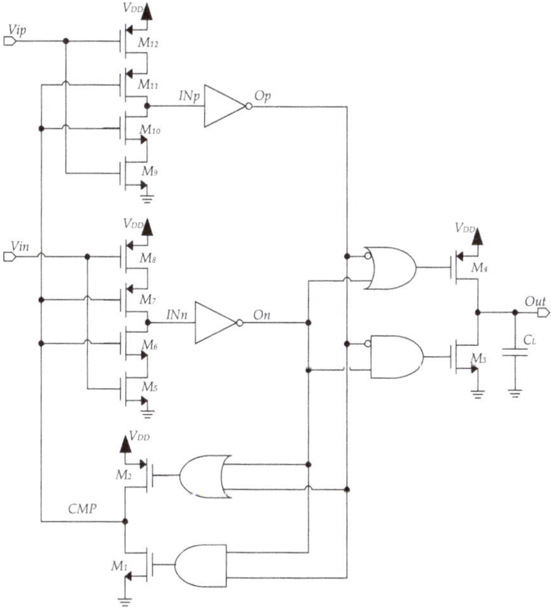

Figure 2. Implementation of the DIGOTA presented in [10].

The Muller C-elements in Figure 2 implement, in a fully digital fashion, the compensation of the common-mode without requiring any calibration circuitry [10,11].

Even if QFG resistors and Muller C-elements can be implemented in CMOS processes, they are usually not available in the standard cell libraries provided by IC manufacturers, and all the previously reported digital OTA implementations are therefore not immediately suitable for automatic place and route within a semi-custom design flow.

The schematic of the proposed pure standard cell implementation of the digital OTA is reported in Figure 3. In particular, the circuit in Figure 3 is based on the following types of logic gates:

- Inverter gates (IV);
- Exclusive OR gates (XOR);
- Three-state Buffer gates (BT);
- Three-state Inverter gates (IT).

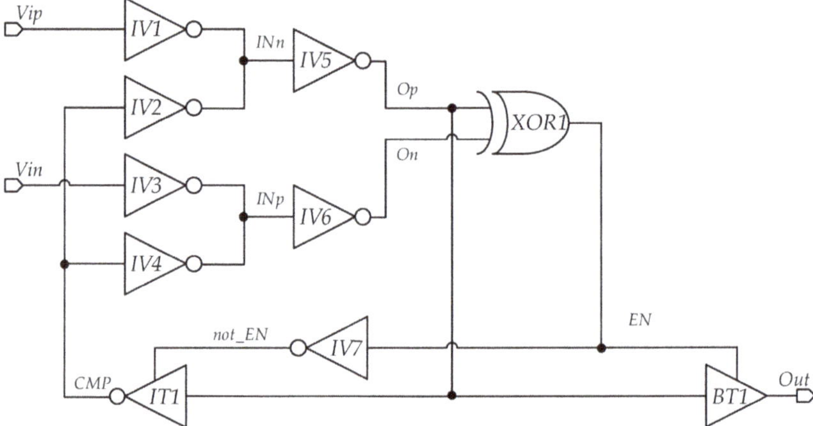

Figure 3. Proposed fully standard cell Implementation of the digital OTA.

Despite its pure standard cell implementation, the circuit operation is very similar to the one of the original implementations of the digital OTA in [9].

The CM extractor is implemented through the XOR1, the IV7 and the IT1 gates in Figure 3 and generates a common mode compensation signal (V_{CMP}) which is then added to the external inputs through the summing network implemented by inverter gates IV1, IV2, IV3 and IV4, thus compensating the common mode as happens in the conventional analog CMOS differential pair.

To explain the summing mechanism, it is sufficient to note that a CMOS inverter acts as a transconductor when its input voltage is close to the logic threshold, therefore, the output current of *IV1* and *IV2* (*IV3* and *IV4*) are summed at their common output node and converted into a voltage through the equivalent resistance at node *INn* (*INp*).

3. Simulation Results

The proposed standard cell-based digital OTA (SC-DIGOTA) has been designed in the 130-nm STMicroelectronics CMOS technology adopting the standard cell library provided by the IC manufacturer. The circuit schematic has been described by using structural Verilog language (see Appendix A), and the layout has been automatically generated within the Cadence Innovus[TM] environment. Transistor level simulations on the post layout netlist have been carried out within the Cadence Virtuoso framework for analog design, exploiting AC and transient simulations. For AC simulations a bias point is established by applying the input signal on a DC level equal to about $V_{DD}/2$ in order to have also the DC output voltage around $V_{DD}/2$.

3.1. Automatic Place and Route

The automatically generated layout of the proposed digital OTA implementation is shown in Figure 4, showing an area footprint of 9.1 μm × 9.7 μm.

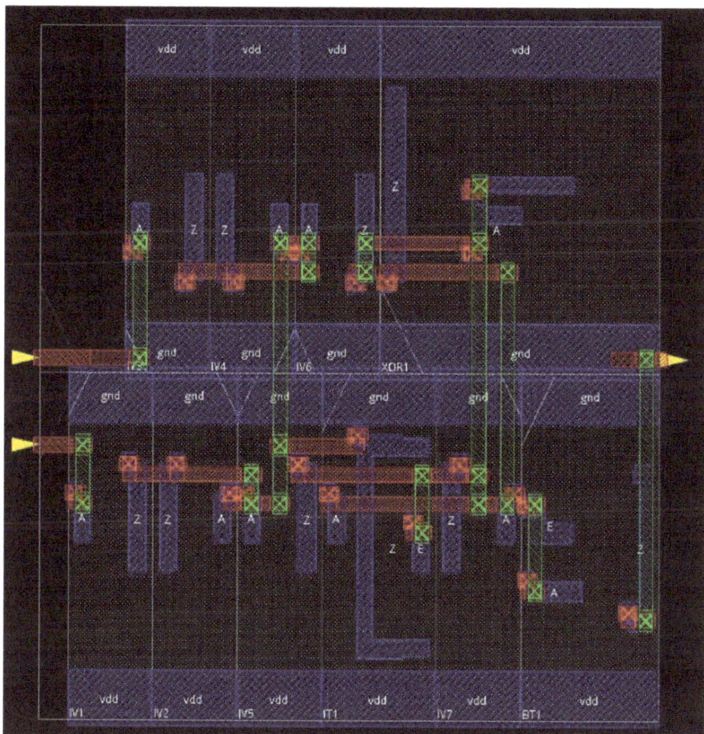

Figure 4. Layout view of the SC-DIGOTA within the Cadence Innovus P&R tool.

3.2. Open Loop Simulations

The amplifier has been simulated assuming a 0.55-V nominal supply voltage and a 250-pF load capacitance. The results of the open loop AC simulations of the proposed OTA are reported in Figure 5, showing that the digital OTA exhibits an overall DC gain and GBW of about 87 dB and 3.15 MHz, respectively. The phase margin of the amplifier results is higher than 65° with all the standard cells sized for minimum area. Figure 6 reports the input-referred noise plot of the proposed digital OTA showing an input-referred Flicker noise of about 4.82 $\mu V/\sqrt{Hz}$ @ 100 Hz and an input-referred white noise of about 175 $nV/\sqrt{Hz}$ @ 100 kHz.

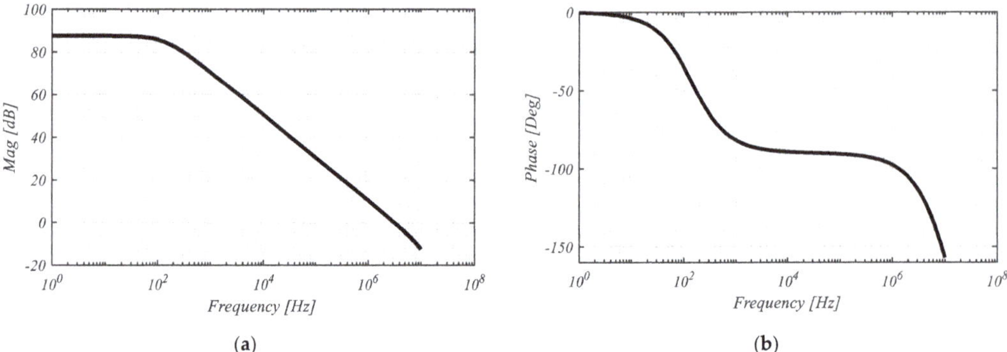

(a)   (b)

Figure 5. Frequency response of the SC-DIGOTA for C$_L$ = 250 pF, magnitude (**a**), phase (**b**).

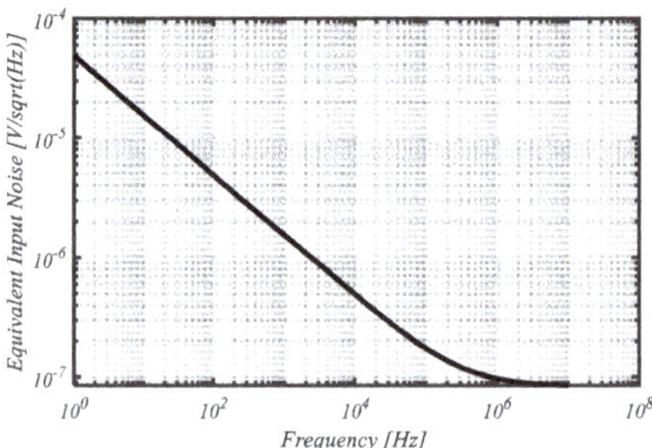

Figure 6. Equivalent input noise of the SC-DIGOTA.

3.3. Closed Loop Simulations

The OTA has then been simulated in a conventional non-inverting unity-gain configuration. The closed loop frequency response of the OTA is depicted in Figure 7, whereas the DC transfer characteristic is shown in Figure 8, highlighting an almost rail-to-rail behavior. Figure 9 shows the time domain response of the circuit to a sinusoidal waveform with a frequency of 10Hz and an amplitude of 200mV. The waveforms of the internal signals INn, INp and CMP of the SC-DIGOTA when processing the sinusoidal signal reported in Figure 9 are reported in Figure 10. Figure 11 shows the time domain response of the circuit to a square wave with a period of 2ms and an amplitude of 200mV. The positive and negative slew-rate have been found to be SR+ = 4.32 V/ms and SR− = 1.03 V/ms respectively.

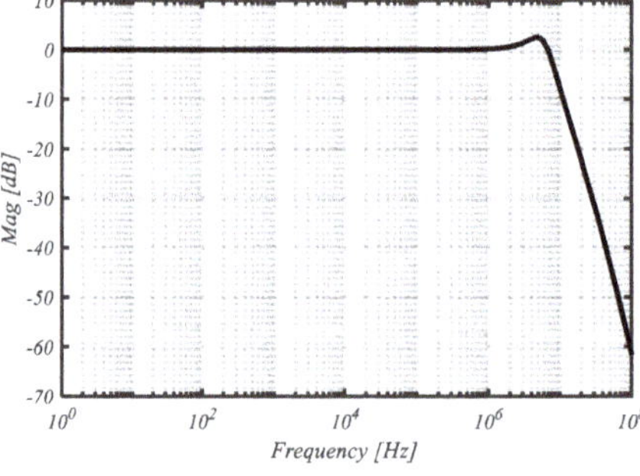

Figure 7. Closed loop frequency response.

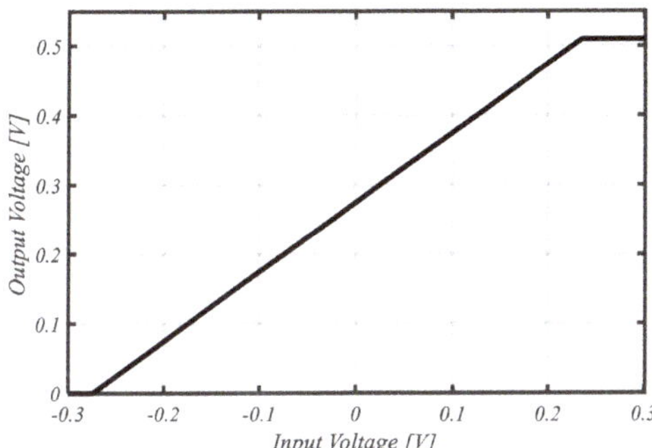

Figure 8. Closed loop dc voltage transfer characteristic.

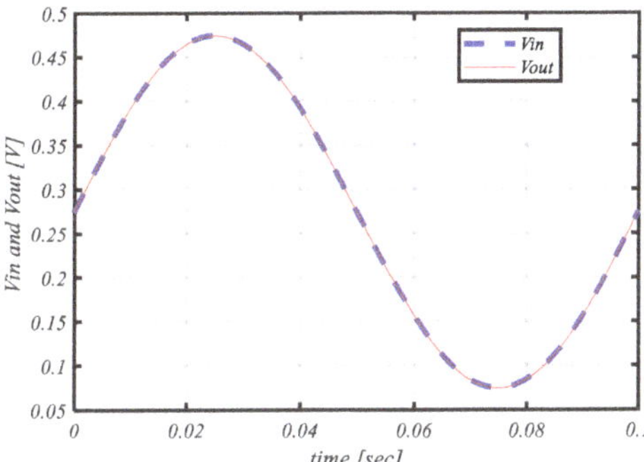

Figure 9. Time domain response to a sinusoidal waveform (frequency = 10 Hz and amplitude = 200 mV) of the SC-DIGOTA in unity gain configuration.

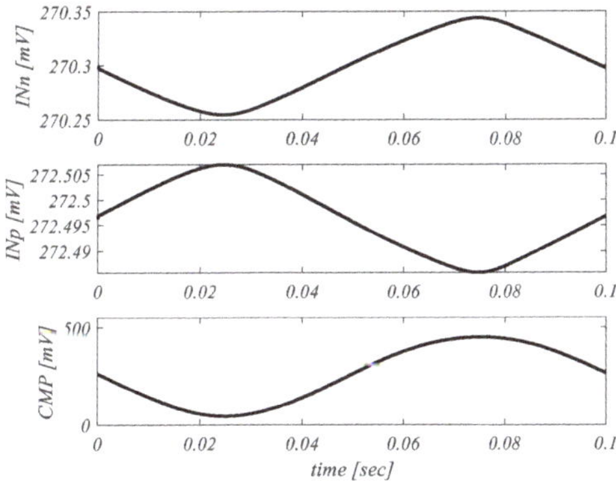

Figure 10. Internal waveforms of the SC-DIGOTA in unity gain configuration in response to a sinusoidal waveform (frequency = 10 Hz and amplitude = 200 mV).

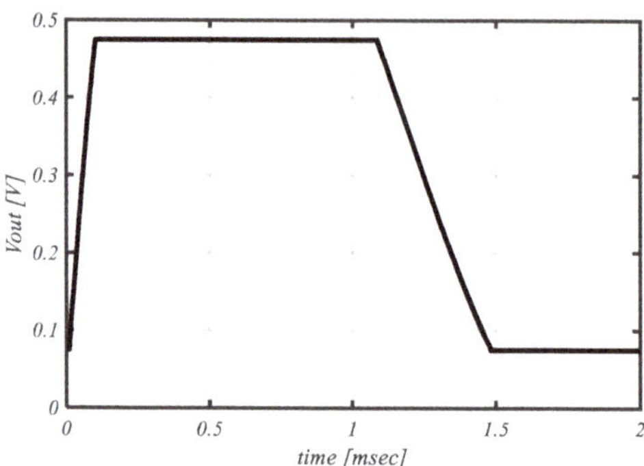

Figure 11. Time domain response to a square wave (period = 2 ms and amplitude = 200 mV) of the SC-DIGOTA in unity gain configuration.

4. Comparison with the Literature

To compare the proposed SC-DIGOTA against state-of-the-art low voltage amplifiers, we refer to the following Figures of Merit:

$$FOMs = \frac{GBWC_L}{P_{diss}} \tag{3a}$$

$$FOM_{S,A} = \frac{GBWC_L}{P_{diss} \cdot Area} \tag{3b}$$

$$FOM_L = \frac{SRC_L}{P_{diss}} \tag{4a}$$

$$FOM_{L,A} = \frac{SRC_L}{P_{diss} \cdot Area} \tag{4b}$$

where GBW is the gain bandwidth product, C_L the load capacitance, SR is the average slew-rate, and P_{diss} is the power consumption. S and L in (3) and (4) denote small-signal and large-signal, respectively, while the $FOM_{S,A}$ and $FOM_{L,A}$ are normalized with respect to the layout area of the OTA.

Table 1 reports the comparison of the SC-DIGOTA against recently published low voltage OTAs taken from the literature, showing how the proposed circuit exhibits very good small signal performance and adequate large signal performance. Due to the very compact layout, the proposed OTA outperforms all other similar designs in terms of $FOM_{S,A}$.

Table 1. Comparison against the state of the art.

	[13]	[14]	[15]	[12]	[12]	**This Work**
supply voltage [V]	0.5	0.3	0.3	0.3	0.5	0.55
OTA architecture	bulk-driven	bulk-driven	bulk-driven	digital	digital	digital
technology [nm]	180	130	130	180	180	130
area (μm^2)	26,000	6400	3600	982	982	88.3
cap load CL [pF]	20	50	40	150	150	250
power [μW]	110	0.0114	0.073	0.0024	0.1075	8.2
DC gain [dB]	52	64.6	41	30	73	87
GBW [kHz]	2500	3.58	18.65	0.250	57.5	3150
average slew rate SR [V/μs]	2.89	0.00093	0.0216	0.000085	0.019	0.0027
in-band input noise [μV]	442.7	-	-	21	122	253
CMRR [dB]	78	61	67.4	41	65	46

Table 1. *Cont.*

	[13]	[14]	[15]	[12]	[12]	This Work
PSRR [dB]	76	28	45	30	50	39
FOMS [MHz·pF/µW]	0.45	15.7	10.2	15.6	80.2	96.6
FOML [(V/µs)·pF/µW]	0.52	4.07	11.8	5.3	26.5	0.58
FOMS,A [MHz·pFµW·mm^2]	17.3	2453	2833	15,885	81,724	1,094,000
FOML,A [V/µs·pFµW·mm^2]	20.2	635.9	3277	5397	27,000	6568

However, it has to be noted that, as pointed out in [16,17], the operation of Digital OTAs is typically strongly sensitive to PVT variations and mismatch, and often requires suitable calibration strategies to achieve high production yield. This also apply to the proposed implementation in which some sort of calibration [16,17] and/or V_{DD} adjustment strategy is required to cope with variations.

5. Conclusions

In this paper, a digital OTA which, unless the others in literature, is realized with only digital gates of a standard cell library is proposed and demonstrated. The post layout circuit, resulting from a fully automatic design process, was simulated in open and closed loop conditions, which completely validate the idea.

In order to compare the performance of the proposed SC-DIGOTA with respect the previously DIGOTAs presented in literature, well-known figures of merits have been used. The comparison shows that added to main strength deriving by the fully standard cell realization, which gives significant advantages in the design step. The SG-DIGOTA results are also very competitive in the small signal domain, especially considering the very small silicon area required, while performing less in the large signal domain.

Author Contributions: Conceptualization, G.S.; methodology, G.S. and G.P.; software, G.S.; validation, G.S. and G.P.; writing—original draft preparation, G.S. and G.P.; writing—review and editing, G.S. and G.P. All authors have read and agreed to the published version of the manuscript.

Funding: This research received no external funding.

Data Availability Statement: The data presented in this study are available in article.

Conflicts of Interest: The authors declare no conflict of interest.

Appendix A

Verilog netlist of the SC-DIGOTA is as follows:

```
module SC-DIGOTA (input Vip, input Vin, output Out);
wire INn;
wire INp;
wire CM;
wire EN;
wire not_EN;
wire OP;
wire ON;
IVLL IV1(.A(Vip),.Z(INn));
IVLL IV2(.A(CM),.Z(INn));
IVLL IV3(.A(Vin),.Z(INp));
IVLL IV4(.A(CM),.Z(INp));
IVLL IV5(.A(INn),.Z(OP));
IVLL IV6(.A(INp),.Z(ON));
IVLL IV7(.A(EN),.Z(not_EN));
EOLL XOR1(.A(OP),.B(ON),.Z(EN));
```

BTSLL BT1(.A(OP),.E(EN),.Z(Out));
ITSLL IT1(.A(OP),.E(not_EN),.Z(CM));
endmodule

References

1. Alioto, M. *Enabling the Internet of Things: From Integrated Circuits to Integrated Systems*; Springer: Berlin/Heidelberg, Germany, 2017.
2. Liu, J.; Park, B.; Guzman, M.; Fahmy, A.; Kim, T.; Maghari, N. A Fully Synthesized 77-dB SFDR Reprogrammable SRMC Filter Using Digital Standard Cells. *IEEE Trans. Very Large Scale Integr. (VLSI) Syst.* **2018**, *26*, 1126–1138. [CrossRef]
3. Aiello, O.; Crovetti, P.; Alioto, M. Standard Cell-Based Ultra-Compact DACs in 40-nm CMOS. *IEEE Access* **2019**, *7*, 126479–126488. [CrossRef]
4. Fick, L.; Fick, D.; Alioto, M.; Blaauw, D.; Sylvester, D. A 346 μm 2 VCO-Based, Reference-Free, Self-Timed Sensor Interface for Cubic-Millimeter Sensor Nodes in 28 nm CMOS. *IEEE J. Solid-State Circuits* **2014**, *49*, 2462–2473. [CrossRef]
5. Richmond, J.; John, M.; Alarcon, L.; Zhou, W.; Li, W.; Liu, T.T.; Alioto, M.; Sanders, S.R.; Rabaey, J.M. Active RFID: Perpetual wireless communications platform for sensors. In Proceedings of the 2012 ESSCIRC (ESSCIRC), Bordeaux, France, 17–21 September 2012; pp. 434–437.
6. Paul, S.; Honkote, V.; Kim, R.G.; Majumder, T.; Aseron, P.A.; Grossnickle, V.; Sankman, R.; Mallik, D.; Wang, T.; Vangal, S.; et al. A Sub-cm3 Energy-Harvesting Stacked Wireless Sensor Node Featuring a Near-Threshold Voltage IA-32 Microcontroller in 14-nm Tri-Gate CMOS for Always-ON Always-Sensing Applications. *IEEE J. Solid-State Circuits* **2017**, *52*, 961–971. [CrossRef]
7. Aiello, O.; Crovetti, P.; Alioto, M. Fully Synthesizable, Rail-to-Rail Dynamic Voltage Comparator for Operation down to 0.3 V. In Proceedings of the 2018 IEEE International Symposium on Circuits and Systems (ISCAS), Florence, Italy, 27–30 May 2018; pp. 1–5.
8. Aiello, O.; Crovetti, P.; Lin, L.; Alioto, M. A pW-Power Hz-Range Oscillator Operating with a 0.3–1.8-V Unregulated Supply. *IEEE J. Solid-State Circuits* **2019**, *54*, 1487–1496. [CrossRef]
9. Crovetti, P.S. A Digital-Based Analog Differential Circuit. *IEEE Trans. Circuits Syst. I Regul. Pap.* **2013**, *60*, 3107–3116. [CrossRef]
10. Toledo, P.; Crovetti, P.; Aiello, O.; Alioto, M. Fully Digital Rail-to-Rail OTA with Sub-1000um2 Area, 250-mV Minimum Supply, and nW Power at 150-pF Load in 180 nm. *IEEE Solid-State Circuits Lett.* **2020**, *3*, 474–477. [CrossRef]
11. Toledo, P.; Crovetti, P.; Klimach, H.; Bampi, S.; Aiello, O.; Alioto, M. A 300mV-Supply, Sub-nW-Power Digital-Based Operational Transconductance Amplifier. *IEEE Trans. Circuits Syst. II Express Briefs* **2021**, *68*, 3073–3077. [CrossRef]
12. Toledo, P.; Crovetti, P.; Aiello, O.; Alioto, M. Design of Digital OTAs with Operation Down to 0.3 V and nW Power for Direct Harvesting. *IEEE Trans. Circuits Syst. I Regul. Pap.* **2021**, *68*, 3693–3706. [CrossRef]
13. Chatterjee, S.; Tsividis, Y.; Kinget, P. 0.5-V analog circuit techniques and their application in OTA and filter design. *IEEE J. Solid-State Circuits* **2005**, *40*, 2373–2387. [CrossRef]
14. Centurelli, F.; Della Sala, R.; Scotti, G.; Trifiletti, A. A 0.3 V, Rail-to-Rail, Ultralow-Power, Non-Tailed, Body-Driven, Sub-Threshold Amplifier. *Appl. Sci.* **2021**, *11*, 2528. [CrossRef]
15. Centurelli, F.; Della Sala, R.; Monsurrò, P.; Scotti, G.; Trifiletti, A. A 0.3 V Rail-to-Rail Ultra-Low-Power OTA with Improved Bandwidth and Slew Rate. *J. Low Power Electron. Appl.* **2021**, *11*, 19. [CrossRef]
16. Toledo, P.; Aiello, O.; Crovetti, P.S. A 300mV-Supply Standard-Cell-Based OTA with Digital PWM Offset Calibration. In Proceedings of the IEEE Nordic Circuits and Systems Conference (NORCAS): NORCHIP and International Symposium of System-on-Chip (SoC), Helsinki, Finland, 29–30 October 2019; pp. 1–5.
17. Toledo, P.; Crovetti, P.; Klimach, H.; Bampi, S. Dynamic and static calibration of ultra-low-voltage, digital-based operational transconductance amplifiers. *Electronics* **2020**, *9*, 983. [CrossRef]

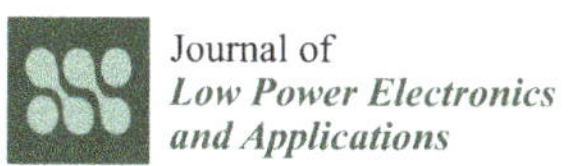

Journal of
*Low Power Electronics
and Applications*

Article

A Tree-Based Architecture for High-Performance Ultra-Low-Voltage Amplifiers

Francesco Centurelli *, Riccardo Della Sala, Pietro Monsurrò, Giuseppe Scotti and Alessandro Trifiletti

Dipartimento di Ingegneria dell'Informazione, Elettronica e Telecomunicazioni (DIET), Università di Roma La Sapienza, 00184 Roma, Italy; riccardo.dellasala@uniroma1.it (R.D.S.); pietro.monsurro@uniroma1.it (P.M.); giuseppe.scotti@uniroma1.it (G.S.); alessandro.trifiletti@uniroma1.it (A.T.)
* Correspondence: francesco.centurelli@uniroma1.it

Abstract: In this paper, we introduce a novel tree-based architecture which allows the implementation of Ultra-Low-Voltage (ULV) amplifiers. The architecture exploits a body-driven input stage to guarantee a rail-to-rail input common mode range and body-diode loading to avoid Miller compensation, thanks to the absence of high-impedance internal nodes. The tree-based structure improves the CMRR of the proposed amplifier with respect to the conventional OTA architectures and allows achievement of a reasonable CMRR even at supply voltages as low as 0.3 V and without tail current generators which cannot be used in ULV circuits. The bias currents and the static output voltages of all the stages implementing the architecture are accurately set through the gate terminals of biasing transistors in order to guarantee good robustness against PVT variations. The proposed architecture and the implementing stages are investigated from an analytical point of view and design equations for the main performance metrics are presented to provide insight into circuit behavior. A 0.3 V supply voltage, subthreshold, ultra-low-power (ULP) OTA, based on the proposed tree-based architecture, was designed in a commercial 130 nm CMOS process. Simulation results show a dc gain higher than 52 dB with a gain-bandwidth product of about 35 kHz and reasonable values of CMRR and PSRR, even at such low supply voltages and considering mismatches. The power consumption is as low as 21.89 nW and state-of-the-art small-signal and large-signal FoMs are achieved. Extensive parametric and Monte Carlo simulations show the robustness of the proposed circuit to PVT variations and mismatch. These results confirm that the proposed OTA is a good candidate to implement ULV, ULP, high performance analog building blocks for directly harvested IoT nodes.

Keywords: body-driven; ultra-low-voltage; ultra-low-power; operational transconductance amplifier; IoT

Citation: Centurelli, F.; Della Sala, R.; Monsurrò, P.; Scotti, G.; Trifiletti, A. A Tree-Based Architecture for High-Performance Ultra-Low-Voltage Amplifiers. *J. Low Power Electron. Appl.* **2022**, 12, 12. https://doi.org/10.3390/jlpea12010012

Academic Editor: Orazio Aiello

Received: 20 January 2022
Accepted: 11 February 2022
Published: 17 February 2022

Publisher's Note: MDPI stays neutral with regard to jurisdictional claims in published maps and institutional affiliations.

1. Introduction

The continuous evolution of electronic systems and the ever increasing symbiotic relationship between humans and electronic devices characterize the era of Internet of Things (IoT) [1,2]. Smart and portable devices, such as laptops, smartphones, smartwatches, fit-trackers and so on, are used more and more often for checking emails, banking management, counter services and the like. Indeed, most of these electronic apparatuses have changed the way we work, study or play.

This IoT revolution has also driven the development of body area networks [3], which exploit implantable and wearable devices, and are widely used in healthcare monitoring and in the study of neurodegenerative diseases such as Parkinson's, Alzheimer's and so on [4–7].

The growing popularity of these electronic devices is also due to their increasing capability to work with low power consumption and low supply voltage in order to maximize battery life or employ energy harvesting techniques.

The stringent requirements in terms of ultra-low-power (ULP) and ultra-low-voltage (ULV) operation set by the above applications have brought about a revolution also in the approach to the design of analog integrated circuits (ICs). In fact, the latter have to be reinvented to enhance the autonomy of smart devices and find a balance between performance, area footprint and power consumption at supply voltages of a few hundreds of millivolts. As such, analog interfaces are among the most challenging building blocks for IoT applications [1,8–11].

The Operational Transconductance Amplifier (OTA) stands out, among the analog building blocks, for its design complexity, especially if ULP and ULV operation are key requirements. In the last few years, there has been a growing trend in the design of ULP OTAs and a plenty of solutions have been proposed in the literature [12–14]. Most of the low voltage OTAs reported in the last decade operate with supply voltages ranging from 0.5 V to about 1 V, and are based on the conventional cascode, folded cascode, multi-stage or gain-boosting approaches, which have been successfully exploited in the past to implement high-performance amplifiers for several application scenarios [15–19]. A novel OTA architecture based on current gain stages to improve bandwidth and slew rate has been recently proposed in [20]. The OTA reported in [20] operates with a supply voltage of 1 V and exhibits state of the art small-signal and large-signal figures of merit. Unfortunately, most of these conventional amplifier topologies are not suited for applications requiring supply voltages lower than 0.5 V, and inverter-based [21–26] and pseudo-differential [27,28] architectures are preferred. However, an aggressive supply voltage scaling severely limits the swing of the control voltage, thus strongly limiting the effectiveness of body bias approaches to set the bias or the common mode current. Therefore, gate-driven amplifiers operating at supply voltages lower than 0.5 V are not able to guarantee either rail-to-rail input common mode range (ICMR) or well-defined bias currents.

The bulk-driven technique [29–31] allows rail-to-rail ICMR in ULV amplifiers at the cost of reduced gain and a resistive input impedance component. Bulk-driven amplifiers are surely one of the best alternatives to attain rail-to-rail input–output swing when a well-defined bias or common mode current is required to increase the robustness against process, supply voltage, and temperature (PVT) variations [32–42]. Indeed, the signal-free gate terminals can be used to accurately set the bias current of the different OTA stages. The bulk-driven technique combined with inverter-based topologies has also been exploited in recent papers to design ULV amplifiers [33,36,39].

A completely novel approach based on fully digital operation to the design of analog differential circuits has been introduced in [43]. Several papers dealing with the fully digital implementation of OTAs for IoT applications have been recently published [44–46]. The digital OTAs in [44,45] are based on the C-Muller element and do not require any passive component. Such digital OTAs are able to operate at supply voltages lower than 0.3 V and are very interesting from the viewpoint of the area footprint and power consumption. However, the operation of this kind of circuits can be sensitive to PVT variations and mismatch and may require suitable calibration strategies to achieve high production yield [47].

Indeed, even if bulk-driven OTAs exhibit some drawbacks with respect to gate-driven ones (higher noise, larger area and lower bandwidth) and to digital OTAs (larger area and power consumption), they can be designed to be robust against PVT and mismatch variations and still represent the best solution to attain rail-to-rail ICMR at supply voltages of the order of 0.3 V.

In this work, we present a novel OTA architecture based on a tree-like structure. This can be viewed as the ULV implementation of the OTA reported in [20], previously proposed by the authors to enhance the bandwidth efficiency. The current gains obtained by means of conventional current mirrors in [20] are not feasible in ULV conditions and have to be implemented by means of other solutions such as the one presented in [48]. In the ULV architeture proposed in this paper, the current gains are implemented by using a different approach which is based on the body-to-gate (B2G) interfaces as will be detailed in the following. The proposed architecture exploits a body-driven input stage to guarantee a rail-to-rail input common mode range and body-diode loading to avoid Miller compensation, thanks to the absence of high-impedance internal nodes. The bias currents and the static output voltages of all the stages implementing the proposed architecture are accurately set through the gate terminals of biasing transistors in order to guarantee a good robustness against PVT variations. However, this biasing strategy results in pseudo-differential stages and therefore has a negative impact on CMRR performance. The proposed tree-like structure improves the CMRR of the OTA with respect to conventional pseudo-differential amplifiers and allows achievement of a reasonable CMRR even in ULV conditions. A 0.3 V supply voltage ULP OTA based on this architecture was designed in a 130 nm CMOS process, and simulation results show state of the art small-signal and large-signal figures of merit (FoMs).

The paper is organized as follows: Section 2 introduces the proposed OTA architecture. Circuit analysis is reported in Section 3. Section 4 deals with design and simulation results and conclusions are drawn in Section 5.

2. Proposed Topology

The block scheme of the proposed OTA architecture is depicted in Figure 1. This architecture of ULV OTA was derived from the OTA introduced by the authors in [20] and is a three stage, tree-like OTA, made up of the cascade of differential-to-single-ended converter stages, to maximize CMRR. Three different topologies are exploited in the three stages of the OTA to optimize the tradeoff between performance and efficiency. Each one of these stages was extensively investigated and their behavior is discussed in the next subsections. It has to be remarked that the proposed ULV OTA makes extensive use of the body terminals of MOS devices and thus it can be implemented only in CMOS technologies (such as triple-well-bulk or FDSOI), where both NMOS and PMOS transistors have available body connections. However, this is not a strong limitation, since most modern processes have available body connections for both PMOS and NMOS transistors.

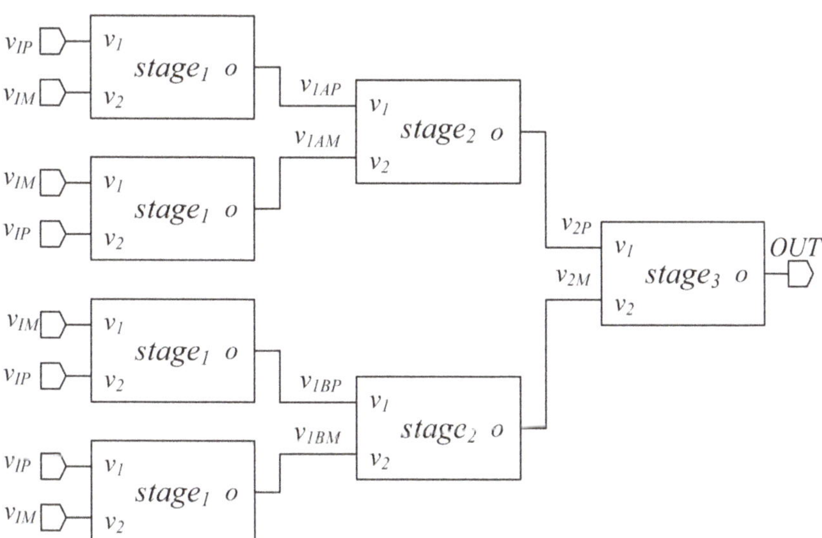

Figure 1. Proposed tree-like architecture of the OTA.

2.1. Stage$_1$

The topology of the blocks denoted as stage$_1$ in Figure 1 is reported in Figure 2, and is made up of transistors M_{1A}, M_{1B} and M_{2A}, M_{2B}. This input stage has the same topology adopted for the OTA in [40]. It is a bulk-driven stage in which the bias current is accurately set through the V_{GN} voltage applied to the gate of transistor M_{2A}. The bias voltage V_{GN} is generated by the biasing circuit reported in Figure 3. The current flowing in M_{2A} is mirrored through M_{1A} and M_{1B}, so that the standby current of all MOS devices is accurately set. The body terminals of transistors M_{1A} and M_{1B} are connected to the input voltages, V_{IP} and V_{IM}, respectively. The output of stage $_1$ is loaded through a body–diode connection on the transistor M_{2B} whose gate voltage is connected to the bias voltage V_{GN}, and results in an output impedance lower than the one of conventional input stages. This stage thus provides limited gain, but allows achievement of a rail-to-rail input common mode range and improvement of the bandwidth. As a consequence, noise and mismatch of the second stage contributes to the total input referred noise and offset. However, even if noise and offset performance are suboptimal, the OTA can still be designed to exhibit acceptable noise and offset, while achieving very good bandwidth efficiency.

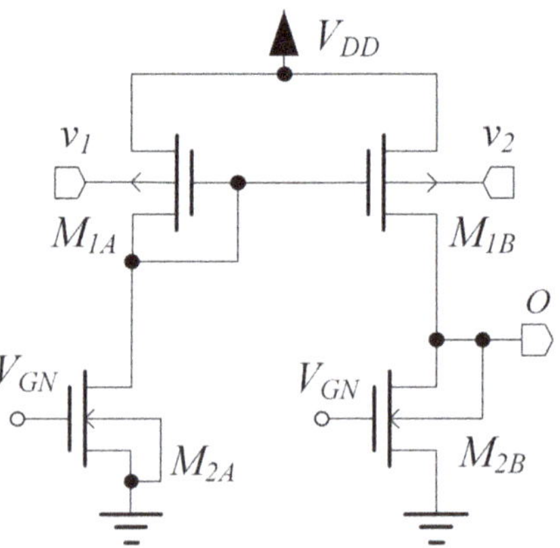

Figure 2. Stage $_1$ used in the proposed OTA architecture.

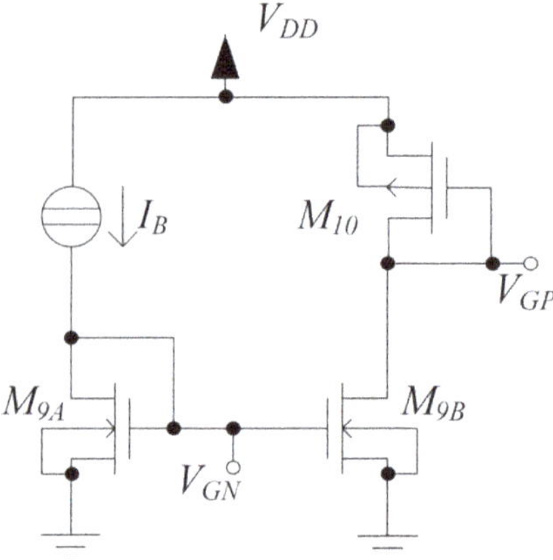

Figure 3. Biasing circuit used in the proposed OTA architecture.

2.2. Stage$_2$

The topology of stage$_2$ is shown in Figure 4. This stage converts the input differential signal to single-ended providing some gain, a well defined bias point and contributing to the overall CMRR. The input signal is applied to the gates of M_{4A} and M_{4B}, and the bias current is set through the gates of M_{3A} and M_{3B} connected to the bias voltage V_{GP} generated by the circuit in Figure 3. The current cancellation given by the body-to-body (B2B) current mirror (Appendix B) M_{4A}, M_{4B} allows to attain good common mode rejection ratio as will be better shown in the next sections. Since the output is body-loaded, also this stage doesn't show any high-impedance internal node and thus does not require any internal compensation.

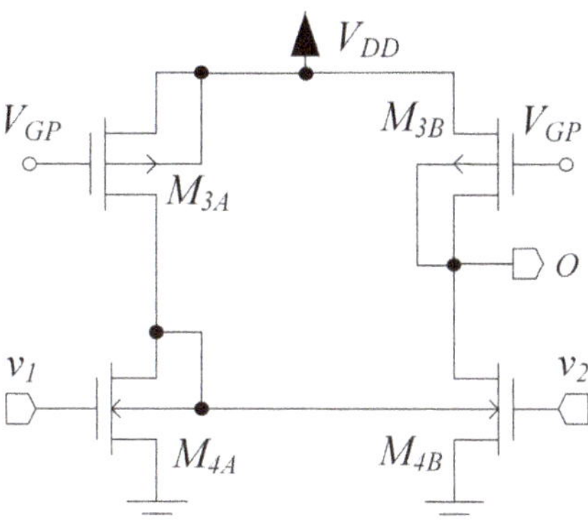

Figure 4. Stage$_2$ used in the proposed OTA architecture.

2.3. Stage$_3$

The topology of stage$_3$ is shown in Figure 5. This stage combines the signal behavior of an inverter-based pseudo-differential pair (Arbel topology) with differential-to-single-ended conversion through the body current mirror and robust biasing, and is composed by an n-input and a p-input stage similar to that of Figure 4, but without diode loading, connected together. The signal is applied to the gates of two PMOS and two NMOS devices, respectively M_{6A}, M_{6B} and M_{8A}, M_{8B}, and the body-diode connections in M_{6A} and M_{7B} implement body-driven current mirrors performing differential-to-single-ended conversion and common mode current cancellation. Transistors M_{5A}, M_{5B} and M_{7A} and M_{7B} act as current sources and are exploited to set the bias current in all the branches of the third stage through V_{GP} and V_{GN}, respectively; thus, each transistor has a well-defined bias point.

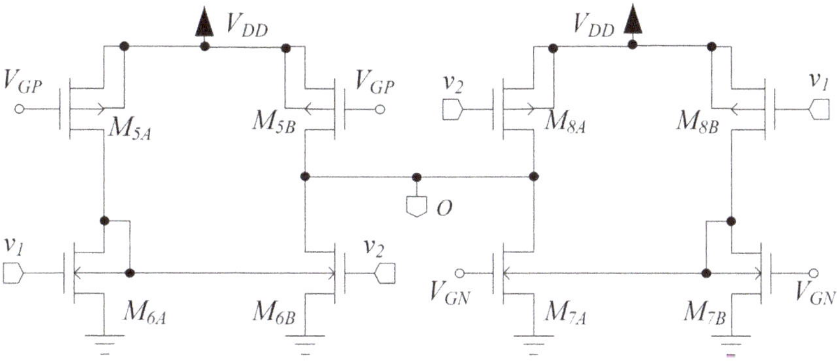

Figure 5. Stage$_3$ used in the proposed OTA architecture.

2.4. Architectural Considerations

It has to be noted that, referring to the proposed architecture, at the interfaces between stage$_1$ and stage$_2$ and between stage$_2$ and stage$_3$, we have a body to-gate (B2G) connection. These B2G connections result in lower voltage gain with respect to the conventional drain-to-gate connections, but the lower gain allows avoidance of high-impedance internal nodes, and therefore compensation capacitors. In fact, even if each B2G interface generates a pole (as shown in Appendix A), it is placed at a much higher frequency than the one given by the output stage, which provides the dominant pole.

3. Circuit Analysis

In this section, the small-signal and large-signal performances of the proposed architecture are analyzed from an analytical point of view, and design equations for the main performance parameters, such as gain, frequency response, slew-rate and noise, are presented to provide insight into circuit behavior.

3.1. Differential Gain

Referring to the small-signal equivalent circuits of stage$_1$, stage$_2$ and stage$_3$, the differential mode gain of the different stages was computed. Using the standard notation for small-signal parameters of MOS devices, the differential gain of the first stage can be expressed as:

$$A_{vd_1} = \frac{g_{mb_1}}{g_{mb_2}} \frac{1 + s\frac{\tau_1}{2}}{(1 + s\tau_1)(1 + s\tau_2)} \tag{1}$$

where:

$$\tau_1 \approx \frac{2C_{gs_1} + C_{gd_1}(1 + \frac{g_{m_1}}{g_{mb_2}}) + C_{gd_2}}{g_{m_1}}$$
$$\tau_2 \approx \frac{C_{gs_4} + C_{gd_4}(1 + \frac{g_{m_4}}{g_{mb_3}}) + C_{gd_2} + C_{gd_1} + C_{bs_2}}{g_{mb_2}} \tag{2}$$

According to usual approximations, the pole-zero doublet in Equation (1) can be neglected.

Thereafter, the differential gain of stage$_2$ can be derived to be:

$$A_{vd_2} = \frac{g_{m_4}}{g_{mb_3}} \frac{1 + s\frac{\tau_3}{2}}{(1 + s\tau_3)(1 + s\tau_4)} \tag{3}$$

where:

$$\tau_3 \approx \frac{2C_{bs_4} + C_{gd_3} + C_{gd_4}}{g_{mb_4}}$$
$$\tau_4 \approx \frac{C_{gs_6} + C_{gs_8} + C_{gd_3} + C_{gd_4} + C_{bs_3} + \frac{g_{m_8}}{g_{out}}C_{gd_8} + \frac{g_{m_6}}{g_{out}}C_{gd_6}}{g_{mb_3}} \tag{4}$$

Moreover, in this case, the pole-zero doublet in Equation (3) can be neglected.

Finally, the stage$_3$ differential gain can be computed by neglecting the pole-zero doublets given by body–diode connections of $M_{6_{A,B}}$ and $M_{7_{A,B}}$; hence, it can be expressed as:

$$A_{vd_3} = \frac{g_{m_8} + g_{m_6}}{g_{out}} \frac{1}{1 + s\frac{C_L}{g_{out}}} \tag{5}$$

where it is denoted with:

$$g_{out} = 2(g_{ds_8} + g_{ds_6}) \tag{6}$$

considering that $M_5 = M_8$ and $M_6 = M_7$.

The overall gain of the amplifier can then be expressed as:

$$A_{vd_{tot}}(s) = 4 \prod_{i=1}^{3} A_{vd_i}(s) \tag{7}$$

and rewritten as:

$$A_{vd_{tot}}(s) = 4 \cdot \frac{g_{m_8} + g_{m_6}}{g_{out}} \cdot \frac{g_{mb_1}}{g_{mb_2}} \cdot \frac{g_{m_4}}{g_{mb_3}} \cdot \frac{1}{(1 + s\frac{C_L}{g_{out}})} \cdot \frac{1}{(1 + s\tau_2)(1 + s\tau_4)} \tag{8}$$

It is evident from Equation (8) that the output capacitance sets the dominant pole since the poles of stage$_1$ and stage$_2$ are at higher frequencies due to the body–diode connected loads and the smaller load capacitances.

Starting from the above results, the gain-bandwidth product (GBW) of the proposed OTA can be computed as:

$$GBW = \frac{g_\alpha}{2\pi \cdot C_L} \tag{9}$$

where:

$$g_\alpha = (g_{m_8} + g_{m_6}) \cdot \frac{g_{mb_1}}{g_{mb_2}} \cdot \frac{g_{m_4}}{g_{mb_3}} \tag{10}$$

The phase margin of the whole OTA can then be expressed as:

$$\varphi_m = \frac{\pi}{2} - \arctan\left(\frac{g_\alpha}{C_L} \cdot \tau_2\right) - \arctan\left(\frac{g_\alpha}{C_L} \cdot \tau_4\right) \tag{11}$$

According to Equation (11), the proposed OTA requires a minimum value of C_L for stability. However, Equation (11) shows also that the desired phase margin can be set by properly designing MOS devices' size for a given load capacitor; a higher C_L results in a smaller GBW and a larger phase margin.

3.2. Common Mode Gain

The common mode gain of stage$_1$ was found to be:

$$A_{vc_1} = -\frac{g_{mb_1}(g_{ds_1} + g_{ds_2})}{g_{mb_2} g_{m_1}} \frac{1 + s\tau_{z_1}}{(1 + s\tau_{p_{1,1}})(1 + s\tau_{p_{2,1}})} \tag{12}$$

where:

$$\tau_{z_1} = \tau_1 \frac{g_{m_1}}{g_{ds_1} + g_{ds_2}} \quad \tau_{p_{1,1}} = \tau_1 \quad \tau_{p_{2,1}} = \tau_2 \tag{13}$$

therefore, the CMRR of stage$_1$ can be expressed as:

$$\text{CMRR}_1 = \frac{g_{m_1}}{g_{ds_1} + g_{ds_2}} \tag{14}$$

The common mode gain of stage$_2$ is:

$$A_{vc_2} = -\frac{g_{m_4}}{g_{mb_4}} \frac{g_{ds_3} + g_{ds_4}}{g_{mb_3}} \frac{1 + s\tau_{z_2}}{(1 + s\tau_{p_{1,2}})(1 + s\tau_{p_{2,2}})} \tag{15}$$

where:

$$\tau_{z_2} = \tau_3 \frac{g_{mb_4}}{g_{ds_3} + g_{ds_4}} \quad \tau_{p_{1,2}} = \tau_3 \quad \tau_{p_{2,2}} = \tau_4 \tag{16}$$

whereas its CMRR amounts to:

$$\text{CMRR}_2 = \frac{g_{mb_4}}{g_{ds_4} + g_{ds_3}} \tag{17}$$

Stage$_3$ shows a common mode gain of:

$$A_{vc_3} - \frac{g_{m_8} + g_{m_6}}{2g_{mb_6}} \frac{1 + s\tau_{z_3}}{(1 + s\tau_{p_{1,3}})(1 + s\tau_{p_{2,3}})} \tag{18}$$

where:

$$\tau_{z_3} = \frac{2C_{bs_6} + C_{gd_8} + C_{gd_6}}{g_{mb_6}} \quad \tau_{p_{1,3}} = 2\frac{2C_{bs_6} + C_{gd_8} + C_{gd_6}}{g_{ds_8} + g_{ds_6}} \quad \tau_{p_{2,3}} = \frac{C_L}{g_{ds_8} + g_{ds_6}} \tag{19}$$

and its CMRR results:

$$\text{CMRR}_3 = \frac{g_{mb_6}}{g_{ds_8} + g_{ds_6}} \tag{20}$$

Due to the body current mirror, the CMRR of these stages is reduced with respect to stage$_1$. Combining the above results, the common mode gain of the proposed tree-like architecture can be derived as:

$$A_{vc_{TOT}} = \prod_{i=1}^{3} A_{vc_i} \tag{21}$$

Finally, the CMRR of the overall OTA can be expressed as:

$$CMRR_{tot} = 4\prod_{i=1}^{3} CMRR_i \tag{22}$$

therefore, the total CMRR is about:

$$CMRR_{tot} = 4 \cdot \frac{g_{m_1}}{g_{ds_1} + g_{ds_2}} \cdot \frac{g_{mb_4}}{g_{ds_4} + g_{ds_3}} \cdot \frac{g_{mb_6}}{g_{ds_8} + g_{ds_6}} \tag{23}$$

By looking at Equation (22), it is evident that the CMRR in typical conditions is high, due both to the cascade of several stages and to the scaling factor of the tree architecture, and that it can be enhanced by further iterating the tree-like structure of the proposed OTA architecture. However, in ULV conditions, PVT variations and mismatch may impact on the stability of the operating point, especially in the presence of a B2G interface, and significantly degrade the CMRR$_{i\text{-th}}$ of the OTA. As a consequence, the CMRR of this architecture is more sensitive to PVT variations and mismatch than other architectures which adopt higher supply voltages and/or a more stable operating point. Anyway, to cope with this problem, design centering techniques are exploited in this work in order to increase the overall CMRR in a given range of PVT and mismatch conditions achieving a reasonable robustness. The above reported frequency analysis shows that the common mode gain presents some zeros that could appear before the unity-gain frequency (depending on the C_L/C_{gs} ratio), thus reducing the CMRR at high frequency. A large load capacitance is usually required to achieve stability, therefore the resulting CMRR reduction is often limited.

3.3. Large-Signal Performances

The large-signal performance of the proposed OTA has been investigated by assuming that the load capacitance C_L is much larger than the other circuit capacitances. The slew-rate is thus determined by the output stage, and it can be assumed that the output voltage v_{O2} of stage$_2$, which drives stage$_3$, is a rail-to-rail signal.

With reference to Figure 5, the output current is given by $I_o = I_{5B} + I_{8A} - I_{6B} - I_{7A}$; positive and negative slew-rates are given by $SR_p = I_{o_{max}}/C_L$ and $SR_m = I_{o_{min}}/C_L$, where $I_{o_{max}}$ and $I_{o_{min}}$ are the maximum positive and negative values of I_o.

For the current, we use the standard relationship for sub-threshold current:

$$I_{n,p} = I_{0_{n,p}}\, e^{\frac{V_{ov} - |V_{th_{n,p}}|}{n_{n,p} U_t}} \tag{24}$$

where $U_t = kT/q$ is the thermal voltage and $|V_{th_{n,p}}| = V_{th_{n,p_0}} - \alpha_{n,p}|V_{bs}|$.

For the positive slew-rate, we have $v_1 = V_{DD}$ and $v_2 = 0$, and we can assume that the body voltages of M_{6_B} and M_{7_A} are approximately 0. By denoting with I_{ref}, the quiescent current of the devices of stage$_3$, we obtain:

$$I_{o_{max}} = I_{ref}\left[1 + e^{\frac{\alpha_n|\Delta V_{BH}|}{n_n U_t}} + e^{\frac{\Delta|V_{GH}|}{n_p U_t}}\right] \tag{25}$$

where: $\Delta V_{BH} = -V_{B0}$, $\Delta V_{GH} = V_{DD} - V_{GP}$ with V_{B0} and V_{GP} the quiescent voltage at body and gate terminals of the NMOS and PMOS devices.

For the negative slew-rate, we have $v_1 = 0$, $v_2 = V_{DD}$ and in this case we derive:

$$I_{0_{min}} = I_{ref}\left[1 - e^{\frac{a_n|\Delta V_{BL}|}{n_n U_t}}\left(1 + e^{\frac{|\Delta V_{GL}|}{n_n U_t}}\right)\right] \tag{26}$$

where: $\Delta V_{BL} = V_{DD} - V_{B0}$ and $\Delta V_{GL} = V_{DD} - V_{GN}$ with V_{GN} as the quiescent voltage at gate terminals of NMOS devices. In this case, we assume that the body terminals of M_{6_B} and M_{7_A} are approximately V_{DD}. Equations (25) and (26) show that, in general, positive and negative slew-rates give different results.

3.4. Noise Analysis

The noise analysis has been carried out assuming that each transistor can be modelled with only one noise current generator, which includes both thermal and flicker noise. The power spectral density of the modelled current generator can be expressed as follows:

$$S_{n_i} = \overline{i^2_{i_w}} + \overline{i^2_{i_f}} \tag{27}$$

where:

$$\overline{i^2_{n(p)_w}} = 4kTn_{n(p)}\gamma g_{m_i} = 2qI_d \tag{28}$$

$$\overline{i^2_{n(p)_f}} = \frac{K_{n(p)}}{fC_{ox}}\frac{g_m^2}{WL} \tag{29}$$

Taking into account that the noise sources due to $stage_3$ can be neglected due to the high gain of the preceding stages (considering also the contribution of the tree structure), the equivalent input noise mainly results from the first two stages and can be expressed as follows:

$$S_{v_{eq}} = \frac{S_{n_1} + S_{n_2}}{2\,g^2_{mb_1}} + \frac{1}{4\,g^2_{m_4}}\cdot\frac{g^2_{mb_2}}{g^2_{mb_1}}(S_{n_3} + S_{n_4}) \tag{30}$$

As it can be observed from Equation (30), the noise performance of the amplifier is worsened by body driving, which shows a transconductance gain (i.e., g_{mb}) which is n-times lower than g_m. Consequently, in order to reduce the equivalent input noise, larger transistors are required. The result in Equation (30) can be written in a less concise form as:

$$S_{v_{eq}} \approx \frac{1}{16}\left(4S_{no1} + \frac{2S_{no2}}{A_V^2}\right) \tag{31}$$

where

$$S_{no1} = \frac{2}{g^2_{mb_1}}(S_{n_1} + S_{n_2}) \tag{32}$$

and

$$S_{no2} = \frac{2}{g^2_{m_4}}(S_{n_3} + S_{n_4}) \tag{33}$$

are the input-referred noise spectra for the first and second stage (contribution of the single cell). Factor 16 in the denominator of (31) accounts for the $2^{(N-1)}$ gain contribution of a N-level tree architecture, whereas the factors 4 and 2 in the numerator consider how many identical cells are present.

4. Amplifier Design and Simulation Results

The proposed OTA has been designed and simulated in a 130 nm CMOS process from STMicroelectronics. Small-signal and large-signal figures of merit (FoMs) were used to compare it against recently published OTAs with supply voltages lower than 0.5 V. Extensive parametric and Monte Carlo simulations were carried out in order to assess the robustness of the amplifier to PVT variations and mismatch referring to both open-loop and closed-loop simulation test benches.

4.1. Sizing

The transistors in the stages implementing the architecture in Figure 1 were sized as reported in Table 1. The bias voltages V_{GN} and V_{GP} in Figures 2, 4 and 5, are generated by the biasing circuit shown in Figure 3. Moreover, the sizing of the NMOS transistors M_{9A} and M_{9B} and of the PMOS transistor (M_{10}) of the biasing circuit are reported in Table 1. The voltages V_{GN} and V_{GP} propagate the bias current, $I_B = 4$ nA, through body-mirroring or gate-mirroring.

Table 1. Transistors' sizing.

Transistor	Stage	Width [μm]	Length [μm]	I_{bias} [nA]
M_{1A}, M_{1B}	1	4.465	1.000	4
M_{2A}, M_{2B}, M_{9A}, M_{9B}	1	0.375	3.000	4
M_{3A}, M_{3B}, M_{10}	2	4.465	1.000	4
M_{4A}, M_{4B}	2	0.375	3.000	4
M_{5A}, M_{5B}, M_{8A}, M_{8B}	3	13.390	1.000	19.67
M_{6A}, M_{6B}, M_{7A}, M_{7B}	3	1.125	3.000	19.67

4.2. Circuit Simulations

The proposed OTA was simulated within the Cadence Virtuoso environment assuming a supply voltage of 0.3 V and an output load capacitance of 50 pF.

Referring to the open-loop simulation test bench the differential gain (magnitude and phase) was evaluated as reported in Figure 6. As can be observed from the figure, the phase margin is about 52.40°, whereas the gain-bandwidth product is about 35.16 kHz. Figure 6 also shows the common mode gain in typical conditions.

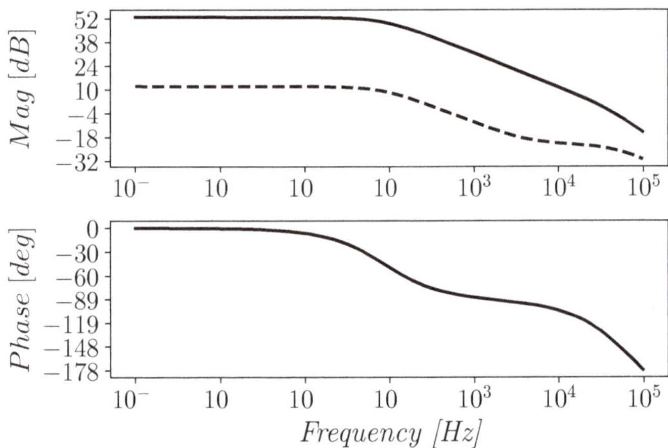

Figure 6. Differential (solid) and common mode (dashed) gain of the proposed OTA.

Figure 7 confirms that the bias currents of all the three stages of the OTA are accurately set and are also very stable for an input signal amplitude going rail-to-rail in closed-loop unity-gain configuration.

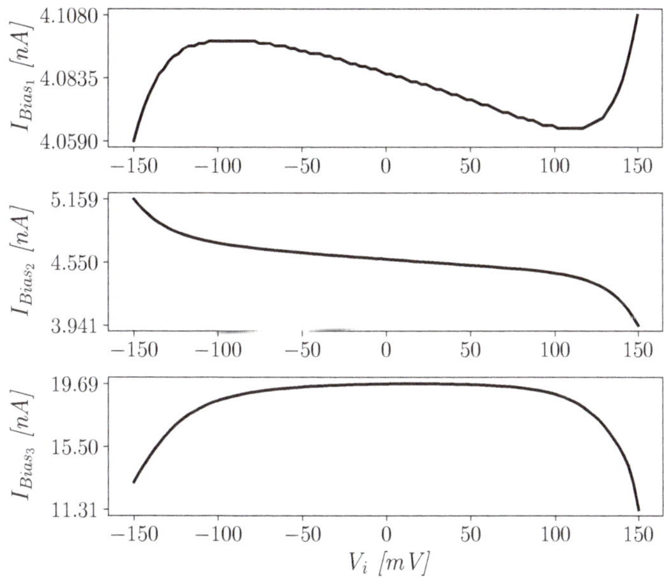

Figure 7. Biasing currents of the three stages vs. input common mode level.

The amplifier was then tested in unity-gain configuration and its transfer characteristic is reported in Figure 8, highlighting the rail-to-rail capabilities of the OTA.

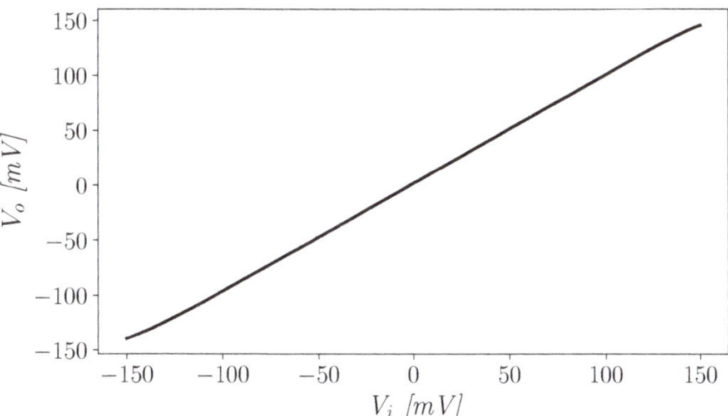

Figure 8. Unity-gain amplifier transcharacteristic.

Sinusoidal waves at different amplitudes and with a frequency of 200 Hz were used to excite the unity-gain amplifier and evaluate distortions. The OTA exhibits very good total harmonic distortion (THD), also with an input signal swing equal to the supply voltage (as depicted in Figure 9). As can be observed from Figure 9, when a 90% signal swing is considered, the THD is about 0.673%, whereas when a full-swing signal is used the THD is still good and equal to about 1.38%. Furthermore, to assess the slew-rate (SR) performance of the amplifier, a full range square wave was used, and results are shown in Figure 10. The amplifier shows positive and negative slew-rate (SR_p and SR_n) equal to 18.61 and 11.51 V/ms, respectively. Though not symmetrical, the worst-case slew-rate is not much worse than the best one, hence large-signal performance is good on both signal edges.

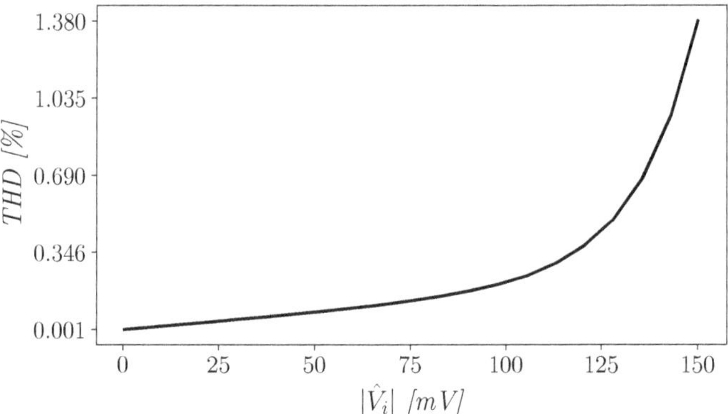

Figure 9. THD vs. amplitude of the input signal in unity-gain configuration.

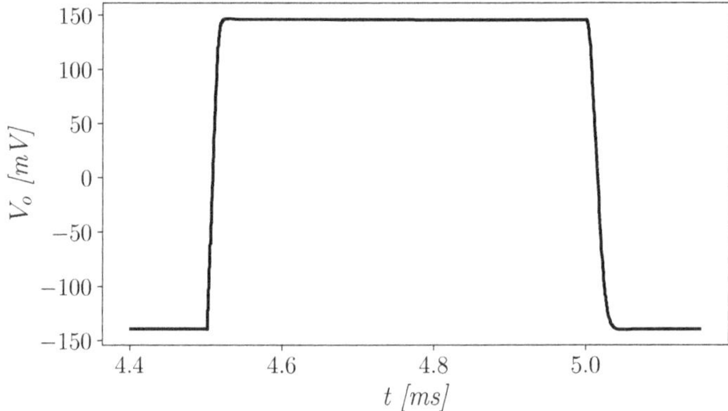

Figure 10. Response to square input wave.

The input-referred noise spectrum of the proposed OTA is reported in Figure 11 and shows a value of about 1.60 μV$/\sqrt{\text{Hz}}$ at 1 kHz.

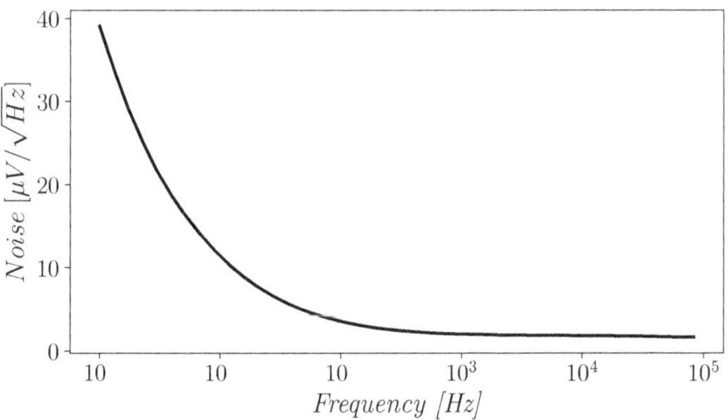

Figure 11. Input-referred noise of the proposed OTA.

4.3. Robustness to Mismatch and PVT Variations

The OTA was then extensively tested by means of parametric and Monte Carlo simulations to demonstrate its robustness to PVT and mismatch variations. Table 2 reports the results of 200 Monte Carlo iterations. Power dissipation (P_D) has a standard deviation lower than the 10% of the mean value. Large-signal performance (i.e., SR_p and SR_m) is close to the nominal value, whereas the attained mean value of the phase margin m_φ is about 53°. The standard deviation of the offset is relatively large, confirming the suboptimal performance in terms of noise and offset of the proposed OTA. Its value is however similar to other ULV OTAs reported in the literature.

Table 2. Performance under mismatch variations.

	Mean	**StdDev**	**Min**	**Max**
P_D (nW)	20.85	1.44	16.6	24.34
Idiss (nA)	69.50	4.80	55.33	81.13
Offset (mV)	3.84	15.46	−30	50
SR_p (V/ms)	18.54	0.30	17.84	19.42
SR_m (V/ms)	11.63	0.34	10.82	12.52
Gain (1 Hz) (dB)	51.48	1.22	49.59	56.49
CMRR (dB)	42.11	10.44	27.84	98.85
PSRR (dB)	56.13	2.12	48.05	56.39
Mphi (deg)	53.08	6.27	38.25	74.98
GBW (kHz)	32.72	8.42	11.54	49.33
THD (%)	0.74	0.57	0.51	2.61

Figure 12 reports the histogram of the CMRR that clearly shows a log-normal distribution, probably due to the sub-threshold operating condition of the circuit. The architecture exhibits a CMRR up to 98dB for some iterations (as expected from theoretical results in Section 3.2), and remains relatively high under mismatch variations, with a mean value of about 42 dB.

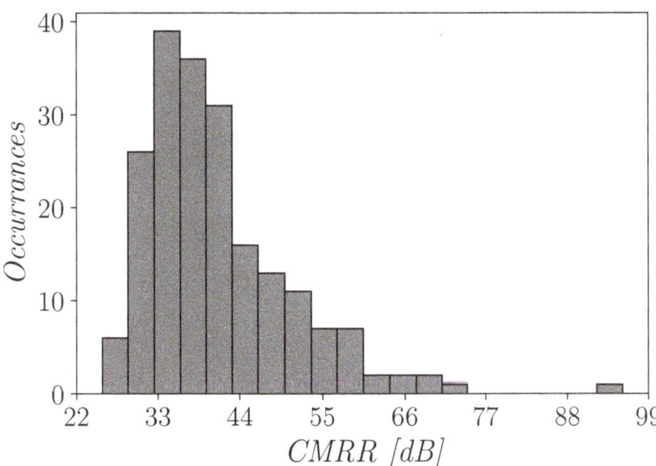

Figure 12. Histogram of the common mode rejection ratio (CMRR) of the proposed OTA for 200 Monte Carlo mismatch iterations.

The power supply rejection ratio (PSRR) of the proposed OTA is also quite good despite the very low supply voltage. Figure 13 reports the histogram of the PSRR, that shows a mean value of about 56.13 dB with a limited variation under mismatch.

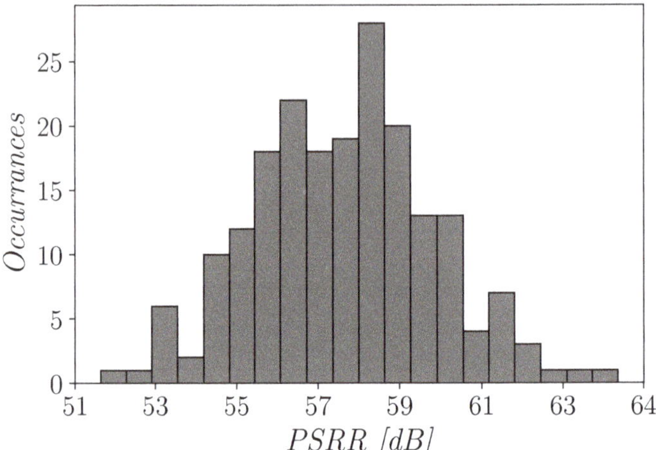

Figure 13. Histogram of the power supply rejection ratio (PSRR) of the proposed OTA for 200 Monte Carlo mismatch iterations.

The performance under PVT variations was investigated taking into account a $\pm 10\%$ supply voltage variation and a $[0, 70]$ °C temperature range. In Table 3, the performance under temperature variations is summarized. Total power consumption, the gain-bandwidth product as well as noise and total harmonic distortion are adequately stable across the considered temperature range. However, it is evident from Table 3 that the differential gain and CMRR degrade at high temperatures; this is probably due to variations in the bias point of $stage_2$ and in particular in transistors M_{4A} and M_{4B} entering the triode region. A temperature-dependent current biasing approach would probably allow achievement of better results, but this has not been considered in this work. Furthermore it has to be noted that an ideal constant current source was considered: while such generator can be devised (e.g., see [49], or using a higher supply voltage for the current reference), this clearly remains a critical issue, dependent on the application environment of the OTA.

Table 3. Performance vs. temperature variations.

Temp (°C)	0.00	16.67	27.00	43.33	50.00	70.00
P_D (nW)	21.48	21.93	21.89	20.40	20.54	21.35
I_D (nW)	71.59	73.10	72.98	68.00	68.46	71.18
SR_p (V/ms)	11.44	15.66	18.61	23.55	25.60	31.76
SR_m (V/ms)	10.11	10.99	11.51	12.47	12.84	13.65
Gain (1Hz) (dB)	58.65	57.61	52	50.07	48.87	46.72
CMRR (dB)	64.45	57.56	44.96	34.31	32.03	26.66
Mphi (deg)	48.63	46.26	52.40	54.54	52.86	48.88
GBW (kHz)	32.85	39.45	35.16	30.80	32.16	37.95
Noise [‡] ($\mu V/\sqrt{Hz}$)	0.60	0.85	1.60	3.42	3.91	4.85
THD (%)	0.45	0.51	0.67	0.72	0.84	1.23

[‡] Computed at 1 kHz.

Table 4 shows that the amplifier is stable under power supply variations, with power dissipation and slew-rate increasing significantly with the supply voltage, whereas CMRR improves at lower supply voltages due to the following design centering approach.

Table 4. Performance vs Voltage Variations.

V_{DD} (mV)	270.0	285.0	300.0	315.0	330.0
P_D (nW)	21.710	21.980	21.890	20.500	20.240
Idiss (nA)	72.370	73.270	72.980	68.350	67.460
SR_p (V/ms)	8.532	12.750	18.610	26.500	36.790
SR_m (V/ms)	7.147	9.161	11.510	14.230	17.210
Gain (1 Hz) (dB)	54.34	53.22	52.93	52.84	53.07
CMRR (dB)	60.340	53.720	44.960	38.740	35.450
Mphi (deg)	47.530	50.230	52.920	53.550	49.570
GBW (kHz)	34.830	35.230	35.160	33.470	36.980
Noise [†] ($\mu V/\sqrt{Hz}$)	0.869	1.011	1.595	2.485	3.161
THD (%)	0.50	0.37	0.29	0.23	0.19

[†] Computed at 1 kHz.

The OTA was then tested under different process corners and results are reported in Table 5. As is evident from Table 5, the proposed OTA shows good performance, even assuming the worst case process conditions.

Table 5. Performance vs. corners.

Corner	TYP	FF	SS	SF	FS
P_D (nW)	21.89	20.32	21.68	21.98	26.60
Idiss (nA)	72.97	67.73	72.27	73.27	88.67
SR_p (V/ms)	18.61	27.32	12.18	28.77	11.63
SR_m (V/ms)	11.51	15.47	8.62	9.00	14.43
Gain (1 Hz)(dB)	52.92	50.41	57.90	55.72	49.93
CMRR (dB)	44.96	33.72	63.31	53.26	35.5
PSRR (dB)	56.40	48.26	73.31	64.93	47.52
Mphi (deg)	52.40	51.37	48.59	42	58.59
GBW (kHz)	35.16	34.43	37.19	49.626	27.55
Noise [‡] ($\mu V/\sqrt{Hz}$)	1.60	3.03	3.03	3.21	5.16
THD (%)	0.67	0.25	0.43	0.95	0.46

[‡] Computed at 1 kHz.

4.4. Discussion and Comparison with the Literature

In order to compare the amplifier with the literature, we employ the two standard figures of merit (FOMs) for small and large-signal performance, namely FOM_S and FOM_L. The FOM_S is defined as:

$$FOM_S = \frac{GBW \cdot C_L}{P_D} \qquad (34)$$

where C_L is the load capacitance; the FOM_L is defined as:

$$FOM_L = \frac{SR_{avg} \cdot C_L}{P_D} \qquad (35)$$

where SR_{avg} is the average (between the positive and negative edge) slew-rate.

However, since most works presented in the literature show an asymmetric slew-rate, it is more meaningful to consider the worst case slew-rate. Consequently, as in [40], we define the $FOM_{L_{WC}}$ as:

$$FOM_{L_{WC}} = \frac{SR_{WC} \cdot C_L}{P_D} \qquad (36)$$

where SR_{WC} is the worst case slew-rate between the positive and negative signal edges.

The proposed amplifier exhibits the largest small-signal FOM among the comparable ULV literature, with a FOM_S approaching 80.29 k against the previously reported record of about 20.16 k attained by [42]. The proposed OTA outperforms gate-driven, body-driven and also digital OTAs. Large-signal performance is also very good, especially if the worst-case *FOM* is considered: the proposed amplifier is the best in the literature. Indeed, the FOM_L is about 34.40 k; furthermore, the worst case $FOM_{L_{WC}}$ also is very good, approximately 26.30 k, which is an awesome result, also given that previous works attained in the best case $FOM_L \approx 21.00$ k and in the worst case $FOM_{L_{WC}} \approx 8.36$ k. The proposed amplifier has a small area occupation with respect to comparable body-driven designs, though the area is larger than digital and gate-driven designs (Table 6).

Table 6. Comparison table.

	This Work *	[42] *	[45] †	[40] *	[39] *	[25] *	[37] †	[50] †	[23] †	[36] *	[51] †
Year	2021	2021	2021	2021	2021	2020	2020	2019	2019	2018	2018
Technology (μm)	0.13	0.13	0.18	0.13	0.13	0.18	0.18	0.18	0.13	0.065	0.18
V_{DD} (V)	0.3	0.3	0.3	0.3	0.3	0.3	0.3	0.3	0.3	0.3	0.3
V_{DD}/V_{TH}	0.86	0.86	0.6	0.86	0.86	0.6	0.6	0.6	0.86	-	0.6
DC_{gain} (dB)	52.92	38.07	30	40.80	64.6	39	98.1	64.7	49.8	60	65.8
C_L (pF)	50	50	150	40	50	10	30	30	2	5	20
GBW (kHz)	35.16	24.14	0.25	18.65	3.58	0.9	3.1	2.96	9100	70	2.78
$m\varphi$ (deg)	52.40	60.15	90	51.93	53.76	90	54	52	76	53	61
$SR_+ \left[\frac{V}{ms}\right]$	18.61	20.02	-	10.83	1.7	-	14	1.9	-	25	6.44
$SR_- \left[\frac{V}{ms}\right]$	11.51	8.44	-	32.37	0.15	-	4.2	6.4	-	25	7.8
$SR_{avg} \left[\frac{V}{ms}\right]$	15.06	14.23	0.085	21.60	0.93	-	9.1	4.15	3.8	25	7.12
THD (%)	0.673	1.635	2	1.4	0.84	1	0.49	1	-	-	1
% of input swing	90	80	90	80	100	23	83.33	85	-	-	93.33
CMRR (dB)	42.11 ‡	54.88	41	67.49	61	30	60	110	-	126	72
PSRR (dB)	56.13 ‡	51.05	30	45	26/28 *	33	61	56	-	90/91 *	62
spot-noise $\left[\frac{\mu V}{\sqrt{Hz}}\right]$	1.60	3.16	-	2.12	2.69	0.81	1.8	1.6	0.035	2.82	1.85
@freq (Hz)	1000	1000	-	1000	100	1000	-	-	100,000	1000	36
Power (nW)	21.89	59.88	2.4	73	11.4	0.6	13	12.6	1800	51	15.4
Mode	BD	BD	DIGITAL	BD	BD	GD	BD	BD	GD	BD	BD
$FOM_S \left[\frac{MHz \cdot pF}{mW}\right]$	80.29 k	20.16 k	15.89 k	10.20 k	15.72 k	15.00 k	7.15 k	7.05 k	10.11 k	6.86 k	3.61 k
$FOM_L \left[\frac{V \cdot pF}{\mu s \cdot mW}\right]$	34.40 k	11.88 k	5.40 k	11.82 k	4.08 k	-	21.00 k	9.88 k	4.67 k	2.45 k	9.25 k
$FOM_{L_{WC}} \left[\frac{V \cdot pF}{\mu s \cdot mW}\right]$	26.30 k	7.04 k	-	5.93 k	4.52 k	-	6.30 k	4.52 k	-	2.45 k	8.36 k
Area $[mm^2]$	0.0052 ★	0.0027	0.000982	0.0036	0.0036	0.00047	0.0098	0.0085	-	0.003	0.0082

* Simulated; † Measured; ‡ Monte Carlo mean-value; * $PSRR_+/PSRR_-$ [dB]; ★ area estimated accounting for the minimum distances due to deep N-Wells for body connections.

5. Conclusions

In this work, we propose a novel tree-based OTA architecture that exploits body-driven stages to achieve rail-to-rail ICMR, and body-diode loads to avoid Miller compensation, improving the bandwidth efficiency. A ULV ULP OTA exploiting this approach was designed in a 130 nm CMOS process from STMicroelectronics. Simulation results show a dc gain higher than 52 dB, a gain-bandwidth product of about 35.16 kHz with nominal CMRR and PSRR, respectively, equal to 42.11 dB and 56.13 dB. Large-signal characteristics are also very good both in terms of THD and slew-rate. Due to the very limited power consumption of about 21.89 nW, the OTA exhibits state-of-the-art small-signal and large-signal FoMs. Summarizing, the overall performance of the proposed OTA shows record-breaking small-signal and large-signal performance, relatively large DC gain and reasonable PSRR and CMRR performance. The OTA exhibits good stability and robustness against PVT and mismatch variations.

Author Contributions: Conceptualization, F.C., R.D.S. and G.S.; data curation, R.D.S.; investigation, F.C., R.D.S., P.M. and G.S.; software, R.D.S.; validation, R.D.S.; supervision, F.C. and G.S.; writing—original draft preparation, R.D.S. and P.M.; writing—review and editing, F.C. and G.S.; funding acquisition, A.T. All authors have read and agreed to the published version of the manuscript.

Funding: This research received no external funding.

Conflicts of Interest: The authors declare no conflict of interest.

Appendix A. Body-to-Gate (B2G) Interface

This section aims to explain the body-to-gate (B2G) interface which is exploited in each stage ith-1, ith interface. Following the notation in Figure A1a, the current gain can be expressed as:

$$\frac{I_{out}}{I_{in}} = \frac{g_{m_B}}{g_{mb_A}} \left(\frac{1}{1 + \frac{1}{g_{mb_A}/g_{ds_A}}} \right) \frac{1}{1 + s\frac{C_{gs_B}+C_{hs_A}+C_{gd_A}+C_{gd_B}\chi_\alpha}{g_{mb_A}+g_{ds_A}}} \tag{A1}$$

where χ_α derives from Miller approximation on C_{gd_B} and can be therefore expressed as:

$$\chi_\alpha = \frac{g_{m_B}}{\left(g_{ds_B} + g_{load}\right)} \tag{A2}$$

where g_{load} load conductance and as a consequence it could be equal to $g_{mb_{load}}$ or $g_{ds_{load}}$ (respectively, for stage$_{1,2}$ and stage$_3$). It is possible thereafter to conclude that the interface behaves as a small signal current-mirror with gain.

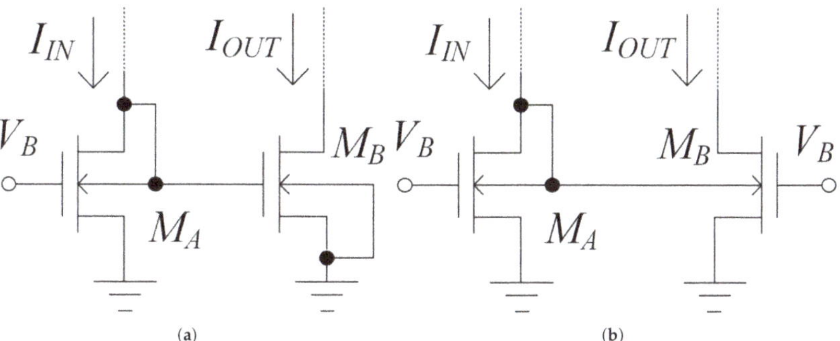

Figure A1. (a) Body-to-gate (B2G) interface; (b) body-to-body (B2B) mirror.

Appendix B. Body-to-Body (B2B) Mirror

This section aims at explaining the body-to-body (B2B) interface which is exploited in each stage. Following the notation in Figure A1b, the current gain can be expressed as:

$$\frac{I_{out}}{I_{in}} = \frac{g_{mb_B}}{g_{mb_A}} \left(\frac{1}{1 + \frac{1}{g_{mb_A}/g_{ds_A}}} \right) \frac{1}{1 + s\frac{C_{gd_A}+C_{bs_A}+C_{bs_B}+C_{bd}\chi_\beta}{g_{ds_A}+g_{mb_A}}} \tag{A3}$$

where also in this case χ_β denotes the Miller approximation and can be derived as:

$$\chi_\beta = \frac{g_{mb_B}}{\left(g_{ds_B} + g_{load}\right)} \tag{A4}$$

Finally, it can be concluded that the interface could be considered as a B2B mirror that enables a small-signal current mirror whose gain is fixed by properly sizing M_A and M_B.

References

1. Alioto, M. *Enabling the Internet of Things—From Integrated Circuits to Integrated Systems*; Springer: Berlin/Heidelberg, Germany, 2017.
2. Sobin, C.C. A survey on architecture, protocols and challenges in IoT. *Wirel. Pers. Commun.* **2020**, *112*, 1383–1429. [CrossRef]
3. Wu, T.; Wu, F.; Redouté, J.M.; Yuce, M.R. An autonomous wireless Body Area Network implementation towards IoT connected healthcare applications. *IEEE Access* **2017**, *5*, 11413–11422. [CrossRef]
4. Lee, J.; Johnson, M.; Kipke, D. A tunable biquad switched-capacitor amplifier-filter for neural recording. *IEEE Trans. Biomed. Circuits Syst.* **2010**, *4*, 295–300. [CrossRef] [PubMed]
5. Della Sala, R.; Monsurrò, P.; Scotti, G.; Trifiletti, A. Area-efficient low-power bandpass Gm-C filter for epileptic seizure detection in 130 nm CMOS. In Proceedings of the 26th IEEE International Conference on Electronics, Circuits and Systems (ICECS), Genoa, Italy, 27–29 November 2019; pp. 298–301.
6. Liu, Z.; Tan, Y.; Li, H.; Jiang, H.; Liu, J.; Liao, H. A 0.5-V 3.69-nW complementary source-follower-C based low-pass filter for wearable biomedical applications. *IEEE Trans. Circuits Syst. I Regul. Pap.* **2020**, *67*, 4370–4381. [CrossRef]
7. Swaroop, K.N.; Chandu, K.; Gorrepotu, R.; Deb, S. A health monitoring system for vital signs using IoT. *Internet Things* **2019**, *5*, 116–129. [CrossRef]
8. Toledo, P.; Rubino, R.; Musolino, F.; Crovetti, P. Re-thinking analog integrated circuits in digital terms: A new design concept for the IoT era. *IEEE Trans. Circuits Syst. II Express Briefs* **2021**, *68*, 816–822. [CrossRef]
9. Aiello, O.; Crovetti, P.; Alioto, M. Ultra-low power and minimal design effort interfaces for the Internet of Thing. In Proceedings of the ICSyS19IEEE International Circuits and Systems Symposium (ICSyS), Kuantan, Malaysia, 18–19 September 2019, pp. 1–4.
10. Harpe, P.; Gao, H.; Dommele, R.; Cantatore, E.; van Roermund, A.H.M. A 0.20 mm^2 3 nW signal acquisition IC for miniature sensor nodes in 65 nm CMOS. *IEEE J. Solid-State Circuits* **2016**, *51*, 240–248. [CrossRef]
11. Chi, Q.; Yan, H.; Zhang, C.; Pang, Z.; Xu, L.D. A reconfigurable smart sensor interface for industrial WSN in IoT environment. *IEEE Trans. Ind. Inform.* **2014**, *10*, 1417–1425. [CrossRef]
12. Grasso, A.D.; Pennisi, S. Ultra-low power amplifiers for IoT nodes. In Proceedings of the ICECS18 IIEEE International Conference on Electronics, Circuits and Systems (ICECS), Bordeaux, France, 9–12 December 2018; pp. 497–500.
13. Richelli, A.; Colalongo, L.; Kovacs-Vajna, Z.; Calvetti, G.; Ferrari, D.; Finanzini, M.; Pinetti, S.; Prevosti, E.; Savoldelli, J.; Scarlassara, S. A survey of low voltage and low power amplifier topologies. *J. Low Power Electron. Appl.* **2018**, *8*, 22. [CrossRef]
14. Khateb, F.; Dabbous, S.B.A.; Vlassis, S. A survey of non-conventional techniques for Low-voltage Low-power analog circuit design. *Radioengineering* **2013**, *22*, 415–427.
15. Cabrera-Bernal, E.; Pennisi, S.; Grasso, A.D.; Torralba, A.; Carvajal, R.G. 0.7-V three-stage class-AB CMOS operational transconductance amplifier. *IEEE Trans. Circuits Syst. I Regul. Pap.* **2016**, *63*, 1807–1815. [CrossRef]
16. Taherzadeh-Sani, M.; Hamoui, A.A. A 1-V process-insensitive current-scalable two-stage opamp with enhanced DC gain and settling behavior in 65-nm digital CMOS. *IEEE J. Solid-State Circuits* **2011**, *46*, 660–668. [CrossRef]
17. Paul, A.; Ramirez-Angulo, J.; Lopez-Martin, A.J.; Carvajal, R.G.; Rocha-Perez, J.M. Pseudo-three-stage Miller op-amp with enhanced small-signal and large-signal performance. *IEEE Trans. Very Large Scale Integr. (VLSI) Syst.* **2019**, *27*, 2246–2259. [CrossRef]
18. Riad, J.; Estrada-López, J.J.; Padilla-Cantoya, I.; Sánchez-Sinencio, E. Power-scaling output-compensated three-stage OTAs for wide load range applications. *IEEE Trans. Circuits Syst. I Regul. Pap.* **2020**, *67*, 2180–2192. [CrossRef]
19. Wang, Y.; Zhang, Q.; Yu, S.S.; Zhao, X.; Trinh, H.; Shi, P. A robust local positive feedback based performance enhancement strategy for non-recycling folded cascode OTA. *IEEE Trans. Circuits Syst. I Regul. Pap.* **2020**, *67*, 2897–2908. [CrossRef]
20. Centurelli, F.; Della Sala, R.; Monsurrò, P.; Scotti, G.; Trifiletti, A. A novel OTA architecture exploiting current gain stages to Boost bandwidth and slew-rate. *Electronics* **2021**, *10*, 1638. [CrossRef]
21. Aguirre, P.C.D.; Susin, A.A. PVT compensated inverter-based OTA for low-voltage CT sigma-delta modulators. *Electron. Lett.* **2018**, *54*, 1264–1266. [CrossRef]
22. Braga, R.A.; Ferreira, L.H.; Coletta, G.D.; Dutra, O.O. A 0.25-V calibration-less inverter-based OTA for low-frequency Gm-C applications. *Microelectron. J.* **2019**, *83*, 62–72. [CrossRef]
23. Lv, L.; Zhou, X.; Qiao, Z.; Li, Q. Inverter-based subthreshold amplifier techniques and their application in 0.3-V $\Sigma\Delta$-modulator. *IEEE J. Solid-State Circuits* **2019**, *54*, 1436–1445. [CrossRef]
24. Manfredini, G.; Catania, A.; Benvenuti, L.; Cicalini, M.; Piotto, M.; Bruschi, P. Ultra-low-voltage inverter-based amplifier with novel common-mode stabilization loop. *Electronics* **2020**, *9*, 1019. [CrossRef]
25. Rodovalho, L.H.; Aiello, O.; Rodrigues, C.R. Ultra-low-voltage inverter-based operational transconductance amplifiers with vVoltage gain enhancement by improved composite transistors. *Electronics* **2020**, *9*, 1410. [CrossRef]
26. Rodovalho, L.H.; Rodrigues, C.R.; Aiello, O. Self-biased and supply-voltage scalable inverter-based operational transconductance amplifier with improved composite transistors. *Electronics* **2021**, *10*, 935. [CrossRef]
27. Baghtash, H.F. A 0.4 V, body-driven, fully differential, tail-less OTA based on current push-pull. *Microelectron. J.* **2020**, *99*, 104768. [CrossRef]
28. Ghosh, S.; Bhadauria, V. An ultra-low-power near rail-to-rail pseudo-differential subthreshold gate-driven OTA with improved small and large signal performances. *Analog. Integr. Circuits Signal Process.* **2021**, *109*, 345–366. [CrossRef]

29. Allen, P.E.; Blalock, B.J.; Rincon, G.A. 1 V CMOS opamp using bulk-driven MOSFETs. In Proceedings of the ISSCC'95-International Solid-State Circuits Conference, San Francisco, CA, USA, 15–17 February 1995; pp. 192–193. [CrossRef]

30. Blalock, B.J.; Allen, P.E.; Rincon-Mora, G.A. Designing 1-V op amps using standard digital CMOS technology. *IEEE Trans. Circuits Syst. II Analog. Digit. Signal Process.* **1998**, *45*, 769–780. [CrossRef]

31. Stockstad, T.; Yoshizawa, H. A 0.9-V 0.5-/spl mu/A rail-to-rail CMOS operational amplifier. *IEEE J. Solid-State Circuits* **2002**, *37*, 286–292. [CrossRef]

32. Ferreira, L.; Sonkusale, S. A 60-dB gain OTA operating at 0.25-V power supply in 130-nm digital CMOS process. *IEEE Trans. Circuits Syst. I Regul. Pap.* **2014**, *61*, 1609–1617. [CrossRef]

33. Colletta, G.D.; Ferreira, L.H.; Pimenta, T.C. A 0.25-V 22-nS symmetrical bulk-driven OTA for low-frequency G_m G m-C applications in 130-nm digital CMOS process. *Analog. Integr. Circuits Signal Process.* **2014**, *81*, 377–383. [CrossRef]

34. Abdelfattah, O.; Roberts, G.W.; Shih, I.; Shih, Y.C. An ultra-low-voltage CMOS process-insensitive self-biased OTA with rail-to-rail input range. *IEEE Trans. Circuits Syst. I Regul. Pap.* **2015**, *62*, 2380–2390. [CrossRef]

35. Akbari, M.; Hashemipour, O. A 63-dB gain OTA operating in subthreshold with 20-nW power consumption. *Int. J. Circuit Theory Appl.* **2017**, *45*, 843–850. [CrossRef]

36. Veldandi, H.; Shaik, R.A. A 0.3-V pseudo-differential bulk-input OTA for low-frequency applications. *Circuits Syst. Signal Process.* **2018**, *37*, 5199–5221. [CrossRef]

37. Kulej, T.; Khateb, F. A 0.3-V 98-dB Rail-to-Rail OTA in 0.18 μm CMOS. *IEEE Access* **2020**, *8*, 27459–27467. [CrossRef]

38. Woo, K.C.; Yang, B.D. A 0.25-V rail-to-rail three-stage OTA with an enhanced DC gain. *IEEE Trans. Circuits Syst. II Express Briefs* **2020**, *67*, 1179–1183. [CrossRef]

39. Centurelli, F.; Della Sala, R.; Scotti, G.; Trifiletti, A. A 0.3 V, rail-to-rail, ultralow-power, non-tailed, body-driven, sub-tThreshold amplifier. *Appl. Sci.* **2021**, *11*, 2528. [CrossRef]

40. Centurelli, F.; Della Sala, R.; Monsurrò, P.; Scotti, G.; Trifiletti, A. A 0.3 V rail-to-rail ultra-low-power OTA with improved bandwidth and slew rate. *J. Low Power Electron. Appl.* **2021**, *11*, 19. [CrossRef]

41. Fortes, A.; Quirino, F.A.; da Silva, L.A.; Girardi, A. Low power bulk-driven OTA design optimization using cuckoo search algorithm. *Analog. Integr. Circuits Signal Process.* **2021**, *106*, 99–109. [CrossRef]

42. Centurelli, F.; Della Sala, R.; Monsurró, P.; Tommasino, P.; Trifiletti, A. An ultra-low-voltage class-AB OTA exploiting local CMFB and body-to-gate interface. *AEU Int. J. Electron. Commun.* **2022**, *145*, 154081. [CrossRef]

43. Crovetti, P.S. A digital-based analog differential circuit. *IEEE Trans. Circuits Syst. I Regul. Pap.* **2013**, *60*, 3107–3116. [CrossRef]

44. Toledo, P.; Crovetti, P.; Aiello, O.; Alioto, M. Fully digital rail-to-rail OTA with sub-1000-μm² area, 250-mV minimum supply, and nW power at 150-pF load in 180 nm. *IEEE Solid-State Circuits Lett.* **2020**, *3*, 474–477. [CrossRef]

45. Toledo, P.; Crovetti, P.; Aiello, O.; Alioto, M. Design of digital OTAs with operation down to 0.3 V and nW power for direct harvesting. *IEEE Trans. Circuits Syst. I Regul. Pap.* **2021**, *68*, 3693–3706. [CrossRef]

46. Toledo, P.; Crovetti, P.; Klimach, H.; Bampi, S.; Aiello, O.; Alioto, M. 300mV-supply, sub-nW-power digital-based operational transconductance amplifier. *IEEE Trans. Circuits Syst. II Express Briefs* **2021**, *68*, 3073–3077. [CrossRef]

47. Toledo, P.; Crovetti, P.; Klimach, H.; Bampi, S. Dynamic and static calibration of ultra-low-voltage, digital-based operational transconductance amplifiers. *Electronics* **2020**, *9*, 983. [CrossRef]

48. Fiorelli, R.; Arnaud, A.; Galup-Montoro, C. Series-parallel association of transistors for the reduction of random offset in non-unity gain current mirrors. In Proceedings of the ISCAS04 IEEE International Symposium on Circuits and Systems, Vancouver, BC, Canada, 23–26 May 2004; Volume 1, pp. 881–884.

49. Narasimman, N.; Kim, T.T. A 0.3 V, 49 fJ/conv.-step VCO-based delta sigma modulator with self-compensated current reference for variation tolerance. In Proceedings of the ESSCIRC Conference 2016: 42nd European Solid-State Circuits Conference, Lausanne, Switzerland, 12–15 September 2016; pp. 237–240. [CrossRef]

50. Kulej, T.; Khateb, F. A compact 0.3-V class AB bulk-driven OTA. *IEEE Trans. Very Large Scale Integr. (VLSI) Syst.* **2020**, *28*, 224–232. [CrossRef]

51. Kulej, T.; Khateb, F. Design and implementation of sub 0.5-V OTAs in 0.18-μm CMOS. *Int. J. Circuit Theory Appl.* **2018**, *46*, 1129–1143. [CrossRef]

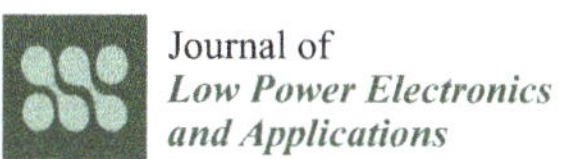

Journal of
*Low Power Electronics
and Applications*

Article

A 0.5 V Sub-Threshold CMOS Current-Controlled Ring Oscillator for IoT and Implantable Devices

Andrea Ballo [1], Salvatore Pennisi [1,*], Giuseppe Scotti [2] and Chiara Venezia [1]

[1] Dipartimento di Ingegneria Elettrica Elettronica e Informatica (DIEEI), University of Catania, 95125 Catania, Italy; andrea.ballo@unict.it (A.B.); chiara.venezia@phd.unict.it (C.V.)
[2] Dipartimento di Ingegneria dell'Informazione Elettronica e Telecomunicazioni (DIET), Sapienza University of Rome, 00184 Rome, Italy; giuseppe.scotti@uniroma1.it
[*] Correspondence: salvatore.pennisi@unict.it; Tel.: +39-095-7382318

Abstract: A current-controlled CMOS ring oscillator topology, which exploits the bulk voltages of the inverter stages as control terminals to tune the oscillation frequency, is proposed and analyzed. The solution can be adopted in sub-1 V applications, as it exploits MOSFETS in the subthreshold regime. Oscillators made up of 3, 5, and 7 stages designed in a standard 28-nm technology and supplied by 0.5 V, were simulated. By exploiting a programmable capacitor array, it allows a very large range of oscillation frequencies to be set, from 1 MHz to about 1 GHz, with a limited current consumption. Considering, for example, the five-stage topology, a nominal oscillation frequency of 516 MHz is obtained with an average power dissipation of about 29 μW. The solution provides a tuneable oscillation frequency, which can be adjusted from 360 to 640 MHz by controlling the bias current with a sensitivity of 0.43 MHz/nA.

Keywords: ring oscillator; body biasing; tuning range

Citation: Ballo, A.; Pennisi, S.; Scotti, G.; Venezia, C. A 0.5 V Sub-Threshold CMOS Current-Controlled Ring Oscillator for IoT and Implantable Devices. *J. Low Power Electron. Appl.* **2022**, *12*, 16. https://doi.org/10.3390/jlpea12010016

Academic Editor: Orazio Aiello

Received: 8 February 2022
Accepted: 3 March 2022
Published: 9 March 2022

Publisher's Note: MDPI stays neutral with regard to jurisdictional claims in published maps and institutional affiliations.

1. Introduction

The Internet of Things (IoT), wireless sensor networks, and the emergence of other energy-harvested microsystems pose continuous challenges and create ever-growing interest in CMOS ultra-low-power analog and mixed-signal system-on-chip solutions. In this framework, applications such as wearable and implantable medical devices, body sensor networks, etc., often require a controlled oscillator (CO) with a minimum power consumption, small layout area, low phase noise, and adequate frequency tuning range to cope with process and/or temperature variations [1–6]. COs are also fundamental blocks of phase-locked loops (PLLs) to provide the timing basis in clock control, clock generator circuits, RFID tags, and systems that use clock-dependent circuits, such as switching power converters and so on [7–9].

CMOS COs can be categorized in two main families. The first includes LC resonant oscillators and, the second, ring oscillators. LC oscillators are mainly used in applications where both a high-phase noise and quality factor (Q) are required. Due to their spiral inductors' large area and high-power dissipation, they cannot be used in ultra-low-power systems on a chip and where physical dimensions must be limited [10]. As is well known, ring oscillators (ROs) consist of an odd number of cascaded delay elements, usually identical to each other, that form a ring where the last stage output is connected to the first stage input. A further categorization is performed based on the control variable, which is often a voltage or a current.

Controlled ring oscillator topologies exhibit a good frequency tuning range, low power dissipation, low design complexity, occupy a small area, and compatibility with CMOS processes. Moreover, ring oscillators are more power efficient compared to relaxation oscillators, although these can achieve a wider tuning range [11].

To improve the frequency-tuning range and phase-noise margin, several design approaches for low-power COs have been reported in the literature. Among these techniques, we mention the combined current-starving technique (i.e., current-controlled oscillator) with a negative skewed-delay approach to improve the power delay product (i.e., product between dissipated power and single gate delay) [12]. A conventional voltage controlled oscillator (VCO) with a negative resistance, multiple-gated circuit and bypass capacitor to suppress high-order harmonics has been reported [13], whereas a digital control circuit to manage oscillation frequency has also been described [1]. An approach that uses positive feedback in each stage to operate with only two stages, instead of three, decreasing the power consumption can be found [14], whereas a frequency tuning cell that consists of one NMOS and one PMOS to form a transmission gate, used to tune the oscillation frequency by varying the gate voltage, has been presented [15]. Finally a dynamic threshold technique (DTMOS) to reduce the threshold voltage of NMOS transistors in the inverting stage with the aim to achieve fast transition and high operating frequency has been presented [3].

The idea of exploiting the bulk terminal to control the oscillation frequency of a RO and the effect of bulk voltage variations on RO phase noise have been analyzed [16], whereas an adaptive body-bias generator for low voltage CMOS VLSI circuits in which a RO was used to estimate the delay of CMOS gates has also been presented [17].

In this work, we exploit a body-biasing technique, originally utilized in the analog domain [18–20], and recently applied to set the quiescent current of the generic inverter stage [21] to design a low-power low-voltage current-controlled ring oscillator (CCO) in 28-nm bulk CMOS technology.

The proposed approach allows to guarantee a static output voltage equal to half the supply voltage, in spite of the value of the bias current, which is tuned in order to control the oscillation frequency. In this way, since any offset in the input output voltage transfer characteristic is removed by the body bias loop, the inverter stages can be reliably cascaded, thus greatly enhancing the tuning range and robustness to PVT variations of the proposed RO.

The manuscript is organized as follows. Section 2 describes the proposed solution. Section 3 reports accurate small- and large-signal analyses of the proposed oscillator. Section 4 includes some simulation results and, finally, the authors' conclusions are summarized in Section 5.

2. The Proposed Solution

Figure 1 shows a circuit schematic of the proposed current-controlled ring oscillator (CCRO). It consists of an N-stage ring oscillator in which the single stage is made up of a CMOS inverter where bulk terminals of both transistors (M_{Pi} and M_{Ni}, with $i = 1, 2, \ldots N$ and N an odd number greater than 3) are made accessible. An output capacitor, C_i in the red-dashed box, is added at the output of each stage with the aim of setting the nominal oscillation frequency (coarse tuning), and to locally make the single stage insensible to parasitic capacitances, as will be clarified in the next section. The body potentials of both transistors, V_{BP} and V_{BN}, are generated from the auxiliary topology depicted in Figure 2, the aim of which is to establish the maximum current flowing in the reference inverter (M_{PR}-M_{NR}), i.e., when the input is at the logic threshold, $V_{DD}/2$. For this purpose, in quiescent conditions, the input terminal of this reference inverter is set to $V_{DD}/2$ and, thanks to the overall negative feedback implemented by error amplifier A_2, such condition is transferred also to the output. Note that also the drain voltage of transistor M_{PA} is kept to $V_{DD}/2$ thanks to A_1. This allows us to set the same nominal operating points for both M_{PR} and M_{PA}.

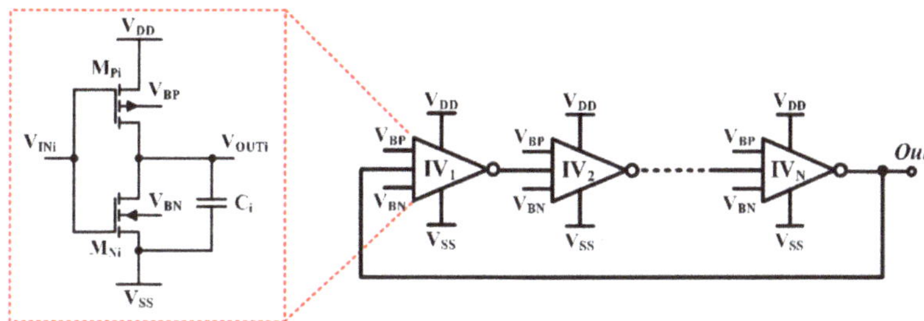

Figure 1. Simplified schematic of the proposed current-controlled ring oscillator.

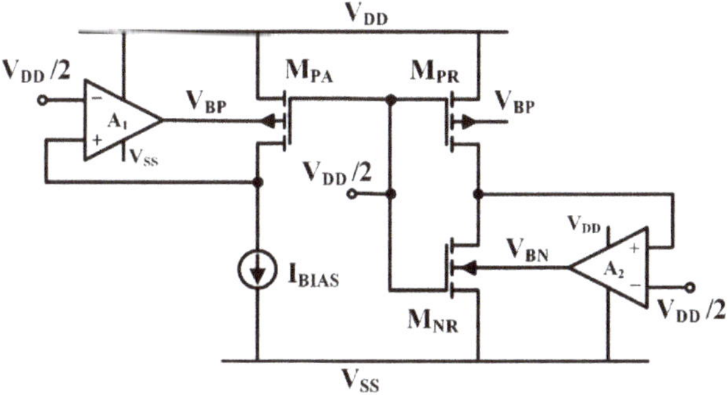

Figure 2. Simplified schematic of the biasing section generating V_{BN} and V_{BP} for the RO in Figure 1.

As far as the quiescent current control is concerned, it is implemented through the bulk terminals via voltage V_{BP} for the p-channel transistors, and V_{BN} for the n-channel ones. These voltages are generated by A_1 and A_2, exploiting a technique proposed in [19] and utilized also in [21].

In brief, starting from the biasing current I_{BIAS}, transistor M_{PA} is forced to generate voltage V_{BP}, which is also applied to M_{PR}. Therefore, current I_{BIAS} in M_{PA} is mirrored by transistor M_{PR} that, as already stated, together with M_{NR}, constitutes the reference inverter. Note also that A_2 generates the required bulk voltages, V_{BN}, for M_{NR} to drive the same current of M_{PR} under the constraints listed in the following:

(a) assigned aspect ratios $(W/L)_{PA}$, $(W/L)_{PR}$ and $(W/L)_{NR}$;
(b) $I_{D,PR/NR} = kI_{BIAS}$, where $k = (W/L)_{PR}/(W/L)_{PA}$;
(c) $V_{SG,PR} = V_{GS,NR} = V_{DD}/2$;
(d) $V_{SD,PR} = V_{DS,NR} = V_{DD}/2$, assuming ideal input virtual short in A_1 and A_2.

Of course, aspect ratios of M_{PR} and M_{NR} must be set so that the required bulk voltages are within V_{DD} and ground. Moreover, the mirroring error between the biasing branch and the reference one is reduced using a careful layout style.

It should be noted that the auxiliary amplifiers A_1 and A_2 should provide a maximum (rail-to-rail) output voltage range, whereas input common mode range is not a concern as input voltage is kept constant to $V_{DD}/2$. Therefore, simple symmetrical OTAs biased in subthreshold, can be effectively used. An example of implementation of this type of amplifier is found in [21,22], and is shown in Figure 3.

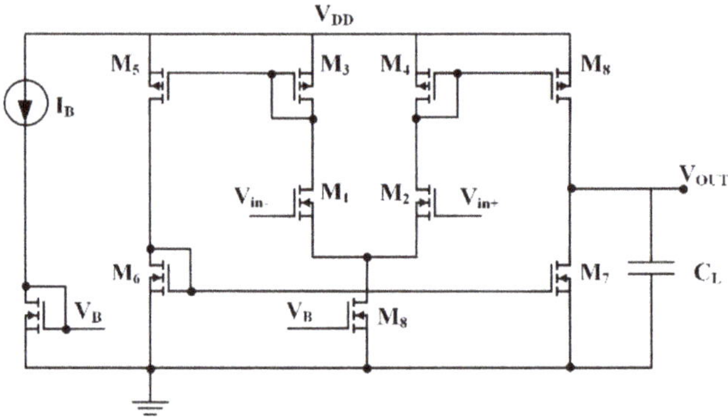

Figure 3. Simplified schematic of simple mirror OTA [21] used in this work.

Voltages V_{BN} and V_{BP} are then applied to the inverters forming the ring oscillator in Figure 1, limiting at the desired value the maximum current flowing when the input voltage is equal to the threshold. Indeed, consider transistor M_{N1} of the first inverter stage, the exploded view of which is depicted in the red-dashed box in Figure 1. Let us remember that, in quiescent conditions, V_{IN1} is equal to $V_{DD}/2$. Consequently, M_{NR} and M_{N1} have respectively the same source, gate, and bulk voltage and hence the drain current of M_{N1} is related to that of M_{NR} in a mirror-like condition:

$$I_{D,N1} = \frac{(W/L)_{N1}}{(W/L)_{NR}} I_{BIAS} \tag{1}$$

where equality is accurately verified because the source-drain voltage of M_{N1} is also equal to $V_{DD}/2$. Similar considerations hold for all the transistors in the ring oscillator, in practice, all p-channel and n-channel devices have their current linked to I_{BIAS} via the current-mirror relations

$$I_{D,Pi} = \frac{(W/L)_{Pi}}{(W/L)_{PR}} I_{BIAS} \tag{2}$$

$$I_{D,Ni} = \frac{(W/L)_{Ni}}{(W/L)_{NR}} I_{BIAS} \tag{3}$$

where $(W/L)_{Pi}$ and $(W/L)_{Ni}$, with $i = 1, 2, \ldots N$, are, respectively, the aspect ratios of the generic p-channel and n-channel MOSFET in the ring oscillator.

3. Small- and Large-Signal Analysis of the Proposed Ring Oscillator

In order to design a conventional ring oscillator, analytical extraction of design equations is carried out by using two main approaches.

The first type of analysis considers small-signal equivalent model of the sub-blocks and Barkhausen stability criterion. In this approach the single gate is seen as working in an operating point (biasing or linearity conditions) and the whole system is analysed in the frequency domain. For this reason, hereinafter we will refer to this approach as an *analog* or *small-signal approach*. As an example, let us consider the conventional CMOS inverter in Figure 4 and its equivalent small-signal circuit.

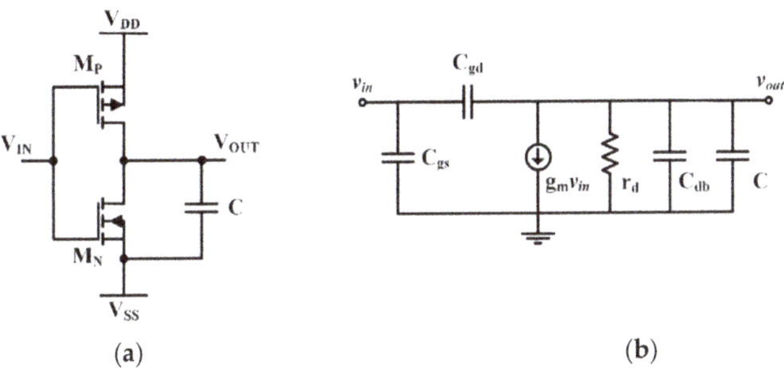

Figure 4. Conventional CMOS inverter (**a**) and its equivalent small-signal circuit (**b**).

When working around an operating point, the inverter behaves as the linear network reported on the right side of Figure 4, the parameters of which are expressed below for the MOS transistors operated in the sub-threshold region.

$$g_m = g_{m,p} + g_{m,n} \simeq 2\frac{I_D}{nV_T} \tag{4}$$

$$r_d = r_{d,n} \parallel r_{d,p} = \frac{nV_T}{2\lambda_{DS}I_D} \tag{5}$$

$$C_{gs} = C_{gs,p} + C_{gs,n} \simeq \frac{2}{3}C_{OX}\left(W_pL_p + W_nL_n\right) + C_{OX}\left(W_pL_{ov} + W_nL_{ov}\right) \tag{6}$$

$$C_{gd} = C_{gd,p} + C_{gd,n} \approx C_{OX}\left(W_pL_{ov} + W_nL_{ov}\right) \tag{7}$$

$$C_{db} = C_{db,p} + C_{db,n} \approx 2C_{Jp/n}\Big|_{V_{DD}/2}\left[1 - \frac{1}{m_j}\frac{V_{DD}/2}{V_{bi}}\left(1 - \frac{V_{DD}/2}{V_{bi}}\right)\right] \tag{8}$$

where parameter n is the sub-threshold slope, $V_T = kT/q$ is the thermal voltage, with k the Boltzmann constant, T is the absolute temperature and C_{OX} is the oxide capacitance for unit of area. In addition, λ_{DS} is the channel modulation coefficient, L_{ov} is the length of the overlap portion, $C_{Jp/n}$ is the capacitance of the S/D junctions (evaluated at the voltage $V_{DD}/2$ in (4e)), m_j is the grading coefficient and V_{bi} is the built-in voltage.

The product between the transconductance g_m (4) and the output resistance r_d (5) yields a constant value, independent of the biasing current I_D and equal to the maximum of the g_m/I_D curves [23]. In such case, only the channel modulation coefficient, λ_{DS}/nV_T, can be changed by sizing the transistors, in order to (slightly) change the inverter intrinsic gain (i.e., $g_m r_d$). Note that the drain-induced barrier lowering (DIBL) effect is included in the channel modulation coefficient through the parameter λ_{DS}. Parasitic capacitance contribution accounts for three capacitances expressed in (6)–(8).

The gate-to-source equivalent capacitance is proportional to C_{OX} and is constituted by a first term that depends on the MOSFET active areas and by the operating condition (assumed with MOSFETs in saturation) and a second term that depends on the overlap capacitance. A similar contribution forms the gate-to-drain equivalent capacitance, C_{gd}, expressed in (7). The drain-to-bulk capacitance, unlike the previous two, is a non-linear capacitance which depends on S/D diffused areas (included in $C_{Jp/n}$) and the applied voltage, i.e., the drain-to-bulk voltage. Referring to Figure 4a, both are evaluated in the quiescent point, i.e., at $V_{DD}/2$, and, from Figure 4b, the output node results to be loaded by the sum of the capacitances (8) and the additional one, C.

The above derivation, when applied to the proposed circuit, yields the same equations except for (4e) that becomes:

$$C_{db} = C_{db,p} + C_{db,n} \approx 2C_{Jp/n}\Big|_{V_{DD}/2-V_B} \left[1 - \frac{1}{m_j} \frac{V_{DD}/2 - V_B}{V_{bi}} \left(1 - \frac{V_{DD}/2 - V_B}{V_{bi}} \right) \right] \tag{9}$$

However, the effect of C_{db} on the oscillation frequency can be neglected if an additional capacitance, C, sufficiently large, is connected in parallel. Analysis of the complete ring oscillator leads to closed-loop gain and phase shift which satisfy Barkhausen's criteria for the common pulsation, ω_p, since the output node electrically coincides with the input one, therefore $|H(j\omega_p)| = 1$, and the a total phase shift of $180°$ is constantly achieved for an odd number of stages N. The result of these concurrent conditions ensures oscillation whose frequency is expressed by:

$$f_{OSC} = \frac{\omega_P}{2\pi} = \frac{1}{2\pi r_d C_{tot}} \tan\left(\frac{\pi}{N} \right) \tag{10}$$

Here c_{tot} gathers all the capacitive contributions (6), (7) doubled for Miller's effect, (9) and C. It can be noted that, being the output small-signal resistance inversely proportional to the biasing current, I_D, a proportional control of the oscillation frequency can be operated by varying the current itself. Various works presented in literature demonstrated that such kind of analysis is inaccurate when the number of stages exceeds 3, hence (10) is rarely used to design a ring oscillator.

On the other hand, the second approach consider the oscillator as the cascade of an odd-number of digital inverting gates where the output of the last gate is fed-back to the input of the first one. In this framework, the single inverter is characterized by its propagation delay, τ_{PD}, and the frequency of the generated signal follows the expression:

$$f_{OSC} = \frac{1}{2N\tau_{PD}} \tag{11}$$

where the factor 2 derives from the fact that each single voltage node switches N-times τ_{PD}, where N is the number of inverters involved. For a digital gate, the propagation delay is defined as the time required to settle the output node to the middle of its dynamic range as referred to the instant of input changing. Henceforth, we call this approach *digital* or *large-signal approach*. While the simple relation in (11) and its scalability assuming general gate implementation are the strengths of this approach, evaluating τ_{PD} could require a great deal of effort. Therefore, designers often adopt a trial-and-error approach.

To better understand the relation between small- and large-signal behavior, propagation delay of the proposed cell should be evaluated. Figure 5 shows the working principle of the inverting gate in response to an input rail-to-rail signal and its static behavior as well. Assuming the inverter symmetrical and working in sub-threshold, which means to size transistors aspect ratios meeting the relationship

$$\left(\frac{W}{L} \right)_P \bigg/ \left(\frac{W}{L} \right)_N = \frac{I_{ST0,N}}{I_{ST0,P}} e^{\frac{|V_{TH,P}| - V_{TH,N}}{nV_T}} \tag{12}$$

where both $I_{ST0,N}$ ($I_{ST0,P}$), defined as the potential sub-threshold current of the NMOS (PMOS) if the threshold voltage are nullified, and n are technology-dependent parameters, and $V_{TH,N}$ ($V_{TH,P}$) are the threshold voltage of the involved transistors. In (12), the tailing effect of drain-to-source voltages is neglected because we assume that transistors are biased in saturation, i.e., $V_{DD}/2 > V_T$. Moreover, $V_{TH,N}$ ($V_{TH,P}$) implicitly depends on $V_{BS,N}$ ($V_{SB,P}$) through the body effect, as well as on $V_{DS,N}$ ($V_{SD,P}$) through the DIBL effect. Their contributions are taken into account by expressing $V_{TH,N} = V_{TH0,N} - \lambda_{BS,N} V_{BS,N} - \lambda_{DS,N} V_{DS,N}$ ($|V_{TH,P}| = |V_{TH0,P}| - \lambda_{BS,P} V_{SB,P} - \lambda_{SD,P} V_{SD,P}$) where $V_{TH0,N}$ ($V_{TH0,P}$) and $\lambda_{BS,N}$ ($\lambda_{BS,P}$) are two technology parameters, while $\lambda_{DS,N}$ ($\lambda_{SD,P}$) coincides with that used in

(5) [24]. It should be noted that, if (12) is fulfilled, the two transistors are equally strong, which means that for the same gate to source voltage they conduct the same current. Under the aforementioned considerations, a good approximation (typical error < 10%) for the propagation delay is given by [25]:

$$\tau_{PD} = \frac{(V_{DD}/2)C_{TOT}}{I_{ST}|_{V_{DD}=0}\, e^{\frac{V_{DD}(1+\lambda_{DS,N}/2+\lambda_{SD,P}/2)+\lambda_{BS,N}V_{BN}+\lambda_{SB,P}(V_{DD}-V_{BP})}{nV_T}}} = \frac{(V_{DD}/2)C_{TOT}}{I_D e^{\frac{V_{DD}/2}{nV_T}}} \tag{13}$$

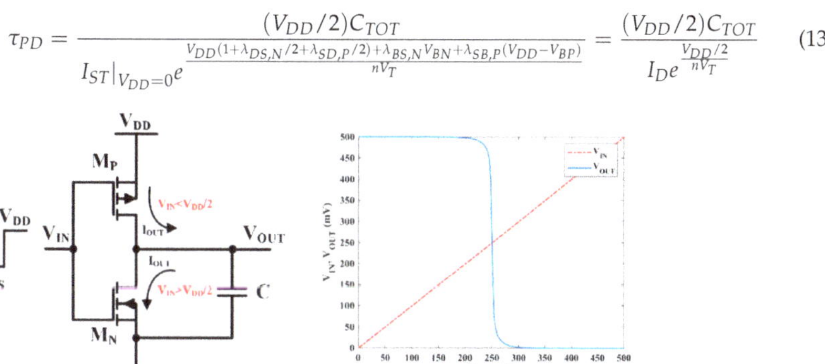

Figure 5. Conventional CMOS inverter (**a**) and its static transfer behavior (**b**).

In the first expression of (13), almost all the technology-dependent characteristics and transistor sizes are gathered in $I_{ST}|_{V_{DD}=0}$ in order to be enucleated from the circuital parameters like voltages V_{DD}, V_{BN}, and V_{BP}. Moreover, the total large-signal capacitance seen at the output node, C_{TOT}, can be assumed to be equal to the small-signal one reported in (10). Finally, (13) has been re-written in the last simple form to highlight the biasing current, I_D.

Replacing (13) in (11), the oscillation frequency is expressed as:

$$f_{OSC} = \frac{I_D}{2N(V_{DD}/2)C_{TOT}} e^{\frac{V_{DD}/2}{nV_T}} \tag{14}$$

As compared with the small-signal counterpart, (14), like (10), shows a linear dependence with the bias current, therefore confirming the possibility to modulate the oscillation frequency of the RO by using the biasing circuit in Figure 2. It should be noted that (14) and (10) give similar information also when the last one loses accuracy. In fact, for $N > 3$ the tangent function can be expanded in Taylor's series, $Tan(\pi/N) \approx \pi/N$ being $\pi/N << 1$. This approximation leads to have:

$$f_{OSC} = \frac{\omega_P}{2\pi} \approx \frac{1}{2Nr_d c_{tot}} = \frac{I_D}{2Nc_{tot}}\left(\frac{2\lambda_{DS}}{nV_T}\right) \tag{15}$$

which differs from (14) only for factor (λ_{DS}/nV_T) that replaces $(e^{\frac{V_{DD}/2}{nV_T}}/V_{DD})$. Thus, it can be claimed that small-signal and large-signal analyses yield results that are similar to those obtained for a conventional topology, such as the current-starved RO [26].

In conclusion, three important metrics for a controlled oscillator are evaluated. Starting from (14), the first is the frequency-to-current first-order slope defined as the derivative function of the frequency versus the control current:

$$\frac{\partial f_{OSC}}{\partial I_D} = \frac{e^{\frac{V_{DD}/2}{nV_T}}}{2N(V_{DD}/2)C_{TOT}} \tag{16}$$

The second one is the total power consumption, made up of a static and a dynamic contribution. While the static part is due only to the leakage current, which coincides with the quiescent one in our case and, as it will be seen it is negligible; the dynamic part represents the major contribution to the power consumption. Consequently, the dynamic power dissipation P_D of a N-stage ring oscillator is given by:

$$P_D = NC_{tot}f_{OSC}(V_{DD})^2 \tag{17}$$

Finally, in a conventional CMOS oscillator, the amount of the phase noise, $L\{\Delta f\}$, (see expression (15) of [27]) is given by the flicker noise and its normalized single-sideband spectral density as given in the following equation

$$L\{\Delta f\} = 10\log\left[\frac{2FkT}{P_{sign}}\left(\frac{f_{OSC}}{2Q\Delta f}\right)^2\right] \tag{18}$$

where Δf is the offset frequency from the nominal one f_{OSC}, Q is the quality factor and F is an empirical fitting parameter that takes the increased noise in Δf into account. The Q factor is typically used in the design of high-order oscillators like LC-type and is defined as the ratio of the energy stored in the oscillating resonator to the energy dissipated per cycle by damping processes. Finally, P_{sign} in (18) is the power of generated signal. Unfortunately, as in the conventional ring oscillator, the quality factor is poor since the energy stored in the node capacitances is reset(discharged) every cycle [27], resulting in a higher phase noise.

Finally, the trade-off between phase noise, power consumption and carrier frequency can be evaluated by using the following figure of merit (*FoM*):

$$FoM = L\{\Delta f\} + 10\log\left(P_{(mW)}\right) - 20\log\left(\frac{f_{OSC}}{\Delta f}\right) \tag{19}$$

where $P_{(mW)}$ is the power consumption expressed in mW, thus normalized to 1 mW.

4. Validation Results

The proposed solution in the version of 3-, 5- and 7-stage CROs was designed in a 28-nm triple-well CMOS technology provided by TSMC and simulated at the schematic level. To set symmetrical behavior of the inverter, of the control bulk voltages ranges and body effect coefficients, MOS transistors with different thresholds were exploited. Specifically, HVT (high threshold) n-channel with 515-mV V_{TH} and SVT (standard threshold) p-channel devices with -460-mV V_{TH}, were adopted. A single power supply of 0.5 V was set and I_{BIAS} was 320 nA. Reference operating temperature was in the range from 0 °C to 60 °C, suitable for implanted and wearable circuits. Transistor dimensions, together with other component values, are summarized in Table 1.

Table 1. Design parameters used in simulations.

Parameter	Value
V_{DD}	0.5 V
I_{BIAS}	320 [a] nA
$(W/L)_{PA}, (W/L)_{PR}, (W/L)_{Pi}$	8.28/0.18 μm/μm
$(W/L)_{NR}, (W/L)_{Ni}$	5.4/0.18 μm/μm
A_1, A_2	30 dB
GBW_{A1}, GBW_{A2}	10 kHz
C_i	10 fF

[a]: This value will be changed to 350 nA after corner analysis.

All p-channel (n-channel) MOSFETS are equal to the reference device 8.28/0.18 (5.4/0.18) μm/μm, where channel length was slightly increased as respect to the minimum one (100 nm) to counteract the short-channel effect. With these design choices the mirroring coefficient, k_i, and the ratios of the transistors' form factors in (1)–(3) are all reduced to the unity. As a consequence of the transistor's dimension, the nominal quiescent current in each branch, which coincides with its short-circuit current, of 320 nA, resulting in a total nominal quiescent current of N-times 320 nA. Coarse tuning capacitor C_i was set to 10 fF for all stages. The DC gain of the auxiliary amplifiers, A_1 and A_2, with transistors in subthreshold, was around 30 dB and the gain-bandwidth product was 10 kHz, while consuming only 50 nA.

The robustness of the quiescent conditions was validated at first. The nominal bulk voltages, V_{BP} and V_{BN}, generated by the circuit in Figure 2 were 251 mV and 249 mV, respectively. The simulated quiescent current in the main ring oscillator in Figure 1 was 961, 1602 and 2243 nA on average, with a standard deviation of 48.5, 78.3, and 107 nA, respectively, for 3-, 5-, and 7-stage topology after running 1000 Monte Carlo iterations. The difference with respect to the expected values is mainly due to the low DC gains of the auxiliary amplifiers, which cause a closed-loop gain error.

Figure 6a shows the body voltages of the transistors involved in the reference inverting gate for a sweeping of the biasing current, I_{BIAS}, in the range 120 nA–820 nA. The voltages fall within the supply rails and, in particular, it is easy to observe that their behaviors are symmetrical, confirming a good sizing of the block and the possibility to exploit the full dynamic range of the control voltages. Currents entering in the body terminals have been also evaluated and reported in Figure 6b to highlight that body junctions are never fully turned on during the control operation. In fact, the values of body currents in the worst case (PMOS), reach around 10 nA, corresponding to less than 2% of the biasing one.

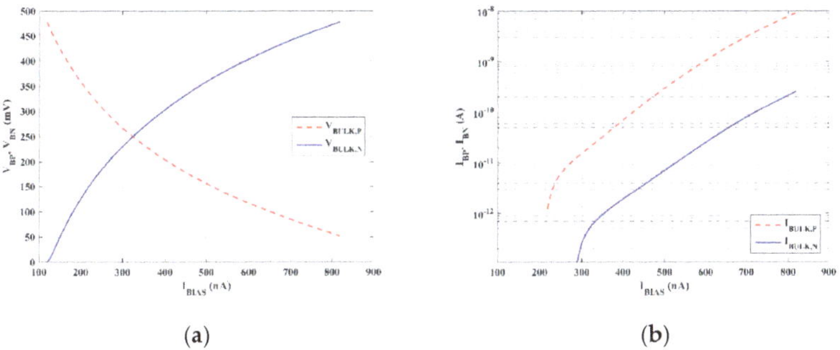

(a)   (b)

Figure 6. Body voltages (**a**) and currents (**b**) in the interested current biasing range at $T = 30\,^{\circ}\mathrm{C}$.

Figure 7a depicts the current flowing in the reference inverting gate when its input is varied from 0 to V_{DD} (500 mV). As expected, the maximum is achieved for $V_{DD}/2$ and accurately follows I_{BIAS} as a validation of the effectiveness of the exploited biasing strategy and the linearity of the relation between the two quantities as well.

Figure 8 illustrates the output signal of the 5-stage CRO with $C_i = 10$ fF for three values of biasing current, 120, 320, and 820 nA, representing the minimum, nominal and maximum value, respectively.

Figure 9 shows the oscillation frequency as a function of biasing current ($C_i = 10$ fF). An oscillation range from 360 MHz to 640 MHz is found with tuning sensitivity, i.e., the ratio between $(f_{MAX} - f_{MIN})/(I_{BIAS,MAX} - I_{BIAS,MIN})$, about equal to 0.43 MHz/nA. Compared with the predicted behavior (linear relationships resulting from (14) and (15)), the obtained one shows a logarithmic relationship with I_{BIAS}. This is confirmed by the inset plot in the same figure, the x-axis of which is logarithmic and slightly extended to cover an entire decade. Such changing in the behavior may be due to the partial operation in moderate inversion region, where subthreshold equations lose accuracy.

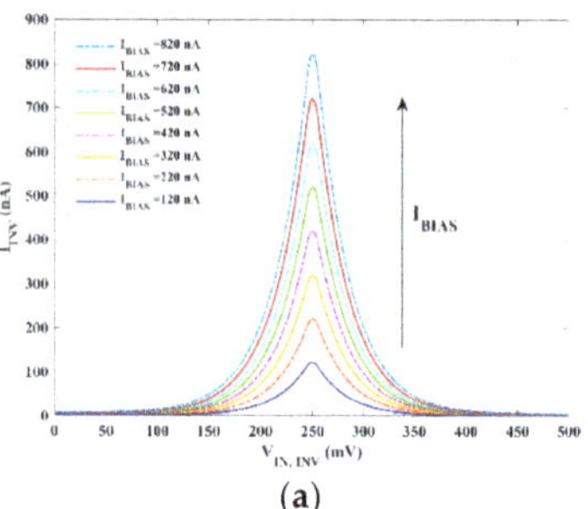
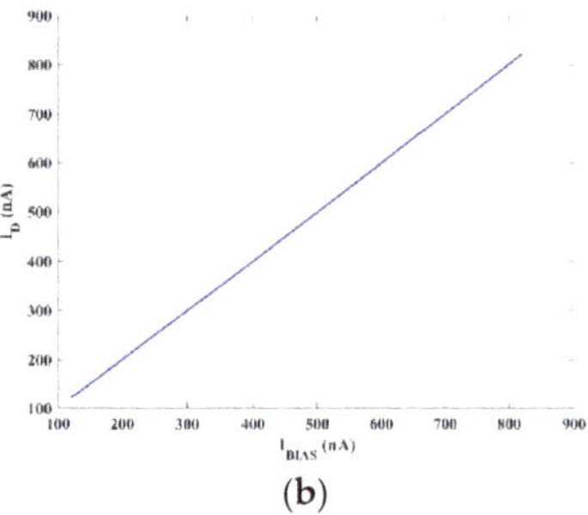

(a)   (b)

Figure 7. Static currents flowing in the reference inverter vs. input voltage for different I_{BIAS} (**a**), and static currents maxima vs. I_{BIAS} ($T = 30\ ^\circ$C) (**b**).

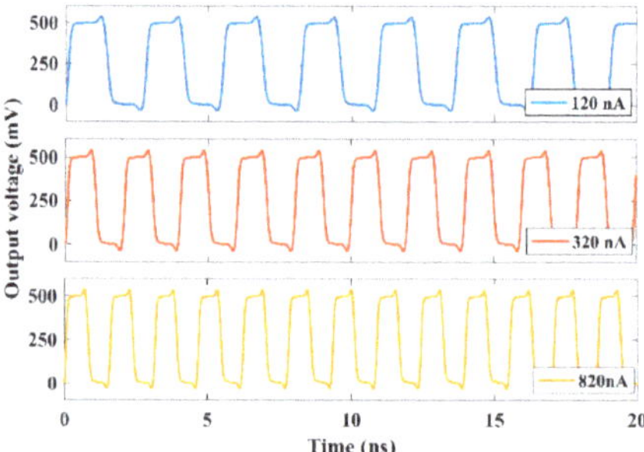

Figure 8. Output signal of 5-stage CRO for three different values of biasing current at $T = 30\ ^\circ$C.

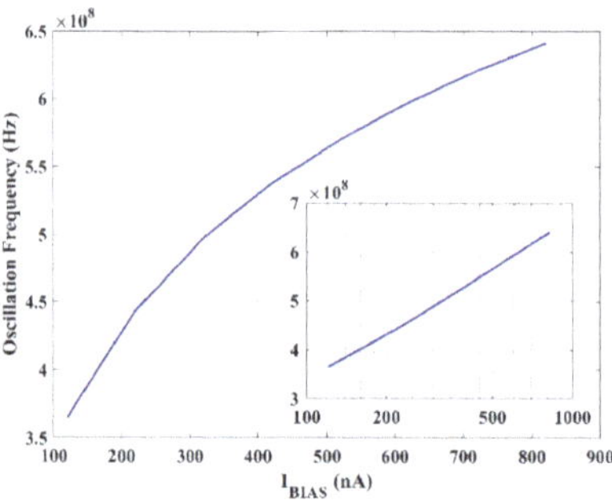

Figure 9. Oscillation frequency of the 5-stage CRO as a function of the biasing current at $T = 30\ ^\circ$C.

Figure 10a,b shows oscillation frequency with number of stages N equal to 3, 5, and 7 and for different coarse tuning capacitances, C_i, in the considered current biasing range (Figure 10a) and for a fixed $I_{BIAS} = 320$ nA (Figure 10b) at $T = 30\ ^\circ$C. It is apparent that frequency varies with the bias current, independently of the number of stages and coarse tuning capacitance. Constant spacing between two adjacent curves shows that the number of stages N acts as a scaling constant factor in the expression of the frequency, as predicted by (14) or, equivalently, (15). Moreover, Figure 10b highlights that the coarse tuning capacitance is comparable, in the range between 10 fF and 100 fF, with the parasitic inverter capacitances, being the oscillation frequency to capacitance relation compressed in this range.

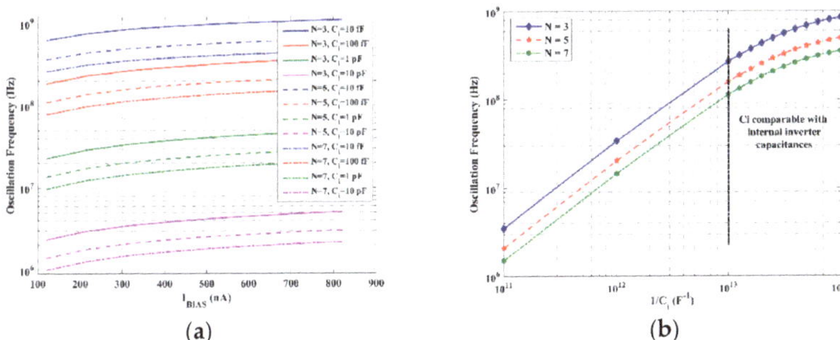

Figure 10. Oscillation frequencies for different coarse tuning capacitance values in the interested current biasing range (**a**) and for a fixed $I_{BIAS} = 320$ nA (**b**) at $T = 30\ ^\circ$C.

Figures 9 and 10 also show that the proposed solution may be used in an automatic design procedure, which, starting from the oscillation frequency specification and the nominal bias current, allows to determine the number of inverter stages (which can be taken from a standard-cell library providing access to the bulk terminals) and the coarse tuning capacitances (which can be taken from a capacitor array). The bias current is then used to perform oscillation frequency tuning to counteract process and temperature variation effects.

At this purpose, corner analyses were carried out for the 5-stage topology as an illustrative example. Figure 11 shows the two bulk-source voltages V_{BN} and V_{BP} as a function of I_{BIAS}, at 30 $^\circ$C. It is apparent that to ensure correct operation (i.e., to maintain bulk voltages within the supply limits) biasing current range must be limited from 240 nA to 460 nA.

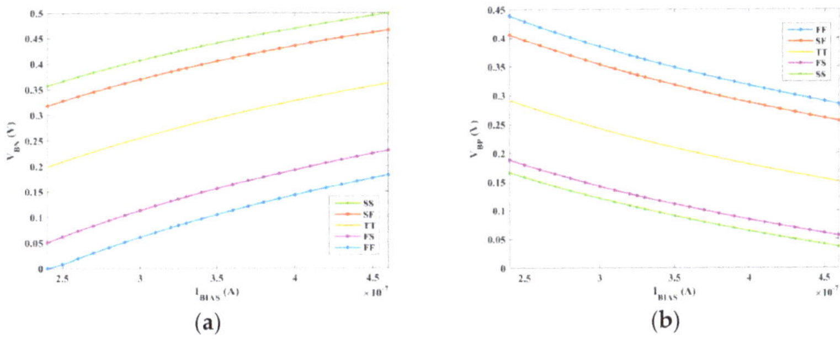

Figure 11. Bulk−source voltage of NMOS (**a**) and PMOS (**b**) transistors vs. biasing current over the 5 basic process corners.

Specifically, NMOS transistors experience the highest process variations. In fact, the upside limit of the current range under slow NMOS corners (SS or SF) must be limited to 460 nA. Vice versa, under fast NMOS corners (FF or FS), I_{BIAS} current must be larger than 240 nA.

Figure 12 shows the simulated oscillation frequency of the 5-stage ring oscillator in this range of I_{BIAS}, for the five basic corners (at 30 $^\circ$C). It can be noted that the tuning frequency range is independent of the process corners. Indeed, the tuning sensitivity is constant regardless the corner and is still approximately equal to 0.43 MHz/nA. The maximum percentage variation between the nominal oscillation frequency and that affected by corners is about 20%.

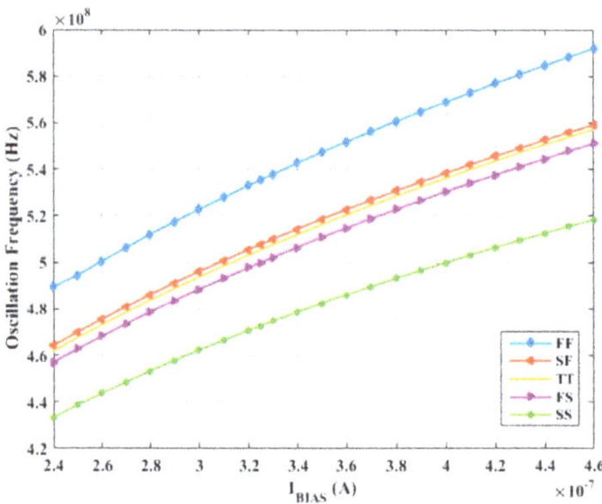

Figure 12. Oscillation frequency vs. biasing current over the 5 basic process corners.

Phase noise versus offset frequency (Δf) for the basic corners is illustrated in Figure 13, which clearly shows the close overlap of the five curves, indicating that also the phase noise is process independent.

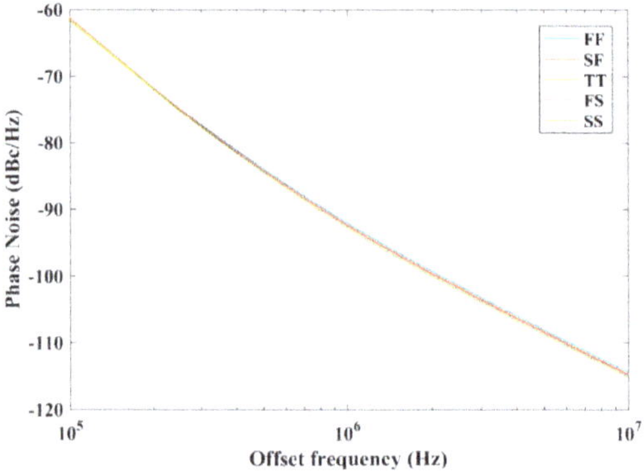

Figure 13. Phase noise vs. offset frequency over the 5 basic process corners.

Phase noise was also simulated in the 0 °C to 60 °C temperature range and across the five corners. The minimum value of the simulated phase noise is −91.86 dBc/Hz when $T = 0$ °C in the FF corner, while the maximum value of the simulated phase noise is −93.72 dBc/Hz when $T = 0$ °C in the SS corner (both values are evaluated at the nominal bias current I_{BIAS} equal to 350 nA).

Tables 2–4 summarize the corner analysis results of the main parameters of the simulated 5-stage current-controlled ring oscillator (nominal I_{BIAS} equal to 350 nA and at 30 °C) for three values of the supply voltage V_{DD}.

Table 2. Corner analysis of the 5-stage current-controlled ring oscillator performed at 30 °C and at V_{DD} = 475 mV.

Corner	TT	FF	FS	SF	SS
Oscillation frequency (MHz)	451.2	481.6	446.2	446.7	397.3
Tuning range (MHz)	87.09	96.38	50.69	59.24	24.12
Phase noise @1 MHz (dBc/Hz)	−92.77	−92.39	−92.90	−92.90	−93.75
Average power consumption (µW)	23.40	24.41	23.31	23.14	20.94

Table 3. Corner analysis of the 5-stage current-controlled ring oscillator performed at 30 °C and at V_{DD} = 500 mV.

Corner	TT	FF	FS	SF	SS
Oscillation frequency (MHz)	516.2	547.3	410.7	418.5	482.5
Tuning range (MHz)	95.44	102.50	94.11	94.94	85.08
Phase noise @1 MHz (dBc/Hz)	−92.47	−92.13	−92.58	−92.40	−92.87
Average power consumption (µW)	28.6	29.8	28.4	28.8	27.5

Table 4. Corner analysis of the 5-stage current-controlled ring oscillator performed at 30 °C and at V_{DD} = 525 mV.

Corner	TT	FF	FS	SF	SS
Oscillation frequency (MHz)	583.3	619.7	577.6	586.5	549.6
Tuning range (MHz)	102.47	59.27	81.37	100.50	93.52
Phase noise @1 MHz (dBc/Hz)	−92.14	−91.98	−92.25	−92.06	−92.48
Average power consumption (µW)	36.16	37.93	35.85	36.29	34.61

Due to process variations, oscillation frequency varies of about 20%. However, tuning range variations across the five corners are limited to 8% and both phase noise and average power consumption variations are 5%. Regarding power consumption, the typical (TT) value is accurately predicted by (14), where C_{tot} can be estimated to be 88.6 fF, including the additional load capacitance of 10 fF. This value agrees with the results shown in Figure 10b.

Mismatch Monte Carlo simulations of the oscillation frequency in typical conditions (V_{DD} = 0.5 V, I_{BIAS} = 350 nA and T = 27 °C) are reported in Figure 14, showing the limited impact of mismatches on the oscillation frequency of the proposed CRO.

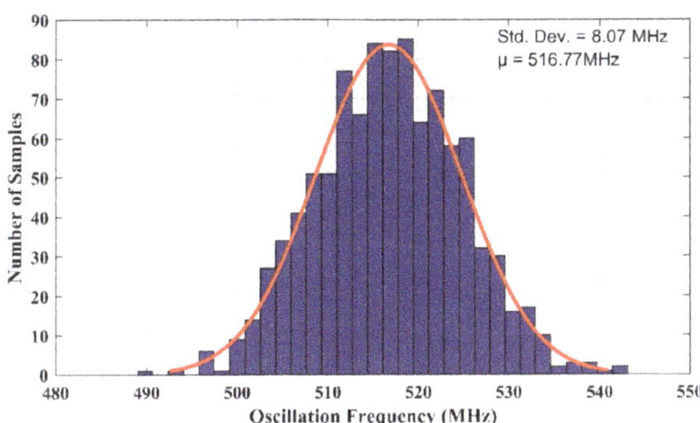

Figure 14. Mismatch Monte Carlo simulations of the oscillation frequency in typical conditions (VDD = 0.5 V, I_{BIAS} = 350 nA and T = 27 °C).

As already mentioned, the tuning capabilities of the proposed CRO can be exploited to compensate for the effects of temperature variations (in a non-exclusive alternative to process variations). To give an example, Figure 15 shows some isofrequency curves (at 557 MHz, 538 MHz, 516 MHz, 491 MHz, and 462 MHz) in I_{BIAS} vs. temperature plot. Each point of the curves establishes the current I_{BIAS} needed to set the target frequency between around 450 MHz and 557 MHz in the operating range from 0 °C to 60 °C.

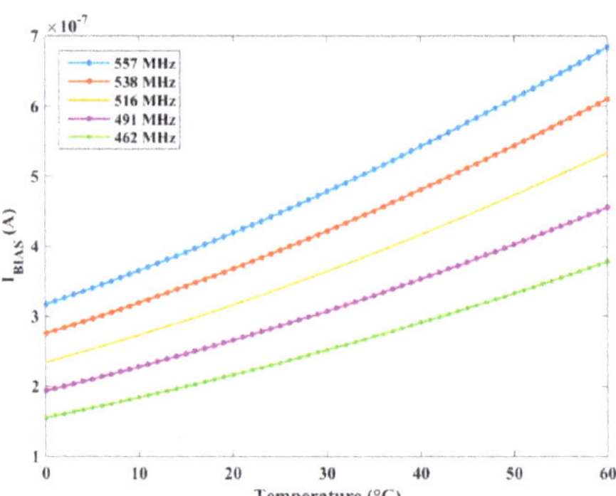

Figure 15. Isofrequency curves in the I_{BIAS} vs. temperature plot (TT).

The tuning capabilities of the proposed CRO can be exploited also to compensate the effects of supply voltage (V_{DD}) variations. To give an example, Figure 16 shows some isofrequency curves (at 557 MHz, 538 MHz, 516 MHz, 491 MHz, and 462 MHz) in the I_{BIAS} vs. V_{DD} plot. Each point of the curves establishes the current I_{BIAS} needed to set the target frequency between around 450 MHz and 557 MHz in the operating range from 475 mV to 525 mV.

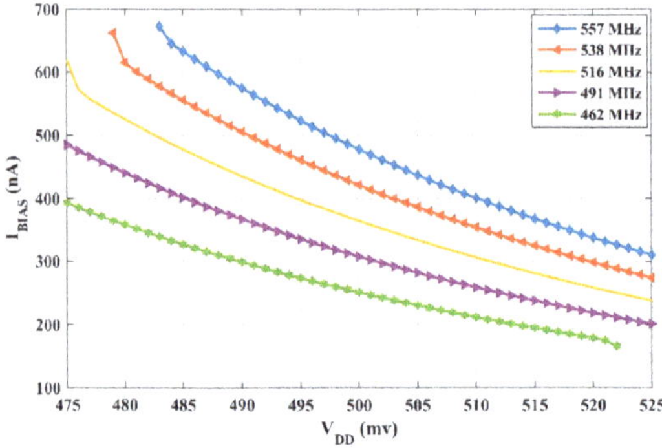

Figure 16. Isofrequency curves in the I_{BIAS} vs. V_{DD}, (TT).

The layout of the five stage CRO is reported in Figure 17, showing an area footprint of 12.4 μm × 7.5 μm. To better assess the reliability of the above results, post layout simulations have been carried. The main effect of the layout resulted in a parasitic capacitance of 6.5 fF at the output node of the CRO. Once reduced the explicit coarse tuning capacitance of an amount equal to the parasitic capacitance, post layout simulations resulted to be in very good agreement with schematic level simulations.

Table 5 compares some recent controlled oscillator topologies presented in the literature with the proposed one, which provides the best performance in terms of phase noise and power consumption, determining the best FoM as defined by (19).

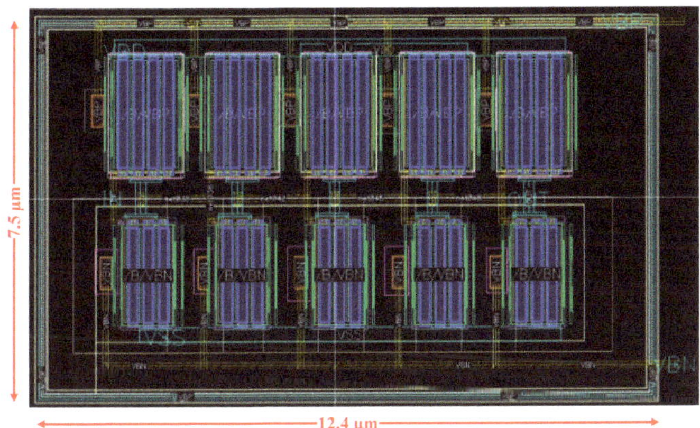

Figure 17. Layout of the five stage CRO.

Table 5. Comparison with the state-of-the-art.

Reference	[28]	[29]	[30]	This Work [b]
Tech. (nm)	180	65	65	28
V_{DD} (mV)	500	600	700	500
N stages	3	3	4	5
Type of control	Voltage	Voltage	Voltage	Current
Osc. frequency (MHz)	82–370	250–800	880–1360	360–640
Phase Noise (dBc/Hz)@1 MHz	−82	−86.38	−90	−92.47
Power consumption (μW)	60	146.2	360	28.6
FoM [a] (dBc/Hz)	−145.6	−153.2	−153.6	−164

[a]: see (16); [b]: simulations.

5. Conclusions

A novel approach to tune the delay of the basic inverter cell of CROs has been presented in this paper. The approach allowed to accurately set the maximum current of all the inverters in the CRO through a body bias loop and to tune the oscillation frequency by controlling the value of a reference current. Small-signal and large-signal analysis of the proposed CRO topology have been carried out to provide insight into circuit behavior and to provide useful design equations.

Current controlled ring oscillators made up of 3, 5, and 7 stages have been designed referring to a commercial 28-nm technology and with a supply voltage of 0.5 V.

Simulation results demonstrated that the proposed approach allows to optimize the tradeoff between tuning range, phase noise and power consumption, as demonstrated by the value of the FoM which outperforms all the similar designs in the literature. Extensive parametric and corner simulations have demonstrated a good robustness of the proposed CROs to PVT variations despite the adoption of a very short channel process node.

Author Contributions: Conceptualization: S.P. and G.S.; data curation: A.B. and C.V.; original draft preparation: A.B. and C.V.; writing—review and editing: all authors; supervision: S.P. and G.S. All authors have read and agreed to the published version of the manuscript.

Funding: This research received no external funding.

Institutional Review Board Statement: Not applicable.

Informed Consent Statement: Not applicable.

Data Availability Statement: The data presented in this study are available in article.

Conflicts of Interest: The authors declare no conflict of interest.

References

1. Corres-Matamoros, A.; Martínez-Guerrero, E.; Rayas-Sanchez, J.E. A programmable CMOS voltage controlled ring oscillator for radio-frequency diathermy on-chip circuit. In Proceedings of the 2017 International Caribbean Conference on Devices, Circuits and Systems (ICCDCS), Cozumel, Mexico, 5–7 June 2017; pp. 65–68. [CrossRef]
2. Ghafari, B.; Koushaeian, L.; Goodarzy, F.; Evans, R.; Skafidas, E. An ultra-low-power and low-noise voltage-controlled ring oscillator for biomedical applications. In Proceedings of the IEEE 2013 Tencon—Spring, Sydney, NSW, Australia, 17–19 April 2013; pp. 20–24. [CrossRef]
3. Ranjan, R.; Raman, A.; Kashyap, N. Low Power and High Frequency Voltage Controlled Oscillator for PLL Application. In Proceedings of the 2019 6th International Conference on Signal Processing and Integrated Networks (SPIN), Noida, India, 7–8 March 2019; pp. 212–214. [CrossRef]
4. Zambrano, B.; Garzón, E.; Strangio, S.; Crupi, F.; Lanuzza, M. A 0.05 mm2, 350 mV, 14 nW Fully-Integrated Temperature Sensor in 180-nm CMOS. *IEEE Trans. Circuits Syst. II Express Briefs* **2021**, 1. [CrossRef]
5. Meng, X.; Li, X.; Cheng, L.; Tsui, C.-Y.; Ki, W.-H. A Low-Power Relaxation Oscillator With Switched-Capacitor Frequency-Locked Loop for Wireless Sensor Node Applications. *IEEE Solid-State Circuits Lett.* **2019**, *2*, 281–284. [CrossRef]
6. Ballo, A.; Bruno, G.; Grasso, A.D.; Vaiana, M.G.G. A Compact Temperature Sensor With a Resolution FoM of 1.82 pJ·K2. *IEEE Trans. Instrum. Meas.* **2020**, *69*, 8571–8579. [CrossRef]
7. Alioto, M. (Ed.) *Enabling the Internet of Things: From Integrated Circuits to Integrated Systems*; Springer International Publishing: Berlin/Heidelberg, Germany, 2017. [CrossRef]
8. Ballo, A.; Bottaro, M.; Grasso, A.D. A Review of Power Management Integrated Circuits for Ultrasound-Based Energy Harvesting in Implantable Medical Devices. *Appl. Sci.* **2021**, *11*, 2487. [CrossRef]
9. Stornelli, V.; Barile, G.; Pantoli, L.; Scarsella, M.; Ferri, G.; Centurelli, F.; Tommasino, P.; Trifiletti, A. A New VCII Application: Sinusoidal Oscillators. *J. Low Power Electron. Appl.* **2021**, *11*, 30. [CrossRef]
10. Razavi, B. A study of phase noise in CMOS oscillators. *IEEE J. Solid-State Circuits* **1996**, *31*, 331–343. [CrossRef]
11. Zaman, K.S.; Reaz, M.I.; Haque, F.; Arsad, N.; Ali, S.H.M. Optimization of WiFi Communication System using Low Power Ring Oscillator Delay Cell. In Proceedings of the 2020 IEEE 8th Conference on Systems, Process and Control (ICSPC), Melaka, Malaysia, 11–12 December 2020; pp. 91–94. [CrossRef]
12. Nayak, R.; Kianpoor, I.; Bahubalindruni, P.G. Low power ring oscillator for IoT applications. *Analog Integr. Circuits Signal Process.* **2017**, *93*, 257–263. [CrossRef]
13. Lee, S.-Y.; Hsieh, J.-Y. Analysis and Implementation of a 0.9-V Voltage-Controlled Oscillator With Low Phase Noise and Low Power Dissipation. *IEEE Trans. Circuits Syst. II Express Briefs* **2008**, *55*, 624–627. [CrossRef]
14. Reddy, N.; Pattanaik, M.; Rajput, S.S. 0.4V CMOS based low power voltage controlled ring oscillator for medical applications. In Proceedings of the TENCON 2008–2008 IEEE Region 10 Conference, Hyderabad, India, 19–21 November 2008; pp. 1–5. [CrossRef]
15. Chuang, Y.-H.; Jang, S.-L.; Lee, J.-F.; Lee, S.-H. A low voltage 900 MHz voltage controlled ring oscillator with wide tuning range. In Proceedings of the 2004 IEEE Asia-Pacific Conference on Circuits and Systems, 2004, Tainan, Taiwan, 6–9 December 2004; Volume 1, pp. 301–304. [CrossRef]
16. Srivastava, A.; Zhang, C. An Adaptive Body-Bias Generator for Low Voltage CMOS VLSI Circuits. *Int. J. Distrib. Sens. Netw.* **2008**, *4*, 213–222. [CrossRef]
17. Deen, M.J.; Kazemeini, M.H.; Naseh, S. Performance characteristics of an ultra-low power VCO. In Proceedings of the 2003 International Symposium on Circuits and Systems, 2003. ISCAS '03, Bangkok, Thailand, 25–28 May 2003; Volume 1, p. I. [CrossRef]
18. Ballo, A.; Grasso, A.D.; Pennisi, S.; Venezia, C. High-Frequency Low-Current Second-Order Bandpass Active Filter Topology and Its Design in 28-nm FD-SOI CMOS. *J. Low Power Electron. Appl.* **2020**, *10*, 27. [CrossRef]
19. Monsurró, P.; Pennisi, S.; Scotti, G.; Trifiletti, A. Exploiting the Body of MOS Devices for High Performance Analog Design. *IEEE Circuits Syst. Mag.* **2011**, *11*, 8–23. [CrossRef]
20. Palumbo, G.; Scotti, G. A Novel Standard-Cell-Based Implementation of the Digital OTA Suitable for Automatic Place and Route. *J. Low Power Electron. Appl.* **2021**, *11*, 42. [CrossRef]
21. Ballo, A.; Pennisi, S.; Scotti, G. 0.5 V CMOS Inverter-Based Transconductance Amplifier with Quiescent Current Control. *J. Low Power Electron. Appl.* **2021**, *11*, 37. [CrossRef]
22. Pérez-Nicoli, P.; Veirano, F.; Rossi-Aicardi, C.; Aguirre, P. Design method for an ultra low power, low offset, symmetric OTA. In Proceedings of the 2013 7th Argentine School of Micro-Nanoelectronics, Technology and Applications, Buenos Aires, Argentina, 15–16 August 2013; pp. 38–43.
23. Silveira, F.; Flandre, D.; Jespers, P.G.A. A g/sub m//I/sub D/ based methodology for the design of CMOS analog circuits and its application to the synthesis of a silicon-on-insulator micropower OTA. *IEEE J. Solid-State Circuits* **1996**, *31*, 1314–1319. [CrossRef]
24. Alioto, M. Understanding DC Behavior of Subthreshold CMOS Logic Through Closed-Form Analysis. *IEEE Trans. Circuits Syst. Regul. Pap.* **2010**, *57*, 1597–1607. [CrossRef]
25. Rabaey, J.M. *Digital Integrated Circuits: A Design Perspective*; Prentice-Hall, Inc.: Hoboken, NJ, USA, 1996.

26. Rajahari, G.; Varshney, Y.A.; Bose, S.C. A Novel Design Methodology for High Tuning Linearity and Wide Tuning Range Ring Voltage Controlled Oscillator. In *VLSI Design and Test*; Springer: Berlin/Heidelberg, Germany, 2013; pp. 10–18. [CrossRef]
27. Lee, T.H.; Hajimiri, A. Oscillator phase noise: A tutorial. *IEEE J. Solid-State Circuits* **2000**, *35*, 326–336. [CrossRef]
28. Tianwang, L.; Jiang, J.; Bo, Y.; Xingcheng, H. Ultra low voltage, wide tuning range voltage controlled ring oscillator. In Proceedings of the 2011 9th IEEE International Conference on ASIC, Xiamen, China, 25–28 October 2011; pp. 824–827. [CrossRef]
29. Saheb, Z.; El-Masry, E.; Bousquet, J.-F. Ultra-low voltage and low power ring oscillator for wireless sensor network using CMOS varactor. In Proceedings of the 2016 IEEE Canadian Conference on Electrical and Computer Engineering (CCECE), Vancouver, BC, Canada, 15–18 May 2016; pp. 1–5. [CrossRef]
30. Abdollahvand, S.; Oliveira, L.B.; Gomes, L.; Goes, J. A low-voltage voltage-controlled ring-oscillator employing dynamic-threshold-MOS and body-biasing techniques. In Proceedings of the 2015 IEEE International Symposium on Circuits and Systems (ISCAS), Lisbon, Portugal, 24–27 May 2015; pp. 1294–1297. [CrossRef]

Journal of
*Low Power Electronics
and Applications*

Article

A Standard-Cell-Based CMFB for Fully Synthesizable OTAs

Francesco Centurelli, Riccardo Della Sala * and Giuseppe Scotti

Department of Information and Communication Technologies, Sapienza University of Rome, 00184 Rome, Italy;
francesco.centurelli@uniroma1.it (F.C.); giuseppe.scotti@uniroma1.it (G.S.)
* Correspondence: riccardo.dellasala@uniroma1.it; Tel.: +39-06-4458-5679

Abstract: In this paper, we propose a fully standard-cell-based common-mode feedback (CMFB) loop with an explicit voltage reference to improve the CMRR of pseudo-differential standard-cell-based amplifiers and to stabilize the dc output voltage. This latter feature allows robust biasing of operational transconductance amplifiers (OTAs) based on a cascade of such stages. A detailed analysis of the CMFB is reported to both provide insight into circuit behavior and to derive useful design guidelines. The proposed CMFB is then exploited to build a fully standard-cell OTA suitable for automatic place and route. Simulation results referring to the standard-cell library of a commercial 130 nm CMOS process illustrated a differential gain of 28.3 dB with a gain-bandwidth product of 15.4 MHz when driving a 1.5 pF load capacitance. The OTA exhibits good robustness under PVT and mismatch variations and achieves state-of-the-art FOMs also thanks to the limited area footprint.

Keywords: OTA; CMFB; low voltage; low power; automatic place and route; standard-cell-based analog circuits

Citation: Centurelli, F.; Della Sala, R.;
Scotti, G. A Standard-Cell-Based
CMFB for Fully Synthesizable OTAs.
J. Low Power Electron. Appl. **2022**, *12*,
27. https://doi.org/10.3390/
jlpea12020027

Academic Editor: Orazio Aiello

Received: 28 February 2022
Accepted: 27 April 2022
Published: 5 May 2022

Publisher's Note: MDPI stays neutral
with regard to jurisdictional claims in
published maps and institutional affiliations.

1. Introduction

Recent years have seen a growing interest in ultra-low-voltage operational transconductance amplifiers (OTAs) [1–18] that are a key building block in many analog and mixed-signal applications such as Internet-of-Things (IoT) and biomedical ones [19–22]. This is a strong incentive to innovate the design flow of analog blocks: even if they often constitute just a small fraction of a mixed-signal system, their design requires a large fraction of the overall effort. Indeed, both the schematic and layout design are typically carried out manually, iterating each step several times until specifications are met, also taking into account the required robustness under process, supply voltage and temperature (PVT) variations. To minimize the analog design effort and hence the cost and time-to-market of such mixed-signal applications, circuit solutions for analog blocks based on digital standard cells were explored in [23,24]. The end goal is to achieve a fully automatic design flow for the analog blocks that is similar to the one adopted for the digital section; as an intermediate step, the use of digital standard cells to design analog functions allows for the automating of the place and route steps of the design flow and, in perspective, the achievement of a fully automatic synthesis flow for both analog and digital blocks.

Recently, different approaches to exploit digital-based architectures to mimic the behavior of analog functions were explored in [25]. In particular, the behavior of OTAs has been mimicked through VCO-based architectures [26–28] and the DIGOTA approach [29–31]. Even if all these innovative techniques are very interesting from a research point of view, the most common approach to implementing analog building blocks suitable for automatic place and route exploits the digital standard cells as basic analog amplifiers [32–37]. In fact, the simplest digital gate (the inverter) behaves as a common source amplifier [38], and several inverter-based OTAs [39–50] have been proposed in the literature. However, differently from custom-designed inverters, the standard-cell inverter is typically optimized for area footprint or symmetrical slew rate, and as a consequence, it exhibits a systematic offset in its input–output dc transfer characteristic which impacts the output static voltage and strongly degrades the performance of standard-cell-based cascaded amplifiers. According to the above considerations, the design of standard-cell-based OTAs must cope with additional issues that make achieving good and robust performance a very critical task. Several authors [37,42,50–52] have pointed out that the performances and even the operation of standard-cell-based analog circuits are severely impaired by PVT and mismatch variations, resulting in incorrect bias, large offsets and significant performance variations.

In such a context, the use of a common-mode feedback (CMFB) loop for each differential gain stage becomes mandatory to ensure a stable bias point [53–55], especially for ultra-low-voltage applications. The need to design a standard-cell-based CMFB greatly restricts the design options; some CMFB solutions have been proposed in the literature, but they usually do not involve an explicit reference voltage, resulting in some sensitivity to process, supply voltage and temperature (PVT) variations.

In this paper, we propose a fully standard-cell-based CMFB loop that exploits an explicit reference voltage to guarantee robust biasing, and we exploit it to design a two-stage OTA. Thanks to the proposed approach, a stable dc output voltage is guaranteed for the first stage, allowing a correct biasing of the second stage. The paper is structured as follows: in Section 2, the proposed CMFB is described and analyzed; in Section 3, the design of the standard-cell-based OTA is presented. Section 4 reports the simulation results, and, finally, conclusions are drawn in Section 5.

2. The Proposed CMFB

Figure 1a shows a CMOS inverter that can be thought of as a common-source amplifier. Its dc output voltage depends on the size of NMOS and PMOS devices and is also affected by PVT and mismatch variations. Figure 1b shows the dc input–output transfer characteristic (blue dashed line) and its derivative (continuous green line) for a typical standard-cell inverter, as a function of the input bias voltage V_i for a supply voltage $V_{DD} = 0.3$ V. Figure 1b clearly shows that an incorrect input dc bias results in a drop of voltage gain, thus making multistage amplifiers very difficult to implement if the dc output voltage of basic inverter stages is not controlled. The plot in Figure 1b also highlights the systematic offset of the inverter from a standard-cell library. In fact, the maximum gain is achieved for an input bias voltage different from $V_{DD}/2 = 150$ mV (value marked as a red dashed line). This systematic offset of the inverter (resulting in a logic threshold different from $V_{DD}/2$) is due to the fact that standard cells are not optimized for analog applications, and a trade-off between area, propagation time and balancing constraints is considered.

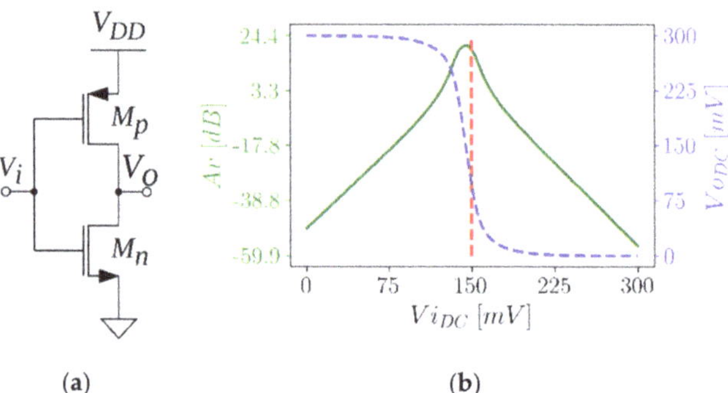

(a)   (b)

Figure 1. CMOS inverter (**a**) and its dc gain vs. input dc voltage (**b**).

A fully differential amplifier can be easily obtained using two inverters; however, it requires a CMFB loop to reduce the common-mode gain and to control the dc output voltage. Furthermore, to ensure correct biasing, multi-stage fully differential amplifiers require a CMFB at each stage. In the absence of accessible terminals to set the bias point of the inverter (e.g., the gate or body terminals of individual devices), the CMFB typically exploits two inverters with shorted outputs to sense the common-mode output voltage and other inverters as current sources to close the loop at the input of the main amplifier [53–55]. However, when applied to the first stage, this technique adds a resistive component to the input impedance of the OTA, and, therefore, the use of common-mode feedforward [33] has been proposed as an alternative. It is worth noting that, typically, this approach is used to reduce the common-mode gain, whereas other techniques such as body biasing [34,53] are exploited to set the output dc common-mode voltage. However, ultra-low voltage applications show low tolerance to biasing errors, and when standard-cell inverters are used, the body terminal is often not available for biasing purposes.

To maintain the advantages of the feedback avoiding this drawback, one option is to use a local common-mode feedback (LCMFB): when applied at transistor level [17], the LCMFB is typically implemented with a pair of common-mode sensing resistors whose central node is connected to the gates of the active load devices. The corresponding standard cell implementation [37] exploits a pair of sensing inverters and a pair of controlling inverters connected to the same output nodes. For the differential mode signal, the load impedance of the LCMFB is the output resistance of the loading inverters, whereas for the common-mode signal the LCMFB provides a low impedance load that reduces the gain, improving the common-mode rejection ratio (CMRR).

In this work, in order to improve the robustness of the dc operating point to PVT variations and to overcome the systematic offset of the standard cell inverters, we propose to add an explicit voltage reference V_{ref} to the standard-cell LCMFB through the inverter I_7, as shown in Figure 2.

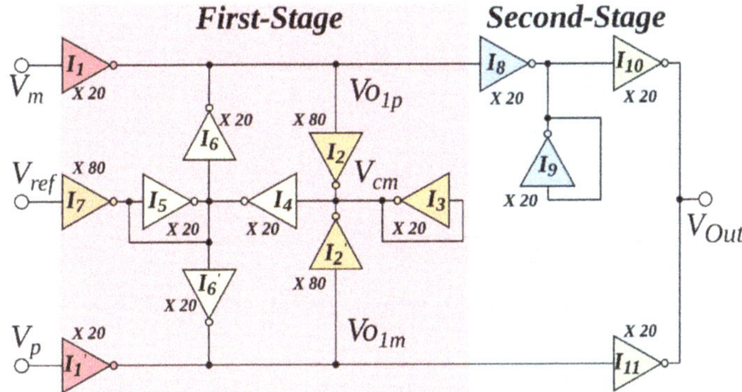

Figure 2. Topology of the standard-cell-based OTA with the proposed CMFB.

The resulting fully differential amplifier topology is depicted in Figure 2 (indicated as *First-Stage*): different colors are used to highlight the inverters constituting the gain stage (I_1 and I_1''), the common-mode estimator (I_2, I_2' and I_3), the reference inverting buffer (I_7, loaded by I_5) and the CMFB auxiliary amplifier (I_4, I_5, I_6 and I_6'). Inverters with their input and output terminals connected together are used as load devices to avoid high impedance nodes in the loop, providing better stability and a degree of freedom to design the circuit, as will be shown following this section. They are equivalent to parallel NMOS and PMOS diode-connected devices; thus, the cascade of an inverter and such a diode-connected inverter is equivalent to the parallel connection of an NMOS and a PMOS diode-loaded common-source stage.

To analyze the proposed topology and obtain design guidelines, we model each inverter I_X ($X = 1, \ldots, 7$) with a transconductance gain G_X and an output conductance G_{oX}, given by

$$G_X = g_{mnX} + g_{mpX} \tag{1}$$

$$G_{oX} = g_{dsnX} + g_{dspX} \tag{2}$$

where g_m and g_{ds} are the small-signal transconductance and output conductance of MOS devices, and n and p subscripts refer to NMOS and PMOS transistors, respectively. We assume that they scale linearly with the size of the devices (hence with the strength of the standard cells, *IV_xN* meaning an inverter whose devices have N times the minimum width), and their ratio is the voltage gain $A_X = G_X/G_{oX}$ that we assume is identical for all the inverters (hence $A_X = A$ for $x = 1, \dots, 7$).

Let α be the ratio of the strengths of inverters I_6 and I_1 (hence $\alpha = G_6/G_1$), $\lambda = G_2/G_3$, $\rho = G_4/G_5$ and $\beta = G_7/G_4$. The differential voltage gain of the first stage in Figure 2 results in

$$A_d = \frac{A}{1+\alpha} \tag{3}$$

where the loading effect of the LCMFB is considered. For the common mode, the analysis provides

$$V_{oc} = A_c V_{ic} + A_R V_{ref} \tag{4}$$

where V_{ic} and V_{oc} are the input and output common-mode components. The gains are

$$A_c = \frac{A}{(1+\alpha)D} = \frac{A_d}{CMRR} \tag{5}$$

$$A_R = \frac{\alpha A \beta \rho \varepsilon_4}{(1+\alpha)D} \tag{6}$$

$$D = 1 + \frac{2\alpha A \lambda \varepsilon_2 \rho \varepsilon_4}{1+\alpha} \tag{7}$$

where the error factors

$$\varepsilon_2 = \frac{1}{1 + \frac{2\lambda+1}{A}} \tag{8}$$

$$\varepsilon_4 = \frac{1}{1 + \frac{\beta\rho+\rho+1}{A}} \tag{9}$$

take into account the effect of the output conductances (ideally, if $G_X \gg G_{oX}$, since A approaches to infinite, the values of ε_2 and ε_4 tend to 1).

Equations (3)–(9) show that the CMRR of the stage is set by D in (7) and allow the ability to derive design guidelines for the choice of the inverter sizes. A trade-off between high CMRR and low gain penalty due to the loading effect involves the factor α: a large value of α maximizes the CMRR, whereas the smaller its value, the lower the reduction of the differential gain (3). A suitable solution is to choose $\alpha = 1$ and maximize the CMRR acting on the other factors.

Correct biasing would require $A_R = 1$, which in the limit of large CMRR implies

$$\beta = 2\lambda\varepsilon_2. \tag{10}$$

CMRR optimization then requires maximizing

$$\beta\rho\varepsilon_4 = \frac{\beta\rho}{1 + \frac{\beta\rho+\rho+1}{A}}. \tag{11}$$

that asymptotically tends to A for increasing β and ρ. Moreover, by (7) and the condition (10), we obtain

$$\lambda = \frac{\beta}{2}\frac{1 + 1/A}{1 - \beta/A} \tag{12}$$

that poses the further design constraints. Equation (11) implies that the size of inverter I_7 must be maximized, and (12) requires $\beta < A$.

In practical situations, the smallest inverter size to be used is limited by matching constraints (the smaller the transistors, the higher the standard deviation of mismatches, hence offset and common mode to differential mode conversion), and the largest inverter size is limited by area and power constraints and by available standard-cells. Hence, there is a limit to the size of I_7 (that is $\beta\rho$ times larger than I_5) and I_2 (that is λ times larger than I_3):

$$\beta\rho \leq R_{max} \qquad \lambda \leq R_{max}. \tag{13}$$

It must also be noted that there is a trade-off between CMRR and the common-mode range: a large value of λ provides a larger CMRR at the cost of the common-mode voltage swing at the output of the stage, since the output of I_2 saturates.

From (12), setting a maximum value of λ poses a more stringent limit on the value of β:

$$\beta_{max} = \frac{2R_{max}}{1 + \frac{2R_{max}+1}{A}}. \tag{14}$$

If $\beta_{max} > R_{max}$, $\beta\rho$ is set to R_{max} and maximizing (11) requires keeping the factor ρ as small as possible ($\rho = 1$). If $\beta_{max} < R_{max}$, which is the case for low values of the gain A, $\beta = \beta_{max}$ must be chosen; (11) then becomes

$$\frac{\beta_{max}\rho}{1 + \frac{\rho(\beta_{max}+1)+1}{A}}. \tag{15}$$

and its optimization involves maximizing ρ, whose maximum value is set by (13) to R_{max}/β_{max}. A flow graph illustrating this design procedure is reported in Figure 3.

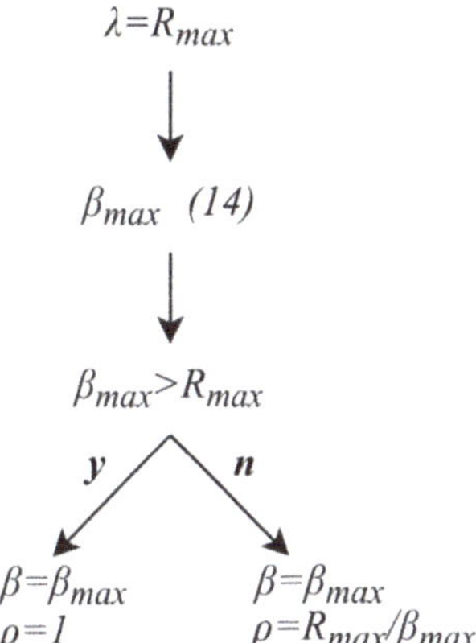

Figure 3. Proposed design approach for the CMFB.

Further insight into the behavior of the differential stage with the proposed CMFB can be gained by adding to the model of each inverter a current source I_{offX} that accounts for the offset of the inverter, i.e., the error in the output voltage with respect to $V_{DD}/2$ when the input voltage is $V_{DD}/2$:

$$I_{offX} = G_{oX} V_o|_{V_i=V_{DD}/2}. \tag{16}$$

This current source accounts for both the systematic offset due to the design of the inverter (PMOS to NMOS size ratio) and the random variation due to mismatches, and we can assume it is proportional to the *strength* of the inverter. The resulting block scheme for the common-mode behavior of the stage is shown in Figure 4; if we consider the contribution to the output common-mode voltage V_{oc} due to these offset current sources, which can be obtained by letting $V_{ic} = V_{ref} = 0$ in Figure 4, we obtain

$$V_{oc}^{(ioff)} = \frac{\alpha A}{D(1+\alpha)}\left[\frac{I_{off1}+I_{off6}}{\alpha G_1} + \frac{\rho\varepsilon_2}{G_3}\left(2I_{off2}+I_{off3}\right) - \frac{\varepsilon_4}{G_5}\left(I_{off4}+I_{off5}+I_{off7}\right)\right]. \tag{17}$$

Equation (17) shows that the residual error on setting $V_{oc} = V_{ref}$ is due also to the offset currents of the inverters and that the CMFB suppresses this error in the limit of its finite loop gain. This is true both for the systematic offset currents and for their random components, thus demonstrating that the circuit provides a stable dc output voltage under process and mismatch variations. It must be noted that a suitable choice of the inverter strengths can lead to minimizing (17) and could be used as a further design constraint for design optimization.

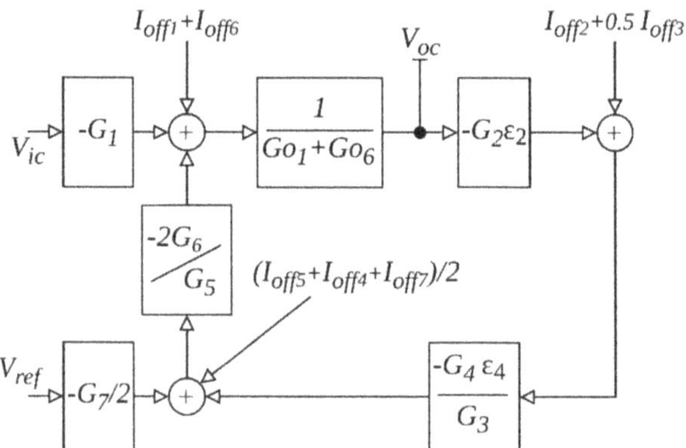

Figure 4. Model for the common-mode half circuit of the proposed CMFB loop considering the offset of the inverters.

3. Standard-Cell-Based OTA

An operational transconductance amplifier must provide high voltage gain, thus requiring, especially in an ultra-low-voltage environment with deep submicron technologies, the cascade of at least two gain stages with a differential-to-single-ended (D2S) conversion.

In the context of a standard-cell-based approach, mimicking the typical analog architecture with an input D2S converter followed by single-ended gain stages would result in a very poor CMRR and in a very high sensitivity to PVT variations. Better performance can be achieved by exploiting differential stages with CMFB loops to improve the CMRR and stabilize the bias point, followed by a final D2S stage.

To illustrate this approach, we propose a two-stage OTA, shown in Figure 2, composed by the fully differential stage described in the previous section (*First-Stage*) followed by a standard-cell-based D2S converter (*Second-Stage*), composed by inverters I_8–I_{11}. Inverters I_8 and I_9 constitute an inverting voltage buffer whose gain is ideally -1, and inverters I_{10} and I_{11} act as transconductance amplifiers driving the same output node.

The voltage gain of the D2S stage is thus ideally

$$\frac{V_{out}}{V_{olp} - V_{olm}} = A_{d2} = \frac{A}{2}. \tag{18}$$

with an infinite CMRR. However, in deep submicron technologies the voltage gain A is limited; hence, the output conductances of I_8 and I_9 cannot be neglected in the analysis. This reduces the gain of the voltage buffer and drastically worsens the CMRR even in typical conditions. Assuming $I_8 = I_9$ and $I_{10} = I_{11}$, differential and common-mode gains of the second stage are

$$A_{d2} = \frac{A}{2} \frac{1 + 1/A}{1 + 2/A} \tag{19}$$

$$A_{c2} = \frac{A}{2} \frac{2/A}{1 + 2/A} = \frac{1}{1 + 2/A} \tag{20}$$

and the overall CMRR results are

$$CMRR_{TOT} = \frac{A_{VD}}{A_{VC}} = \frac{A_d A_{d2}}{A_c A_{c2}} = D \frac{A + 1}{2} \tag{21}$$

where D, defined in (7), is the CMRR of the first stage.

The D2S stage presents an output pole set by the load capacitance C_L; moreover, its dual path nature results in a pole zero doublet similar to that provided by the current mirror load of a differential pair. With reference to Figure 2, assuming $I_8 = I_9$ and $I_{10} - I_{11}$ and considering a differential input to the D2S (i.e., $V_{o1p} = -V_{o1m}$), the frequency response of the D2S can be written as

$$A_{d2} = \frac{A}{2} \frac{2(1 + 1/A)G_8 + sC_X}{(1 + 2/A)G_8 + sC_X} \frac{1}{1 + sC_L/2G_{o10}} \tag{22}$$

where C_X is the total capacitance seen at the output of I_8. Equation (22) shows that the pole and zero due to the inverting buffer I_8–I_9 are spaced by an octave; thus, their effect can be neglected. It must further be noted that Equation (22) poses no constraint on the sizing of inverters I_8 and I_9 with respect to I_{10} and I_{11}; regardless, it could be convenient to use inverters of the same size to provide a symmetric loading to the first stage.

The OTA is stable when driving a sufficiently large load capacitance that makes the output pole dominant; for small load capacitors, a compensation is needed. More in detail, neglecting the pole–zero doublet due to I_8 and I_9, the internal pole of the OTA is given by

$$p_{int} = \frac{G_{o1}(1 + \alpha)}{C_{in2} + C_{in8}} \tag{23}$$

where C_{in2} and C_{in8} are the input capacitances of I_2 and I_8, and the output pole is

$$p_{out} = \frac{2G_{o10}}{C_L} \tag{24}$$

(assuming $I_{10} = I_{11}$). By imposing that the second pole is γ times the unity-gain frequency (where γ is set by the required phase margin), the minimum load capacitance required to have stability with the dominant pole at the output is

$$C_{Lmin} = \gamma \frac{2G_{o10}A_{VD}}{p_{int}} = \gamma \frac{A^2}{4} \frac{G_{o10}}{G_{o1}} (C_{in2} + C_{in8}) \tag{25}$$

where $\alpha = 1$ has been considered. Equation (25) shows that the standard-cell-based OTA has the capacity to drive small capacitors, as required in most on-chip applications, without the need of compensation capacitors thanks to the fact that the intrinsic gain of inverters is low and small-size inverters are used in the output stage. The latter condition, however, limits the slew rate of the OTA.

4. Simulation Results

The OTA in Figure 2 was designed using the standard-cell library of the STMicroelectronics 130 nm CMOS technology. Supply voltage was set to 0.3 V. Taking into account the design guidelines in Section 2 and the mismatch requirements that impose the use of non-minimum size cells, the inverters were designed as specified in Figure 2 (inverter *IV_xN* has devices with N times the minimum width). It must be noted that the low voltage gain of the inverters (A = 19 dB) result in the error factors (8) and (9) that are significantly below 1: in particular, we obtain $\varepsilon_2 = 0.5$ and $\varepsilon_4 = 0.6$. With reference to the analysis in Section 2, the smallest inverters were chosen as 20 times the minimum size inverter (*IV_x20* for all the inverters except I_2 and I_8) and the design factors were set to $\alpha = 1$, $\lambda = 4$, $\rho = 1$ and $\beta = 4$. The resulting static offset is therefore

$$V_{oc}^{(ioff)} = 0.57 \frac{I_{off1}}{G_{o1}}. \tag{26}$$

To assess the effectiveness of the proposed approach in stabilizing the dc output voltage, the pseudo-differential stage with the CMFB was tested under PVT variations. The reference voltage was set to $V_{DD}/2$, and the error of the output dc common-mode voltage with respect to this reference was evaluated. The LCMFB without I_7 and the reference input were also tested for comparison.

Figure 5 shows the relative error on the dc output common-mode voltage for the proposed CMFB and for the LCMFB without the reference input, under variations of temperature and supply voltage. An error is present in typical conditions (300 mV V_{DD}, 27 °C) due to the finite loop gain of the CMFB. When the reference input is present, the output common-mode voltage presents a limited variation when the temperature ranges from 0° to 80 °C, whereas the voltage drifts with the temperature if the reference input is not used. For what concerns the variation of the supply voltage, the dc common-mode output voltage tracks $V_{DD}/2$ with an error, due to the finite loop gain, that presents little variation and is lower than the error achieved by the design without the reference input.

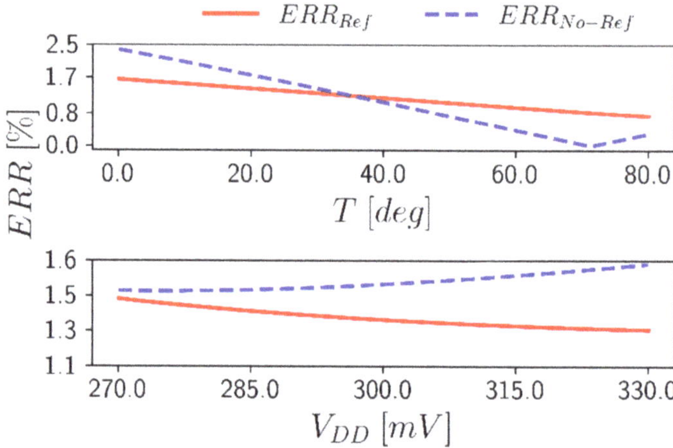

Figure 5. Relative error of the dc output common-mode voltage vs. temperature and supply voltage.

The advantage provided by the reference input is even more evident when the effect of process variations is considered. Figure 6 shows the relative error on the dc output common-mode voltage with respect to the reference input for the extreme process corners. A residual error of about 1.34% is reported in typical (TT) conditions due to the finite loop gain of the CMFB. This error increases in the corners where NMOS and PMOS devices present opposite variations (corners FS and SF); however, the values are below 10% and are one third of the errors obtained if the reference input is not used.

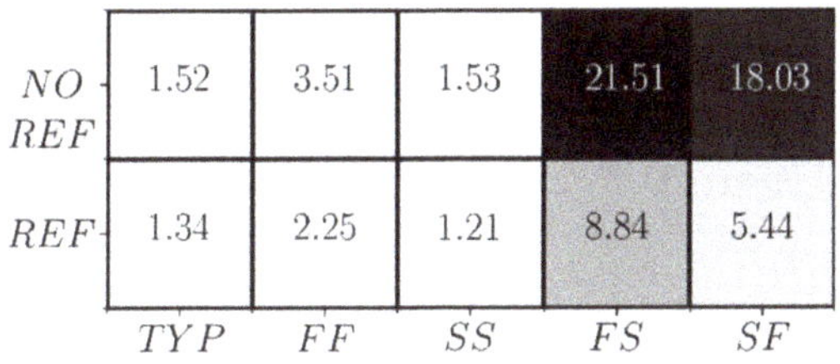

	TYP	FF	SS	FS	SF
NO REF	1.52	3.51	1.53	21.51	18.03
REF	1.34	2.25	1.21	8.84	5.44

Figure 6. Relative error (%) of the dc output common-mode voltage vs. process corners; darker colors correspond to higher relative errors.

The proposed pseudo-differential stage with CMFB thus results in suitable-to-design multi-stage amplifiers, and the OTA in Figure 2 was designed and simulated. The amplifier dissipates 4.4 μW from a 0.3 V supply; this relatively high power consumption is due to the use of large inverters to minimize the mismatches. Figure 7 shows the differential (A_{VD}) and common-mode (A_{VC}) gains of the OTA loaded by a 1.5 pF capacitor. The differential dc gain is 28.27 dB with a 15.42 MHz unity-gain frequency and 54.18° phase margin; the common-mode rejection ratio (CMRR) is about 41.07 dB and is constant across all the bandwidth.

The amplifier was tested in a unity-gain buffer configuration to assess its large-signal performance. The response to an input pulse from 45 to 255 mV (Figure 8) shows identical values for positive and negative slew rates equal to 9.75 V/μs. We also simulated the unity-gain buffer configuration with a 1 MHz sinusoidal input applied, and Figure 9 reports the total harmonic distortion (THD) as a function of the input amplitude. Distortions below 1% (−40 dB THD) are obtained for an input amplitude up to 115 mV, which is about 75% of the rail-to-rail swing.

Figure 10 shows the input-referred noise spectrum of the OTA: a noise corner frequency lower than 1 kHz with a white noise spectral density of 0.497 μV/$\sqrt{\text{Hz}}$ measured at 10 kHz was obtained, resulting in 1.445 mV rms noise when integrated over the whole closed-loop bandwidth.

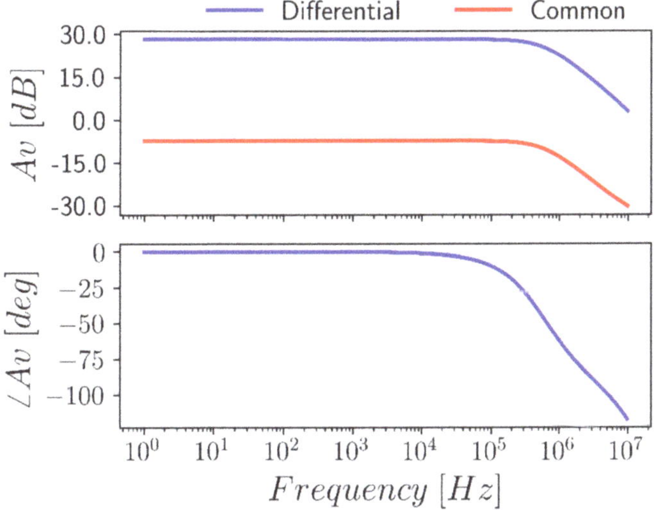

Figure 7. Differential (red) and common-mode (blue) gain of the proposed OTA.

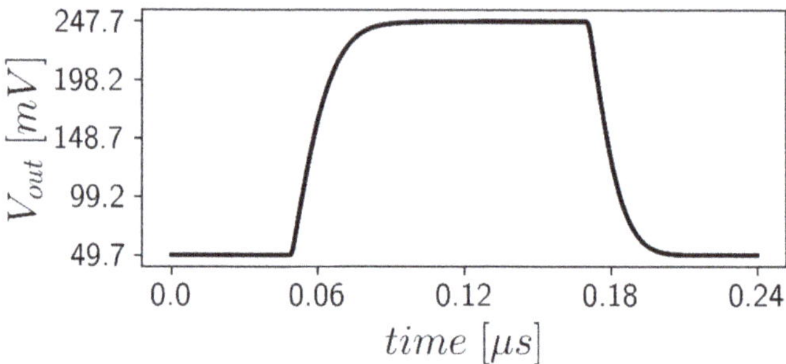

Figure 8. Response of the OTA in unity-gain configuration to a 45-to-255 mV input step.

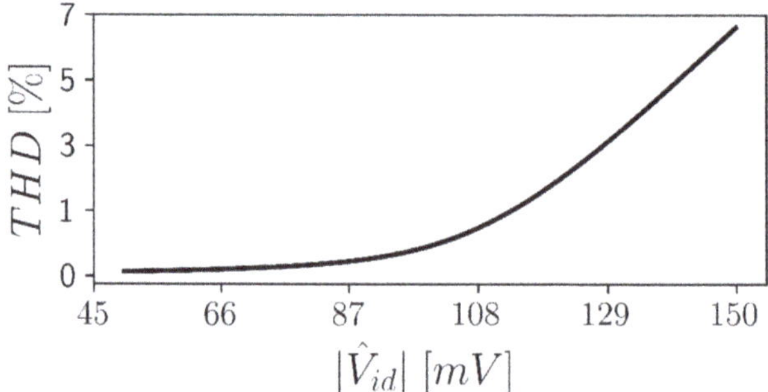

Figure 9. Total Harmonic Distortion vs. input amplitude for a 1 MHz sinusoidal input signal.

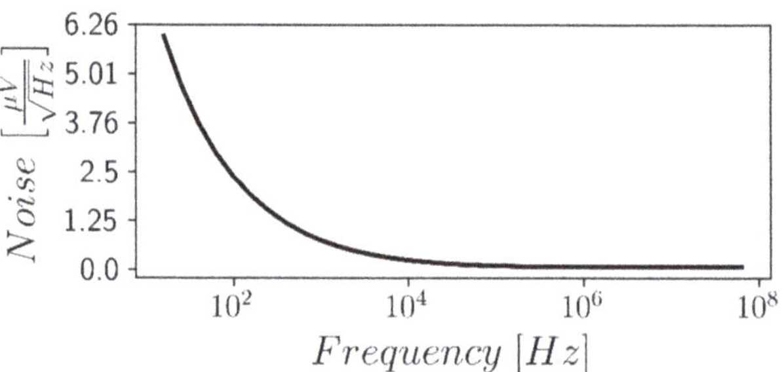

Figure 10. Input-referred noise spectrum of the OTA.

The performance under PVT variations and mismatches was evaluated to assess the robustness of the design. Table 1 reports the main performance parameters of the OTA in five different process corners highlighting how the proposed circuit exhibits a relatively low sensitivity to process variations. The effect of supply voltage and temperature variations is reported in Table 2: power consumption and gain-bandwidth product exhibit a non-negligible variation as expected since there is no bias loop setting the currents. However, the voltage gain A_{VD} and the output dc voltage (measured through the parameter V_{OS}) are extremely stable, confirming the effectiveness of the proposed approach.

Table 1. Variations under process corners.

Corner	FF	SS	FS	SF
A_{VD} [dB]	25.22	31.43	28.04	27.37
GBW [MHz]	26.4	10.24	15.28	14.57
m_φ [°]	63.38	49.18	60.59	60.28
Pd [μW]	7.832	1.89	3.97	3.74
V_{OS} [mV]	0.45	0.18	0.93	0.45
SR [V/μs]	15.15	5.85	12.47	6.33

Table 2. Variations under supply voltage and temperature.

	Voltage Variations		Temperature Variations	
T [°C]	27	27	0	80
V_{DD} [mV]	270	330	300	300
A_{VD} [dB]	27.28	29.03	29.25	26.32
GBW [MHz]	10.03	23.03	9.80	29.41
m_φ [°]	59.94	52.04	52.10	58.50
Pd [μW]	2.65	7.15	2.38	11.21
V_{OS} [mV]	−0.29	−0.29	−0.29	0.17
SR [V/μs]	3.61	12.52	6.91	12.88

Table 3 reports the results of 200 Monte Carlo mismatch simulations that show a good robustness of the proposed OTA, with limited variation of all the parameters. As can be observed in Table 3, under mismatch variations, the standard deviation of the output offset voltage is 9.2 mV, in line with other ULV OTAs taken from the literature. In order to further reduce the standard deviation of V_{OS} under mismatch variations, we can place multiple gates in parallel or exploit standard cells with larger driving capability, at the cost, however, of increased area and power consumption. We consider the proposed design as a good tradeoff between area, power consumption and output offset voltage standard deviation. Figure 11 shows the histogram of the CMRR, which is always higher than 10 dB and presents a log-normal distribution. The histogram shows that, even under mismatch conditions, acceptable values of CMRR are obtained, taking also into account the low value of the differential gain.

Table 3. Results of Monte Carlo mismatch analysis.

	Mean	Std
A_{VD} [dB]	28.2	0.88
GBW [MHz]	15.78	1.91
m_ϕ [°]	54.47	3.12
$CMRR$ [dB]	24.68	8.56
Pd [μW]	4.49	0.11
V_{OS} [mV]	0.002	9.2
SR [V/μs]	9.12	1.02

The layout of the proposed standard-cell OTA was generated by means of an automatic place and route flow by using the Cadence Innovus tool and is shown in Figure 12. The OTA occupies an area of $16.4 \times 10 \ \mu m^2$ that is very limited, notwithstanding the use of large inverters to minimize the mismatches. The layout has been generated automatically starting from a Verilog netlist of the circuit, which is reported in the Appendix A.

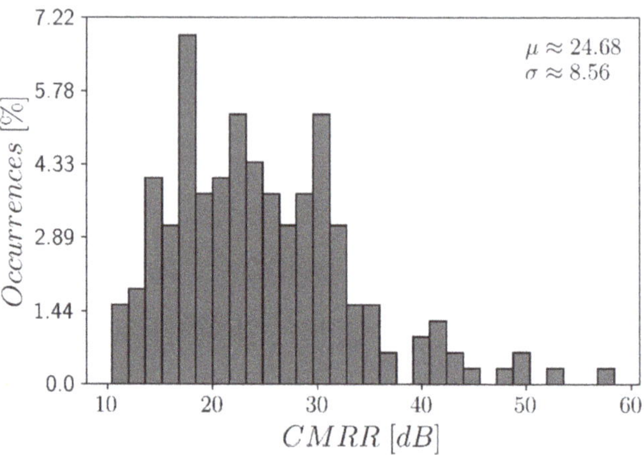

Figure 11. Histogram of the CMRR for 200 Monte Carlo mismatch iterations.

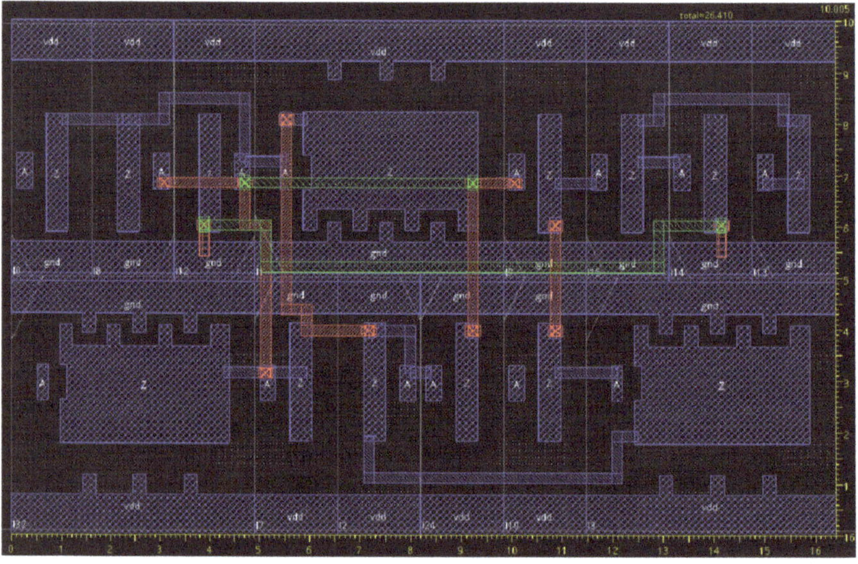

Figure 12. Layout of the proposed OTA generated by using the Cadence Innovus automatic place and route flow.

Table 4 reports the comparison of the proposed OTA with other ULV OTAs from the literature. To compare the proposed OTA against state-of-the-art low-voltage amplifiers, we refer to the following commonly used figures of merit:

$$FOM_S = \frac{GBW\ C_L}{Pd} \tag{27}$$

$$FOM_L = \frac{SR_{AVG}\ C_L}{Pd} \tag{28}$$

$$FOM_{S,A} = \frac{GBW\ C_L}{Pd\ Area} \tag{29}$$

$$FOM_{L,A} = \frac{SR_{AVG}\, C_L}{Pd\ Area} \tag{30}$$

where GBW is the gain-bandwidth product, C_L the load capacitance, SR_{AVG} is the average slew rate and Pd is the power consumption. Subscripts S and L in (27) and (28) denote small signal and large signal, respectively, while the figures of merit (29) and (30) are normalized with respect to the layout area of the OTA. The comparison shows that the proposed circuit exhibits very good small signal performance and adequate large signal performance. Due to the very compact layout, the proposed OTA outperforms all other similar designs in terms of $FOM_{L,A}$. The proposed OTA also outperforms almost all other designs in terms of $FOM_{S,A}$. Only [30] exhibits a higher $FOM_{S,A}$; however, the OTA in [30] is made up of minimum-sized standard cells that result in high sensitivity to process variations and mismatches.

Table 4. Comparison with the literature.

Ref	This Work	[30]	[31]	[17]	[18]	[46]	[10]	[12]	[35]	[36]	[3]
Year	2022	2021	2021	2022	2022	2020	2020	2020	2019	2019	2015
Tech. [nm]	130	180	180	130	130	180	65	180	130	130	65
V_{DD} [V]	0.3	0.55	0.3	0.3	0.3	0.3	0.25	0.5	0.3	0.25	0.35
A_{VD} [dB]	28.3	87	30	38.1	52.9	39	70	79.5	49.8	25	43
GBW [kHz]	15,420	3150	0.25	24.14	35.16	0.9	9.5	36	9100	7.23	3600
m_φ [°]	54	65	90	60	52	90	89	65	76	90	56
C_L [pF]	1.5	250	150	50	50	10	15	15	2	30	3
SR_{AVG} [V/ms]	9075	2.7	0.085	14.23	15.06	–	2	9.7	3800	–	5600
Pd [nW]	4406	8200	2.4	59.9	21.9	0.6	26	60	1800	55	17,000
Area [μm^2]	164	88.3	982	2700	5200	472	2000	3395	–	52,000	5000
Type	STD	STD	DIG	BD	BD	IB	BD	BD	IB	IB	BD
FOM_S [MHz pF/μW]	5.25	96.04	15.62	20.15	80.31	15	5.48	9	10.11	3.94	0.63
FOM_L [V pF/μs μW]	3.09	0.08	5.31	11.88	34.40	–	1.15	2.42	4.22	–	0.99
$FOM_{S,A}$ [MHz pF/μW μm^2]	32.01	1088	15.9	7.46	15.44	31.78	2.74	2.65	–	0.07	0.13
$FOM_{L,A}$ [V pF/ms μW μm^2]	18.84	0.93	5.4	4.4	6.61	–	0.57	0.71	–	–	0.19

STD = standard-cell-based; DIG s= DIGOTA; BD = body-driven; IB = inverter-based.

5. Conclusions

In this paper, we have presented a fully standard-cell-based common-mode feedback (CMFB) loop with an explicit voltage reference that allows the improvement of the CMRR of fully synthesizable standard-cell-based amplifiers and the stabilization of the dc output voltage with respect to PVT variations. A complete analysis of the circuit was presented to derive design guidelines. Simulations displayed that the use of an explicit reference input enhances the robustness of the CMFB to PVT variations.

The proposed CMFB was exploited to design a standard-cell-based OTA made up of only digital gates taken from a standard-cell library. The layout of the OTA was implemented by using a fully automated place and route flow by using the Cadence Innovus tool and starting from the Verilog netlist of the circuit. Simulation results illustrated very good values of both the small signal and large signal FOMs normalized to the area footprint of the circuits with a very good robustness of all the main performance parameters to PVT variations.

We remark that, due to the adoption of the proposed CMFB and to the design equations derived in this paper, the proposed standard-cell-based OTA results are more robust to PVT variations with respect to DIGITAL and standard-cell-based OTAs previously reported in the literature. However, it is worth noting that the performance attained by standard-cell-based OTAs is still less robust with respect to PVT and mismatch variations than that of OTAs designed with a custom analog design approach, which exhibit a well-defined bias current. Moreover, one of the main drawbacks of standard-cell single-ended OTAs is that the D2S converter (the last stage of the proposed OTA) cannot achieve good performance under PVT and mismatch variations, resulting in low and variable CMRR, thus reducing the ICMR of the whole architecture. Therefore, one of the goals of future works will be to achieve better ICMR and CMRR performance by standard-cell D2S converters in order to enhance the performance of standard-cell-based OTAs.

Author Contributions: Conceptualization, F.C., R.D.S. and G.S.; methodology, F.C., R.D.S. and G.S.; software, R.D.S.; validation, F.C., R.D.S. and G.S.; formal analysis, F.C. and R.D.S.; investigation, F.C., R.D.S. and G.S.; data curation, R.D.S.; writing—original draft preparation, F.C.; writing—review and editing, F.C., R.D.S. and G.S.; visualization, R.D.S.; supervision, F.C. and G.S. All authors have read and agreed to the published version of the manuscript.

Funding: This research received no external funding.

Conflicts of Interest: The authors declare no conflict of interest.

Appendix A

Verilog netlist of the proposed OTA is as follows:

```verilog
'timescale 1 ns/1 ns

module OTA (
inout REF,
inout Vim,
inout Vip,
inout Vout );

IV_X20 I24 ( .Z(feed), .A(CM));
IV_X20 I15 ( .Z(net8), .A(Vop1));
IV_X20 I14 ( .Z(Vout), .A(net8));
IV_X20 I13 ( .Z(net8), .A(net8));
IV_X20 I12 ( .Z(Vout), .A(Vom1));
IV_X20 I10 ( .Z(Vop1), .A(Vim));
IV_X20 I9 ( .Z(Vop1), .A(feed));
IV_X20 I8 ( .Z(Vom1), .A(feed));
IV_X20 I7 ( .Z(feed), .A(feed));
IV_X20 I2 ( .Z(CM), .A(CM));
IV_X20 I0 ( .Z(Vom1), .A(Vip));
IV_X80 I32 ( .Z(feed), .A(REF));
IV_X80 I3 ( .Z(CM), .A(Vop1));
IV_X80 I1 ( .Z(CM), .A(Vom1));

endmodule
```

References

1. Chatterjee, S.; Tsividis, Y.; Kinget, P. 0.5-V analog circuit techniques and their application in OTA and filter design. *IEEE J. Solid State Circuits* **2005**, *40*, 2373–2387. [CrossRef]
2. Ferreira, L.; Sonkusale, S.R. A 60-dB Gain OTA Operating at 0.25-V Power Supply in 130-nm Digital CMOS Process. *IEEE Trans. Circuits Syst. I Regul. Pap.* **2014**, *61*, 1609–1617. [CrossRef]
3. Abdelfattah, O.; Roberts, G.W.; Shih, I.; Shih, Y.-C. An Ultra-Low-Voltage CMOS Process-Insensitive Self-Biased OTA with Rail-to-Rail Input Range. *IEEE Trans. Circuits Syst. I Regul. Pap.* **2015**, *62*, 2380–2390. [CrossRef]
4. Akbari, M.; Hashemipour, O. A 63-dB gain OTA operating in subthreshold with 20-nW power consumption. *Int. J. Circuit Theory Appl.* **2017**, *45*, 843–850. [CrossRef]
5. Richelli, A.; Colalongo, L.; Kovacs-Vajna, Z.; Calvetti, G.; Ferrari, D.; Finanzini, M.; Pinetti, S.; Prevosti, E.; Savoldelli, J.; Scarlassara, S. A Survey of Low Voltage and Low Power Amplifier Topologies. *J. Low Power Electron. Appl.* **2018**, *8*, 22. [CrossRef]
6. Veldandi, H.; Shaik, R.A. A 0.3-V Pseudo-Differential Bulk-Input OTA for Low-Frequency Applications. *Circuits Syst. Signal Process.* **2018**, *37*, 5199–5221. [CrossRef]
7. Grasso, A.D.; Pennisi, S. Ultra-Low Power Amplifiers for IoT Nodes. In Proceedings of the 2018 25th IEEE International Conference on Electronics, Circuits and Systems (ICECS), Bordeaux, France, 9–12 December 2018.

8. Kulej, T.; Khateb, F. A Compact 0.3-V Class AB Bulk-Driven OTA. *IEEE Trans. Very Large Scale Integr. Syst.* **2020**, *28*, 224–232. [CrossRef]

9. Renteria-Pinon, M.; Ramirez-Angulo, J.; Diaz-Sanchez, A. Simple Scheme for the Implementation of Low Voltage Fully Differential Amplifiers without Output Common-Mode Feedback Network. *J. Low Power Electron. Appl.* **2020**, *10*, 34. [CrossRef]

10. Woo, K.-C.; Yang, B.-D. A 0.25V Rail-to-Rail Three-Stage OTA with an Enhanced DC Gain. *IEEE Trans. Circuits Syst. II Express Briefs* **2020**, *67*, 1179–1183. [CrossRef]

11. Kulej, T.; Khateb, F. A 0.3-V 98-dB Rail-to-Rail OTA in 0.18 μm CMOS. *IEEE Access* **2020**, *8*, 27459–27467. [CrossRef]

12. Deo, N.; Sharan, T.; Dubey, T. Subthreshold biased enhanced bulk-driven double recycling current mirror OTA. *Analog Integr. Circuits Signal Process.* **2020**, *105*, 229–242. [CrossRef]

13. Centurelli, F.; Sala, R.D.; Scotti, G.; Trifiletti, A. A 0.3 V, Rail-to-Rail, Ultralow-Power, Non-Tailed, Body-Driven, Sub-Threshold Amplifier. *Appl. Sci.* **2021**, *11*, 2528. [CrossRef]

14. Centurelli, F.; Sala, R.D.; Monsurrò, P.; Scotti, G.; Trifiletti, A. A 0.3 V rail-to-rail ultra-low-power OTA with improved bandwidth and slew rate. *J. Low Power Electron. Appl.* **2021**, *11*, 19. [CrossRef]

15. Dong, S.; Wang, Y.; Tong, X.; Wang, Y.; Liu, C. A 0.3-V 8.72-nW OTA with Bulk-Driven Low-Impedance Compensation for Ultra-Low Power Applications. *Circuits Syst. Signal Process.* **2021**, *40*, 2209–2227. [CrossRef]

16. Ghosh, S.; Bhadauria, V. An ultra-low-power near rail-to-rail pseudo-differential subthreshold gate-driven OTA with improved small and large signal performances. *Analog Integr. Circuits Signal Process.* **2021**, *109*, 345–366. [CrossRef]

17. Centurelli, F.; Sala, R.D.; Monsurrò, P.; Tommasino, P.; Trifiletti, A. An ultra-low-voltage class-AB OTA exploiting local CMFB and body-to-gate interface. *AEU Int. J. Electron. Commun.* **2022**, *145*, 154081. [CrossRef]

18. Centurelli, F.; Sala, R.D.; Monsurrò, P.; Scotti, G.; Trifiletti, A. A tree-based architecture for high-performance ultra-low-voltage amplifiers. *J. Low Power Electron. Appl.* **2022**, *12*, 12. [CrossRef]

19. Ng, K.A.; Xu, Y.P. A Low-Power, High CMRR Neural Amplifier System Employing CMOS Inverter-Based OTAs with CMFB Through Supply Rails. *IEEE J. Solid State Circuits* **2016**, *51*, 724–737.

20. Giustolisi, G.; Palumbo, G. A g_m/I_D-based design strategy for IoT and ultra-low-power OTAs with fast-settling and large capacitive loads. *J. Low Power Electron. Appl.* **2021**, *11*, 21. [CrossRef]

21. Silva, R.S.; Rodovalho, L.H.; Aiello, O.; Rodrigues, C.R. A 1.9 nW, Sub-1 V, 542 pA/V Linear Bulk-Driven OTA with 154 dB CMRR for Bio-Sensing Applications. *J. Low Power Electron. Appl.* **2021**, *11*, 40. [CrossRef]

22. Vafaei, M.; Parhizgar, A.; Abiri, E.; Salehi, M.R. A low power and ultra-high input impedance analog front end based on fully differential difference inverter-based amplifier for biomedical applications. *AEU Int. J. Electron. Commun.* **2021**, *142*, 154005. [CrossRef]

23. Newton, S.M.; Kinget, P.R. A 4th-order analog continuous-time filter designed using standard cells and automatic digital logic design tools. In Proceedings of the 2016 IEEE International Symposium on Circuits and Systems (ISCAS), Montreal, QC, Canada, 22–25 May 2016.

24. Liu, J.; Park, B.; Guzman, M.; Fahmy, A.; Kim, T.; Maghari, N. A Fully Synthesized 77-dB SFDR Reprogrammable SRMC Filter Using Digital Standard Cells. *IEEE Trans. Very Large Scale Integr. Syst.* **2018**, *26*, 1126–1138. [CrossRef]

25. Toledo, P.; Rubino, G.R.; Musolino, F.; Crovetti, P. Re-Thinking Analog Integrated Circuits in Digital Terms: A New Design Concept for the IoT Era. *IEEE Trans. Circuits Syst. II Express Briefs* **2021**, *68*, 816–822. [CrossRef]

26. Drost, B.; Talegaonkar, M.; Hanumolu, P.K. Analog Filter Design Using Ring Oscillator Integrators. *IEEE J. Solid State Circuits* **2012**, *47*, 3120–3129. [CrossRef]

27. Hsu, C.-W.; Kinget, P.R. A 40MHz 4th-order active-UGB-RC filter using VCO-based amplifiers with zero compensation. In Proceedings of the ESSCIRC 2014—40th European Solid State Circuits Conference (ESSCIRC), Venice, Italy, 22–26 September 2014; pp. 359–362.

28. Kalani, S.; Bertolini, A.; Richelli, A.; Kinget, P.R. A 0.2V 492nW VCO-based OTA with 60 kHz UGB and 207 μVrms noise. In Proceedings of the 2017 IEEE International Symposium on Circuits and Systems (ISCAS), Baltimore, MD, USA, 28–31 May 2017.

29. Crovetti, P. A Digital-Based Analog Differential Circuit. *IEEE Trans. Circuits Syst. I Regul. Pap.* **2013**, *60*, 3107–3116. [CrossRef]

30. Palumbo, G.; Scotti, G. A Novel Standard-Cell-Based Implementation of the Digital OTA Suitable for Automatic Place and Route. *J. Low Power Electron. Appl.* **2021**, *11*, 42. [CrossRef]

31. Toledo, P.; Crovetti, P.; Aiello, O.; Alioto, M. Design of Digital OTAs with Operation Down to 0.3 V and nW Power for Direct Harvesting. *IEEE Trans. Circuits Syst. I Regul. Pap.* **2021**, *68*, 3693–3706. [CrossRef]

32. Nauta, B. A CMOS transconductance-C filter technique for very high frequencies. *IEEE J. Solid-State Circuits* **1992**, *27*, 142–153. [CrossRef]

33. Barthélemy, H.; Meillère, S.; Gaubert, J.; Dehaese, N.; Bourdel, S. OTA based on CMOS inverters and application in the design of tunable bandpass filter. *Analog Integr. Circuits Signal Process.* **2008**, *57*, 169–178. [CrossRef]

34. Vlassis, S. 0.5 V CMOS inverter-based tunable transconductor. *Analog Integr. Circuits Signal Process.* **2012**, *72*, 289–292. [CrossRef]

35. Lv, L.; Zhou, X.; Qiao, Z.; Li, Q. Inverter-Based Subthreshold Amplifier Techniques and Their Application in 0.3-V ΔΣ-Modulators. *IEEE J. Solid-State Circuits* **2019**, *54*, 1436–1445. [CrossRef]

36. Braga, R.A.S.; Ferreira, L.H.C.; Coletta, G.D.; Dutra, O.O. A 0.25-V calibration-less inverter-based OTA for low-frequency G_m-C applications. *Microelectron. J.* **2019**, *83*, 62–72. [CrossRef]

37. Manfredini, G.; Catania, A.; Benvenuti, L.; Cicalini, M.; Piotto, M.; Bruschi, P. Ultra-Low-Voltage Inverter-Based Amplifier with Novel Common-Mode Stabilization Loop. *Electronics* **2020**, *9*, 1019. [CrossRef]
38. Bae, W. CMOS Inverter as Analog Circuit: An Overview. *J. Low Power Electron. Appl.* **2019**, *9*, 26. [CrossRef]
39. Michel, F.; Steyaert, M.S.J. A 250 mV 7.5 µW 61 dB SNDR SC ΔΣ modulator using near-threshold-voltage-biased inverter amplifiers in 130 nm CMOS. *IEEE J. Solid State Circuits* **2012**, *47*, 709–721. [CrossRef]
40. Yang, Z.; Yao, L.; Lian, Y. A 0.5-V 35-µW 85-dB DR double-sampled ΔΣ modulator for audio applications. *IEEE J. Solid State Circuits* **2012**, *47*, 722–735. [CrossRef]
41. Suadet, A.; Kasemsuwan, V. A CMOS inverter-based class-AB pseudo-differential amplifier with current-mode common-mode feedback (CMFB). *Analog Integr. Circuits Signal Process.* **2013**, *74*, 387–398. [CrossRef]
42. Ismail, A.; Mostafa, I. A process-tolerant, low-voltage, inverter-based OTA for continuous-time Σ-Δ ADC. *IEEE Trans. Very Large Scale Integr. Syst.* **2016**, *24*, 2911–2917. [CrossRef]
43. Yeknami, A.F. A 300-mV ΔΣ Modulator Using a Gain-Enhanced, Inverter-Based Amplifier for Medical Implant Devices. *J. Low Power Electron. Appl.* **2016**, *6*, 4. [CrossRef]
44. De Aguirre, P.C.C.; Susin, A.A. PVT compensated inverter-based OTA for low-voltage CT sigma-delta modulators. *Electron. Lett.* **2018**, *54*, 1264–1266. [CrossRef]
45. Baghtash, H.F. A 0.4 V, tail-less, fully differential trans-conductance amplifier: An all inverter-based structure. *Analog Integr. Circuits Signal Process.* **2020**, *104*, 1–15. [CrossRef]
46. Rodovalho, L.; Aiello, O.; Rodrigues, C. Ultra-Low-Voltage Inverter-Based Operational Transconductance Amplifiers with Voltage Gain Enhancement by Improved Composite Transistors. *Electronics* **2020**, *9*, 1410. [CrossRef]
47. Baghtash, H.F. Bias-stabilized inverter-amplifier: An inspiring solution for low-voltage and low-power applications. *Analog Integr. Circuits Signal Process.* **2020**, *105*, 243–248. [CrossRef]
48. Rodovalho, L.H. Push-pull based operational transconductance amplifier topologies for ultra low voltage supplies. *Analog Integr. Circuits Signal Process.* **2021**, *106*, 111–124. [CrossRef]
49. Lee, S.; Park, S.; Kim, Y.; Kim, Y.; Lee, J.; Lee, J.; Chae, Y. A 0.6-V 86.5-dB DR 40-kHz BW inverter-based continuous-time delta-sigma modulator with PVT-robust body-biasing. *IEEE Solid State Circuits Lett.* **2021**, *4*, 178–181. [CrossRef]
50. Rodovalho, L.; Rodrigues, C.R.; Aiello, O. Self-Biased and Supply-Voltage Scalable Inverter-Based Operational Transconductance Amplifier with Improved Composite Transistors. *Electronics* **2021**, *10*, 935. [CrossRef]
51. Baltolu, A.; Albinet, X.; Chalet, F.; Dallet, D.; Begueret, J.-B. A robust inverter-based amplifier versus PVT for discrete-time integrators. *Int. J. Circuit Theory Appl.* **2018**, *46*, 2160–2169. [CrossRef]
52. Pradeep, R.; Siddharth, R.K.; Kumar, Y.B.N.; Vasantha, M.H. Process corner calibration for standard cell based flash ADC. In Proceedings of the 2019 IEEE International Symposium on Smart Electronic Systems (iSES), Rourkela, India, 16–18 December 2019.
53. Vieru, R.G.; Ghinea, R. Inverter-based ultra low voltage differential amplifiers. In Proceedings of the 2011 International Semiconductor Conference, Sinaia, Romania, 17–19 October 2011.
54. Tanimoto, H.; Yazawa, K.; Haraguchi, M. A fully-differential OTA based on CMOS cascode inverters operating from 1-V power supply. *Analog Integr. Circuits Signal Process.* **2014**, *78*, 23–31. [CrossRef]
55. Ruscio, D.; Centurelli, F.; Monsurrò, P.; Trifiletti, A. Reconfigurable low voltage inverter-based sample-and-hold amplifier. In Proceedings of the 2017 13th Conference on Ph.D. Research in Microelectronics and Electronics (PRIME), Giardini Naxos–Taormina, Italy, 12–15 June 2017.

Journal of
*Low Power Electronics
and Applications*

Article

Bridging the Gap between Design and Simulation of Low-Voltage CMOS Circuits [†]

Cristina Missel Adornes *, Deni Germano Alves Neto, Márcio Cherem Schneider and Carlos Galup-Montoro

Department of Electrical and Electronics Engineering, Federal University of Santa Catarina,
Florianópolis 88040-900, Brazil; deni.alves@posgrad.ufsc.br (D.G.A.N.); m.c.schneider@ufsc.br (M.C.S.);
carlos@eel.ufsc.br (C.G.-M.)
* Correspondence: cristina.m.adornes@posgrad.ufsc.br
† This paper is an extended version of our paper published in 2021 IEEE Nordic Circuits and Systems
  Conference (NorCAS).

Abstract: This work proposes a truly compact MOSFET model that contains only four parameters to assist an integrated circuits (IC) designer in a design by hand. The four-parameter model (4PM) is based on the advanced compact MOSFET (ACM) model and was implemented in Verilog-A to simulate different circuits designed with the ACM model in Verilog-compatible simulators. Being able to simulate MOS circuits through the same model used in a hand design benefits designers in understanding how the main MOSFET parameters affect the design. Herein, the classic CMOS inverter, a ring oscillator, a self-biased current source and a common source amplifier were designed and simulated using either the 4PM or the BSIM model. The four-parameter model was simulated in many sorts of circuits with very satisfactory results in the low-voltage cases. As the ultra-low-voltage (ULV) domain is expanding due to applications, such as the internet of things and wearable circuits, so is the use of a simplified ULV MOSFET model.

Keywords: ACM model; MOSFET modeling; circuit simulation; ultra-low voltage

Citation: Adornes, C.M.; Alves Neto, D.G.; Schneider, M.C.; Galup-Montoro, C. Bridging the Gap between Design and Simulation of Low-Voltage CMOS Circuits. *J. Low Power Electron. Appl.* **2022**, *12*, 34. https://doi.org/10.3390/jlpea12020034

Academic Editors: Orazio Aiello and Andrea Acquaviva

Received: 16 February 2022
Accepted: 5 May 2022
Published: 16 June 2022

Publisher's Note: MDPI stays neutral with regard to jurisdictional claims in published maps and institutional affiliations.

1. Introduction

The design and simulation of integrated circuits (IC) are assisted by compact MOSFET models, which started to be developed in the 1960s [1] for long-channel devices. The technological progress promoted the down-scaling of semiconductor devices, giving rise to short-channel effects and their interference in circuit performance; thereby, these short-channel effects were incorporated into the existing long-channel based models to improve circuit-design efficiency.

Although BSIM [2,3] has been broadly used as the main MOSFET model to simulate MOS circuits in EDA tools, the complexity of its calculations and numerous parameters have opened a gap between circuit simulations and designs by hand [4,5], which has complicated the understandings of how the main MOSFET parameters relate to simulation results. Therefore, it is in designers' interest to have models founded on physics available in the simulator, such as those based on the inversion charge.

In the fast expanding ultra-low-voltage domain [6], some short-channel effects, such as velocity saturation, are not relevant; thus, a simplified MOSFET model can be satisfactory for circuit design. Targeting the increasing number of ultra-low-voltage designs [7–12], this work proposes a four-parameter model (4PM) based on the all-region advanced compact MOSFET model (ACM) [13].

In this work, the 4PM was carried out with the description language Verilog-A to easily simulate circuits in the commercial Cadence® Virtuoso® simulator, which implements BSIM 4.5 through a private propriety interface [14]. Hardware description languages

(HDLs), such as Verilog-A, are interchangeable with different simulators and assist designers in describing circuits and systems in a variety of behavioral modeling levels. We chose Verilog-A because it combines simplicity, functionality and portability [14].

The paper is structured as described in the following lines. Section 2 briefly introduces the four-parameter model (4PM). Section 3 describes the methods employed to extract the model's parameters and describes the extraction results of the parameters with temperature and process variations. Section 4 describes how to carry out the 4PM in Verilog-A for its inclusion in Cadence. Section 5 presents the results of the simulations carried out by using BSIM or the 4PM in Verilog-A. Four circuits designed according to the ACM model were simulated: a CMOS inverter, a ring oscillator, a self-biased current source and a common source amplifier. Conclusions are drawn in Section 6.

2. The Four-Parameter Model (4PM)

The advanced compact MOSFET (ACM) model describes static and small-signal low-frequency characteristics of MOS transistors in all regions of operation [13]. ACM employs three main transistor parameters: the specific current I_S, the threshold voltage V_{T0} and the slope factor n, which are usually sufficient to design a broad amount of circuits.

Nevertheless, the four-parameter model herein also employs drain-induced barrier lowering (DIBL), a secondary effect [13]. In spite of being a very pronounced effect for short-channel transistors, the DIBL cannot be ignored for long-channel transistors in weak inversions. For long-channel transistors in strong inversions (out of the scope of this work), DIBL is overshadowed by channel-length modulation.

In the long-channel ACM model [13], the drain current I_D in Figure 1 is split into the forward term I_F and the reverse term I_R, both of them dependent on the voltage V_{GB}. The component I_F also depends on V_{SB}, while I_R depends on V_{DB}. This source-drain symmetry is given by using (1).

$$I_D = I_F - I_R = I_S \left(i_f - i_r \right) \tag{1}$$

The specific current I_S, depicted in (2), is influenced by the device's geometry and technological parameters, such as the carrier mobility μ, C_{ox}, the slope factor n and temperature through the thermal voltage ϕ_t.

$$I_S = \mu C_{ox} n \frac{\phi_t^2}{2} \frac{W}{L} \tag{2}$$

The relationship between the voltages at the device terminals and the normalized inversion charge density at the source (drain) $q_{IS(D)}$ is established by using the normalized form of the unified charge-control model (UCCM) in (3).

$$\frac{V_P - V_{S(D)B}}{\phi_t} = q_{IS(D)} - 1 + \ln q_{IS(D)} \tag{3}$$

The pinch-off voltage V_P can be approximated by using (4), where V_{T0} is the equilibrium threshold voltage that corresponds to the gate voltage for which $V_P = 0$ and for which σ is the magnitude of the DIBL coefficient. In the four-parameter model, the DIBL effect must comply with the MOSFET symmetry.

$$V_P = \frac{V_{GB} - V_{T0} + \sigma V_{DB} + \sigma V_{SB}}{n} \tag{4}$$

Equation (5) gives the definition of the normalized inversion charge, which is the inversion charge (Q_I) normalized to the pinch-off charge ($-nC_{ox}\phi_t$).

$$q_{IS(D)} = \frac{Q_I}{-nC_{ox}\phi_t} \tag{5}$$

$$q_{IS(D)} = \sqrt{1 + i_{f(r)}} - 1 \tag{6}$$

The voltage-to-inversion level relationship is established by applying (6) to (3), which results in (7a,b), also known as the unified current-control model (UICM). For design purposes, $i_f < 1$ characterizes an operation in a weak inversion (WI), while for $i_f > 100$, it is assumed there is an operation in a strong inversion (SI). For inversion levels between 1 and 100, it is said that the transistors operate in moderate inversion (MI).

$$I_{F(R)} = I_S F\left[\frac{V_P - V_{S(D)}}{\phi_t}\right] \tag{7a}$$

$$F^{-1} = \sqrt{1 + i_{f(r)}} - 2 + \ln\left(\sqrt{1 + i_{f(r)}} - 1\right) \tag{7b}$$

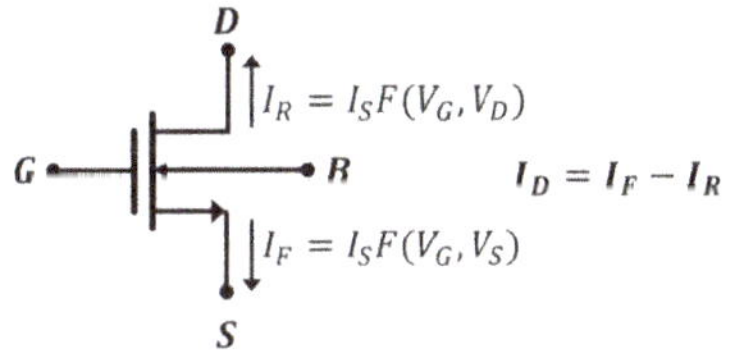

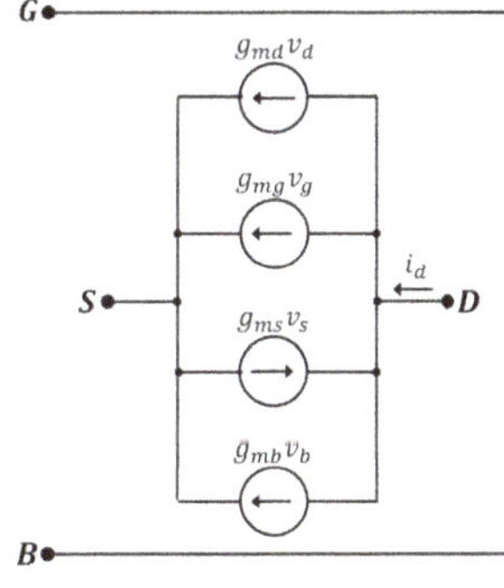

Figure 1. Symbol of an n-channel MOSFET transistor and its four terminals: gate (G), source (S), drain (D) and bulk (B). Source-drain symmetry illustrated by using currents.

2.1. Small-Signal Transconductances

Small-signal transconductances are essential for both the design of integrated circuits and the extraction of the four transistor parameters. Figure 2 presents the low-frequency small-signal model for MOSFET in which the variation of the drain current is expressed by using (8), where g_{mg}, g_{ms}, g_{md} and g_{mb} are, respectively, the gate, source, drain and bulk transconductances given by using (9); v_g, v_s, v_d and v_b represent small variations in the gate, source, drain and bulk voltages, respectively.

$$i_d = g_{mg}v_g - g_{ms}v_s + g_{md}v_d + g_{mb}v_b \tag{8}$$

$$g_{mg} = \frac{\partial I_D}{\partial V_G}; g_{ms} = -\frac{\partial I_D}{\partial V_S}; g_{md} = \frac{\partial I_D}{\partial V_D}; g_{mb} = \frac{\partial I_D}{\partial V_B} \tag{9}$$

Figure 2. Low-frequency small-signal model of the MOSFET.

The relationships between the transconductances and the inversion levels are obtained by applying the partial derivatives of (9) to the UICM along with (1).

The transconductance-to-current ratios, in terms of inversion level, are given by using expressions (10)–(12) in which $I_{D,sat}$ stands for the approximation of the drain current in the saturation region, where $i_r << i_f$ [13].

$$\phi_t \frac{g_{ms}}{I_{D,sat}} = \left(1 - \frac{\sigma}{n}\right) \frac{2}{\sqrt{1 + i_f} + 1} \tag{10}$$

$$\phi_t \frac{g_{md}}{I_{D,sat}} = \left(\frac{\sigma}{n}\right) \frac{2}{\sqrt{1 + i_f} + 1} \tag{11}$$

$$\phi_t \frac{g_m}{I_{D,sat}} = \left(\frac{1}{n}\right) \frac{2}{\sqrt{1 + i_f} + 1} \tag{12}$$

2.2. Dynamic Model

The dynamic model of MOS transistors includes intrinsic and extrinsic capacitances. Figure 3 presents a simplified dynamic model that includes both the intrinsic and extrinsic parts.

In Figure 3a, the extrinsic capacitance $C_{gse(de)}$ includes an unavoidable overlap between the gate and the source (drain) diffusion and fringing capacitances, while the substrate-source (drain) junctions modeled by (nonlinear) diode capacitances correspond to $C_{bse(de)}$. A more complete model for the extrinsic part should include parasitic resistances as well [15].

The field effect of MOS transistors occurs in the intrinsic part between the source and drain. The classical MOSFET model in Figure 3b contains five capacitances added to the small-signal model of Figure 2.

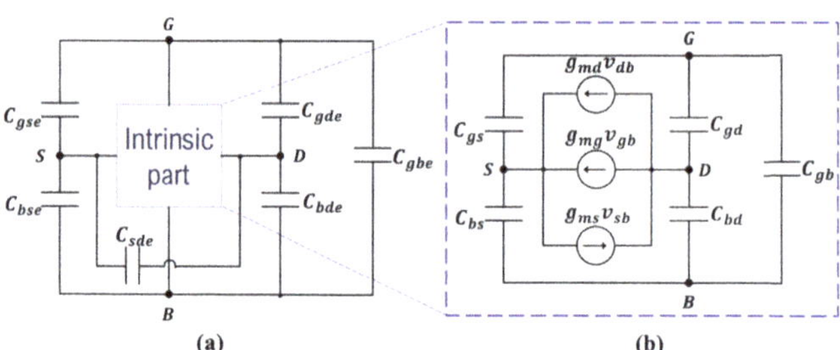

Figure 3. MOSFET dynamic model with (**a**) extrinsic and (**b**) intrinsic parts [13].

The calculation of the intrinsic capacitance coefficients is based on the unified charge-control model (UCCM) and on the quasi-static charge conserving model [13]. The effect of the DIBL parameter on the five intrinsic capacitances is summarized in expressions (15)–(19) in which C_{gs0} and C_{gd0} are the gate-source and gate-drain capacitances of the long-channel model, respectively. In (13) and (14), $\alpha = \frac{1 + q_{iD}}{1 + q_{iS}}$ is the channel linearity factor.

$$C_{gs0} = \frac{2}{3} W L C_{ox} \frac{1 + 2\alpha}{(1 + \alpha)^2} \frac{q_{iS}}{1 + q_{iS}} \tag{13}$$

$$C_{gd0} = \frac{2}{3} W L C_{ox} \frac{\alpha^2 + 2\alpha}{(1 + \alpha)^2} \frac{q_{iD}}{1 + q_{iD}} \tag{14}$$

$$C_{gs} = \left(1 - \frac{\sigma}{n}\right) C_{gs0} - \frac{\sigma}{n} C_{gd0} \tag{15}$$

$$C_{gd} = \left(1 - \frac{\sigma}{n}\right) C_{gd0} - \frac{\sigma}{n} C_{gs0} \tag{16}$$

$$C_{gb} = \left(1 - \frac{1}{n}\right)(WLC_{ox} - C_{gs0} - C_{gd0}) + \frac{2\sigma}{n}[(n-1)WLC_{ox} - C_{gs0} - C_{gd0}] \qquad (17)$$

$$C_{bs} = (n-1)C_{gs} \qquad (18)$$

$$C_{bd} = (n-1)C_{gd} \qquad (19)$$

Figure 4 presents plots of the five intrinsic capacitances normalized to C_{ox} as functions of the pinch-off voltage. The curves were obtained for an NMOS transistor with $\frac{W}{L} = \frac{0.6\ \mu m}{0.3\ \mu m}$ and $V_{DS} = 1$ V.

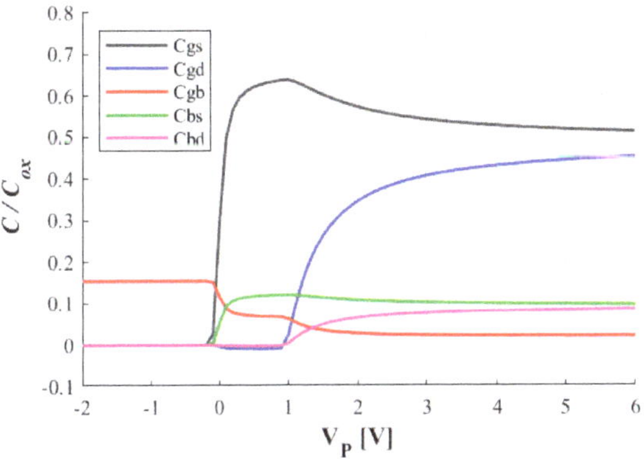

Figure 4. Capacitances (15)–(19) normalized to C_{ox} versus the pinch-off voltage for $V_{DS} = 1$ V.

3. Parameter Extraction

The accuracy of the transistor's characteristics depends on both the model and the accuracy of the parameters. The model's parameters should be easily and accurately extracted; otherwise, the model will not be successful [14]. Thus, this section presents the methods to extract the four transistor parameters.

3.1. Extraction of Threshold Voltage (V_{T0}), Specific Current (I_S) and Slope Factor (n)

The values of the threshold voltage (V_{T0}), the specific current (I_S) and slope factor (n) were extracted from the g_m/I_D curve [16] illustrated in Figure 5b, which was measured with the circuit configuration in Figure 5a.

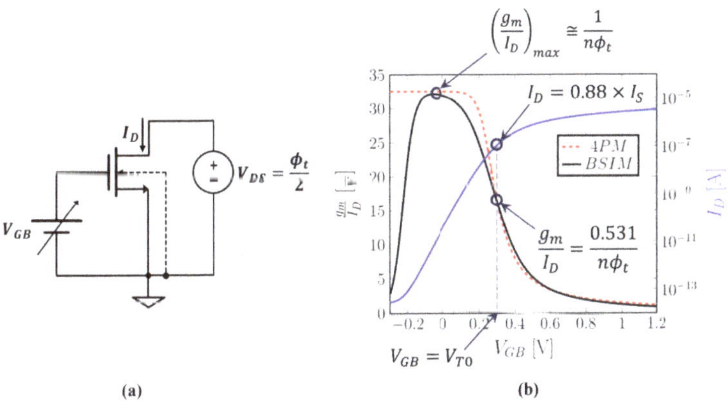

Figure 5. (a) Circuit to extract parameters from (b) the g_m/I_D and I_D curves.

Based on the method described in [16], the values of the threshold voltage and the specific current were determined through the g_m/I_D characteristic written in (20), which was valid for all regions of operation.

$$\frac{g_m}{I_D} = \frac{1}{I_D}\frac{\partial I_D}{\partial V_G} = \frac{2}{n\phi_t\left(\sqrt{1+i_f}+\sqrt{1+i_r}\right)} \tag{20}$$

$$\frac{V_{DS}}{\phi_t} = \sqrt{1+i_f} - \sqrt{1+i_r} + ln\left(\frac{\sqrt{1+i_f}-1}{\sqrt{1+i_r}-1}\right) \tag{21}$$

Expression (21) is obtained by applying the UICM to the drain and source terminals. For $i_f = 3$ and $V_{DS} = \frac{\phi_t}{2}$, expression (21) results in $i_r = 2.12$; under these conditions, V_{T0} corresponds to the gate voltage at which $g_m/I_D = 0.531(g_m/I_D)_{max}$, while I_S corresponds to $I_D/0.88$, where I_D is the drain current at $V_{GB} = V_{T0}$. The method described for the extraction of the values of V_{T0} and I_S assumes that the variation of the slope factor with the gate voltage is negligible. The slope factor (n) can be extracted from (22), which is the asymptotic value of the g_m/I_D curve in a weak inversion. The points used to determine V_{T0}, I_S and n are shown in Figure 5b.

$$\left(\frac{g_m}{I_D}\right)_{max} \approx \frac{1}{n\phi_t} \tag{22}$$

The DIBL factor (σ) does not appear in (20) because the short-channel effects, namely DIBL, velocity saturation and channel length modulation are not relevant in the linear region. Consequently, the extraction of V_{T0}, I_S and n in the linear region is also valid for short-channel devices.

3.2. Extraction of Drain-Induced Barrier-Lowering Factor (σ)

The DIBL factor σ is a small-signal parameter that affects the intrinsic voltage gain of the common source amplifier. Figure 6 presents a schematic to determine the common-source intrinsic gain (CSIG) and the equivalent small-signal model [17] of the amplifier.

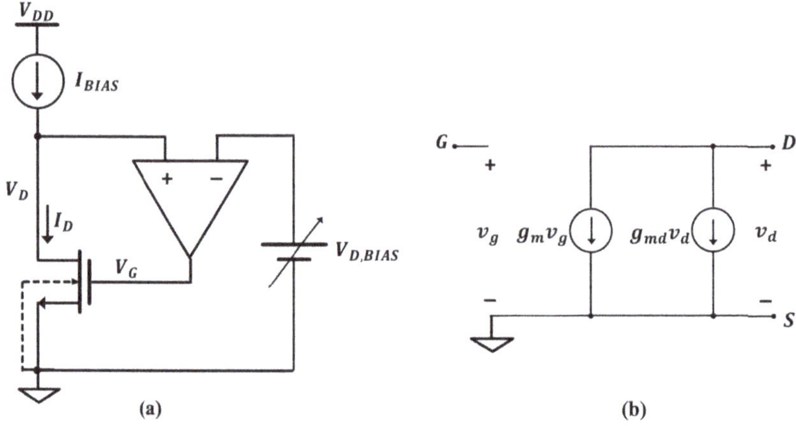

Figure 6. (**a**) Circuit to determine the CSIG and (**b**) its equivalent small-signal model.

In saturation, the use of the transconductance-to-current characteristics (11) and (12) yields the CSIG in (23).

$$A_{V,CS} = \frac{v_d}{v_g} = -\frac{g_m}{g_{md}} = -\frac{1}{\sigma} \tag{23}$$

To determine the common-source intrinsic gain through a simulation, an ideal operational amplifier was included, as shown in Figure 6a, to set the DC operating point required for the small-signal measurement.

3.3. Extraction Results

The g_m/I_D and CSIG methods presented herein were used to extract the four parameters of each transistor used throughout this work. The four parameters were also extracted for various temperatures and corners of process variation.

Figure 7 shows the dependence of the parameters of the 4PM on the temperature of an NMOS transistor with $\frac{W}{L} = \frac{1\ \mu m}{0.3\ \mu m}$. As expected, the threshold voltage is a linearly decreasing function of the temperature [18], whereas the DIBL factor increases linearly with temperature [19,20]. The slope factor is, for practical purposes, independent of the temperature. The dependence of the specific current on the temperature is, in general, not predictable due to uncertainty in the variation of the mobility with the temperature.

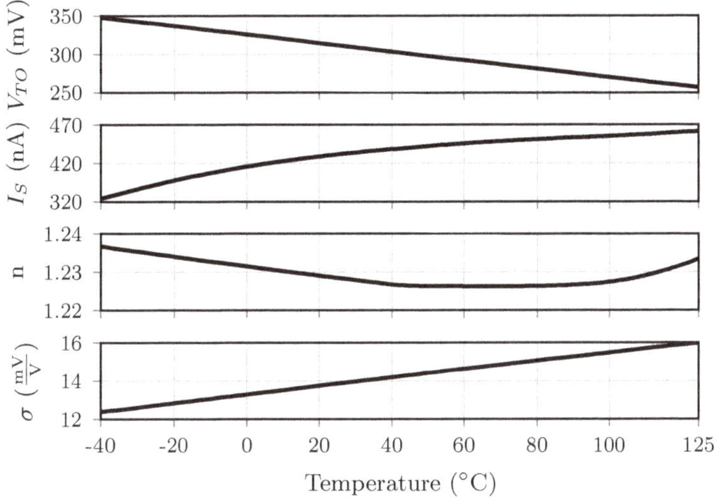

Figure 7. Parameters of the 4PM vs. temperature of a medium (nominal) V_T NMOS transistor with $W/L = 1\ \mu m/0.3\ \mu m$.

Tables 1 and 2 summarize the extracted values for NMOS and PMOS long-channel ($\frac{W}{L} = \frac{1\ \mu m}{1\ \mu m}$) and short-channel ($\frac{W}{L} = \frac{1\ \mu m}{0.3\ \mu m}$) transistors, respectively, from a 0.18 µm technology. The four parameters were extracted at room temperature for extreme corners (slow and fast) and for the typical (nominal) condition.

Table 1. Extracted parameters for medium-V_T NMOS/PMOS transistors with $\frac{W}{L} = \frac{1\ \mu m}{1\ \mu m}$.

Transistor	Slow		Nominal		Fast	
	NMOS	**PMOS**	**NMOS**	**PMOS**	**NMOS**	**PMOS**
V_{TO} [mV]	316	-239	291	-211	266	-183
I_S [nA]	99	35	111	40	124	45
n	1.19	1.18	1.20	1.18	1.22	1.17
$\sigma\left[\frac{mV}{V}\right]$	5.9	18	5.9	18	5.9	19

Table 2. Extracted parameters for medium-V_T NMOS/PMOS transistors with $\frac{W}{L} = \frac{1\,\mu\text{m}}{0.3\,\mu\text{m}}$.

Transistor	Slow		Nominal		Fast	
	NMOS	**PMOS**	**NMOS**	**PMOS**	**NMOS**	**PMOS**
V_{TO} [mV]	338	−272	311	−239	283	−206
I_S [nA]	313	81	420	106	543	137
n	1.24	1.17	1.23	1.18	1.22	1.17
$\sigma\left[\frac{\text{mV}}{\text{V}}\right]$	14	19	14	20	14	20

As expected, the parameters that varied the most were threshold voltage and specific current. The threshold voltage varied 8.6% in relation to the nominal value of the NMOS transistors and 13.3% in relation to the PMOS transistors. The specific current varied around 13% in relation to the nominal value in long-channel transistors and up to 29.3% in short-channel transistors. The effects of these variations in a circuit are presented in Section 5.

4. Including the 4PM in Cadence

To simulate MOS circuits through the 4PM in a commercial simulator, the model was carried out in Verilog-A, an HDL that describes the electrical behavior of analog devices, circuits and systems. The Verilog-A compiler handles every required interaction between the model and the simulation software. Furthermore, Verilog-A supports various functions to assist in descriptions, such as standard mathematical functions, transcendental and hyperbolic functions as well as a set of statistical functions [14].

The inversion levels in the UICM (7) simplify the design of various MOS circuits; however, for a simulator, the voltages at the device's terminals are the inputs, while the current flowing through the device is the output.

When solving (7) for the drain current, a transcendental equation arises, which can be solved numerically. Nonetheless, the simulator solves the equations point by point; thus, iterative calculations to find the solution of one single point waste time and processing power.

Siebel [21] explored some algorithms to improve the implementation of (7) in simulators, reaching the conclusion that algorithm 443 [22] finds an accurate solution for the drain current in only one iteration.

Algorithm 443 solves transcendental equations of the form $x = we^w$. To resemble such a form, the UCCM in (3) can easily be rewritten as (24).

$$e^{\left(\frac{V_P - V_{S(D)B}}{\phi_t} + 1\right)} = q_{IS(D)}e^{q_{IS(D)}} \tag{24}$$

Owing to the similarity of (24) to $x = we^w$, algorithm 443 is employed to determine the drain current by following a few steps: first, the normalized forward and reverse charge densities $q_{IS(D)}$ are determined; then, by applying their values in (6), we obtain the respective inversion levels $i_{f(r)}$, which, at last, are applied in (1), resulting in the drain current I_D. A sample of the Verilog-A description is presented in Appendix A to clarify how algorithm 443 was implemented to solve (24).

For the dynamic model, expressions (13)–(19) were implemented in Verilog-A just after the drain current was calculated. The overlap capacitances were also included as extrinsic capacitances. The transconductances were used as design parameters that could easily be derived from the current–voltage relation, namely UICM.

Model Results

For the sake of comparisons with BSIM 4.5 results, the four-parameter model described in Verilog-A was simulated employing single transistors in typical conditions at room temperature. Figures 8 and 9 present the $I_D \times V_{GS}@V_{DS} = 200$ mV and $I_D \times V_{DS}@V_{GS} = 200$ mV, respectively, for long-channel ($\frac{W}{L} = \frac{1\,\mu m}{1\,\mu m}$) and short-channel ($\frac{W}{L} = \frac{1\,\mu m}{0.3\,\mu m}$) transistors. Note that in both figures, ACM refers to the 4PM.

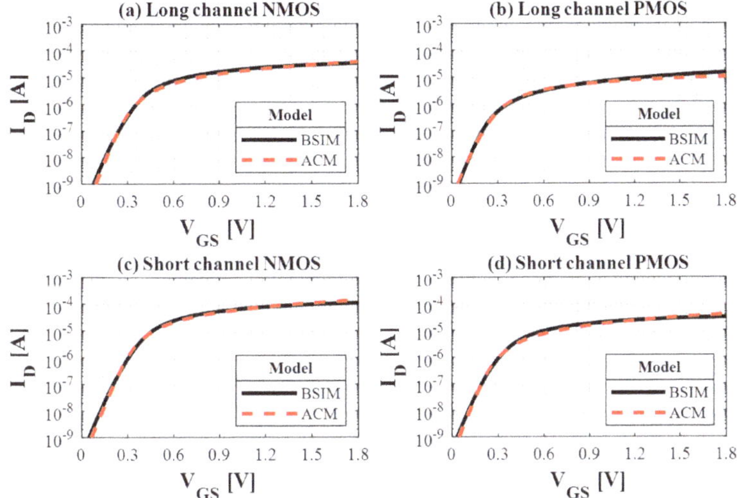

Figure 8. $I_D \times V_{GS}$ @ $V_{DS} = 200$ mV for (**a**) medium (nominal) V_T long-channel NMOS and (**b**) PMOS transistors and for (**c**) medium (nominal) V_T short-channel NMOS and (**d**) PMOS transistors.

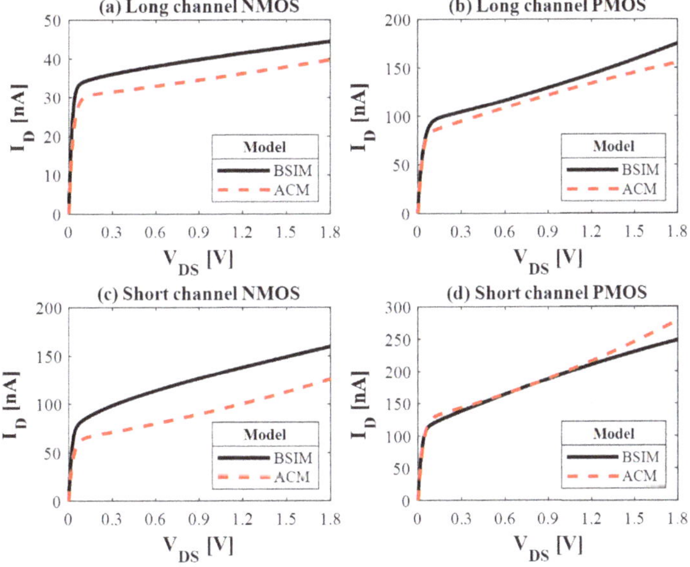

Figure 9. $I_D \times V_{DS}$ @ $V_{GS} = 200$ mV for (**a**) medium (nominal) V_T long-channel NMOS and (**b**) PMOS transistors and for (**c**) medium (nominal) V_T short-channel NMOS and (**d**) PMOS transistors.

Simulations carried out for V_{DS} and V_{GS} with 100 mV, 500 mV and 1 V led to current–voltage characteristics similar to those in Figures 8 and 9; therefore, they were not included herein.

Overall, the Verilog-A simulation for long- and short-channel transistors provided results close to BSIM's. Notably, for high values of V_{DS}, the drain current of the 4PM drifts away from BSIM's due to effects that are not taken into account in the ACM model used herein.

5. Circuit Examples and Simulation Results

Four circuits were simulated through either the 4PM in Verilog-A descriptions or BSIM 4.5 [23]: the classic CMOS inverter, an 11-stage ring oscillator, a self-biased current source (SBCS) and a common-source amplifier.

5.1. CMOS Inverter

The CMOS inverter in Figure 10 is a versatile and simple circuit employed in various ULV digital circuits [6,8] and analog building blocks, such as amplifiers and oscillators [10,24].

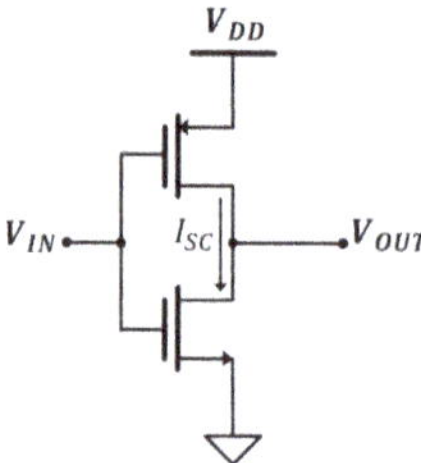

Figure 10. The classic CMOS inverter.

Well-designed CMOS inverters usually present a perfect balance between the N and P networks, which means that in the voltage transfer curve, the mid-point voltage corresponds to $V_{OUT} = V_{IN} = V_{DD}/2$. The CMOS inverter herein was designed to be balanced for the supply voltage V_{DD} = 100 mV, room temperature and typical process parameters.

For this particular design, we chose transistors with threshold voltages lower than those of the standard transistor, which favors them in the design of ULV circuits. They are called medium-V_T transistors, and their minimum channel length is 300 nm in this 0.18 μm technology. The PMOS and NMOS transistors were designed with channel lengths of $L_P = L_N = 300$ nm and widths of $W_P = W_N = 600$ nm. The values in Table 3 correspond to the extracted parameters of these medium-V_T transistors for the simulation through the 4PM in Verilog-A.

Table 3. Corner-extracted parameters for medium-V_T NMOS/PMOS transistors with $\frac{W}{L} = \frac{600\ \text{nm}}{300\ \text{nm}}$.

Transistor	Slow		Nominal		Fast	
	NMOS	**PMOS**	**NMOS**	**PMOS**	**NMOS**	**PMOS**
V_{TO} [mV]	339	−308	309	−269	280	−230
I_S [nA]	206	70	280	89	366	111
n	1.25	1.25	1.24	1.25	1.23	1.24
$\sigma\left[\frac{\text{mV}}{\text{V}}\right]$	15	22	15	23	15	23

The design was validated through a DC analysis in Cadence® by using each model (4PM and BSIM 4.5) separately. The results of the DC simulations for five different supply voltages V_{DD}s at room temperature and typical conditions are depicted in Figure 11, which includes the voltage transfer characteristic (VTC), small-signal gain and short-circuit current (I_{SC}). From Figure 11, it can be verified that the ACM model with only four parameters is sufficient to properly describe the electronic behavior of the classic CMOS inverter in the ULV domain.

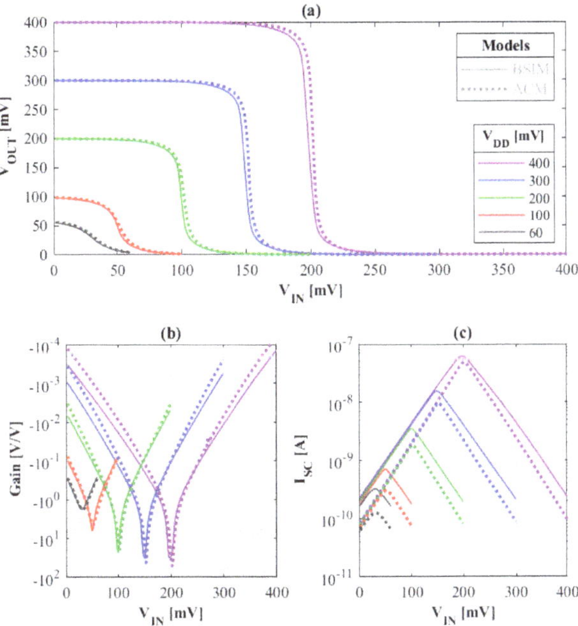

Figure 11. CMOS inverter results of (**a**) voltage-transfer characteristic (VTC), (**b**) small-signal gain and (**c**) short-circuit current.

Figure 12 presents the VTC curves for the CMOS inverter across the corners of process variation for both BSIM and the 4PM at supply voltages of 100 mV and 300 mV and a temperature of 300 K. Even with variations of up to 15% and 30% in the threshold voltage and specific current, respectively, the 4PM clearly adapts to the corners and follows BSIM since the four parameters were extracted for each corner.

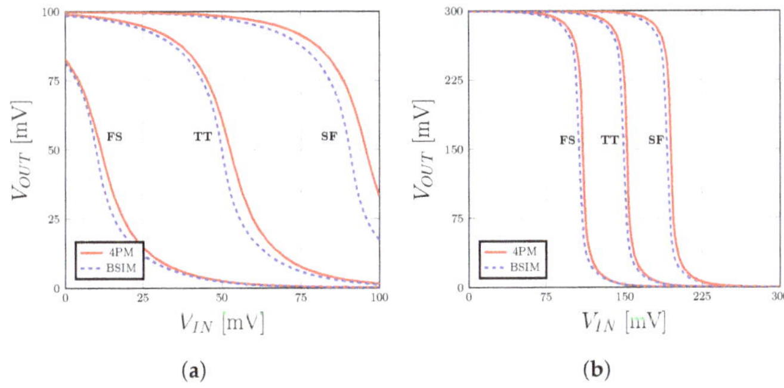

Figure 12. Voltage-transfer characteristics of the CMOS inverter using BSIM and the 4PM across the corners of process variation. (**a**) V_{DD} = 100 mV. (**b**) V_{DD} = 300 mV.

5.2. Ring Oscillator

In Figure 13, the ring oscillator comprises N CMOS inverters in a loop and the load capacitance C_L in between stages, which includes external capacitors that load each node, along with the transistors' intrinsic and extrinsic capacitances presented in Section 2.2. The load capacitance is crucial to set the frequency of oscillation and is critical for the successful start-up of the oscillator.

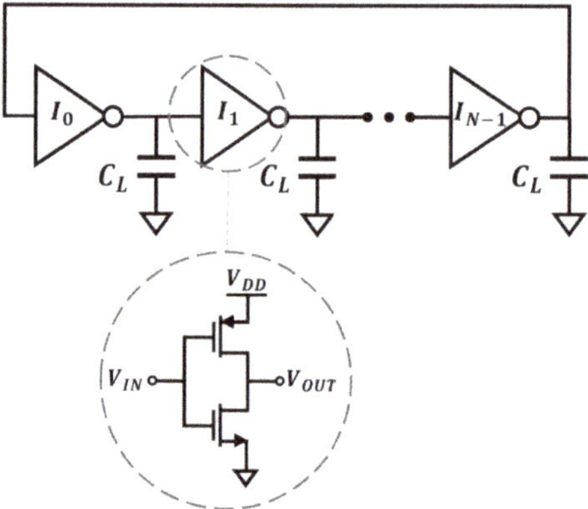

Figure 13. Ring oscillator.

According to [24], in order to facilitate the start-up of the ring oscillator in the ultra-low-voltage domain, the minimum gain required to establish a condition of oscillation can be reduced by increasing the number of stages in the ring oscillator. We chose the number of stages $N = 11$, which corresponds to a minimum voltage gain of 1.04 V/V for the start-up of oscillations.

Figure 14 presents the voltage signal at one of the stages of the ring oscillator for the supply voltage V_{DD} of 100 mV. Table 4 summarizes the frequencies obtained through the use of either ACM or BSIM for various V_{DD} values without the inclusion of any external capacitor.

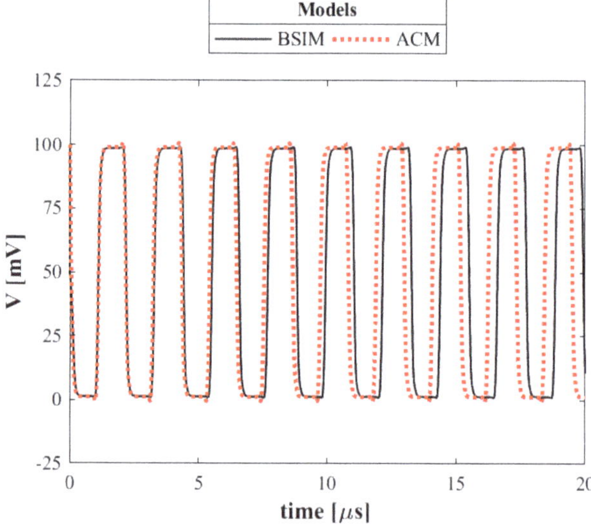

Figure 14. Voltage signal in the time domain at one of the stages of the oscillator. Results for BSIM and 4PM simulations at 100 mV of supply voltage.

Table 4. Oscillation frequency at various VDDs obtained through time-domain simulations of the 11-stage ring oscillator without external C_L.

V_{DD}	BSIM	4PM	$\dfrac{f_{4PM}}{f_{BSIM}}$
400 mV	81.3 MHz	187.2 MHz	2.30
300 mV	23.7 MHz	52.1 MHz	2.20
200 mV	3.79 MHz	5.87 MHz	1.55
100 mV	452 kHz	463 kHz	1.02
60 mV	198 kHz	177 kHz	0.89

As expected, due to a lack of extrinsic capacitances associated with fringing and diode junctions [15] in the implemented dynamic model, the frequency of oscillation using the 4PM was higher than BSIM's overall. Table 4 shows that the oscillation frequency obtained through the 4PM diverged from the frequency obtained through BSIM at $V_{DD} = 300$ mV and 400 mV for more than 200%, which suggests that the implemented dynamic model lacks sufficient information to provide frequency results closer to BSIM's in these voltages.

To further evaluate the difference in the oscillation frequency, we added the external capacitor $C_L = 1\ pF$ between stages. Figure 15 presents the oscillation frequency at supply voltages from 60 mV to 400 mV.

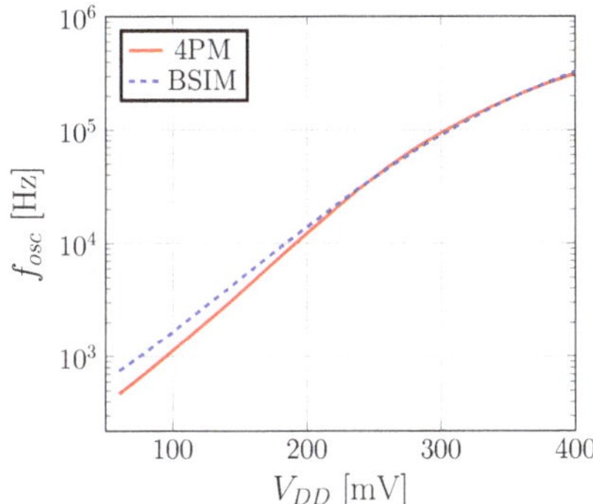

Figure 15. Oscillation frequency vs. the supply voltage V_{DD}.

From Figure 15 and Table 5, it can be seen that the inclusion of high-value external capacitors attenuated the effect of the capacitances inherent to the ring oscillator on the frequency response and, consequently, improved the ACM's accuracy in relation to BSIM for voltages from 200 mV to 400 mV. However, it deteriorated the results for voltages below 100 mV. Overall, the 4PM delivers a time/frequency domain result that closely matches BSIM's.

Table 5. Oscillation frequency at various VDDs obtained through time-domain simulations of the 11-stage ring oscillator with external $C_L = 1\ pF$.

V_{DD}	BSIM	4PM	$\dfrac{f_{4PM}}{f_{BSIM}}$
400 mV	329.4 kHz	316.2 kHz	0.96
300 mV	91.0 kHz	94.4 kHz	1.04
200 mV	14.1 kHz	12.3 kHz	0.87
100 mV	1.67 kHz	1.12 kHz	0.67
60 mV	753 Hz	469 Hz	0.62

These results suggest the capacitances in BSIM have a strong dependence on the supply voltage, a dependence which was not incorporated in the implemented extrinsic dynamic model, hence the observed difference in the oscillation frequency at various supply voltages.

In addition, the computational efficiency was verified by comparing the CPU transient simulation time required to simulate the oscillator with the external $C_L = 1\ pF$ at the supply voltage $V_{DD} = 300$ mV, which provides signals with similar frequencies for BSIM and ACM ($f_{ACM}/f_{BSIM} = 1.04$). The total CPU time required to run the transient analysis with BSIM was 76.25 s, while the same simulation required a total CPU time of 55.64 s using the 4PM in Verilog-A, representing 73% of the time BSIM used, which is very significant when it comes to several long simulation runs.

5.3. Self-Biased Current Source (SBCS)

The design of the self-biased current source (SBCS) in Figure 16, for the output current $I_{OUT} = 100$ nA and supply voltage $V_{DD} = 1.8$ V, was based on the ACM model [25–27].

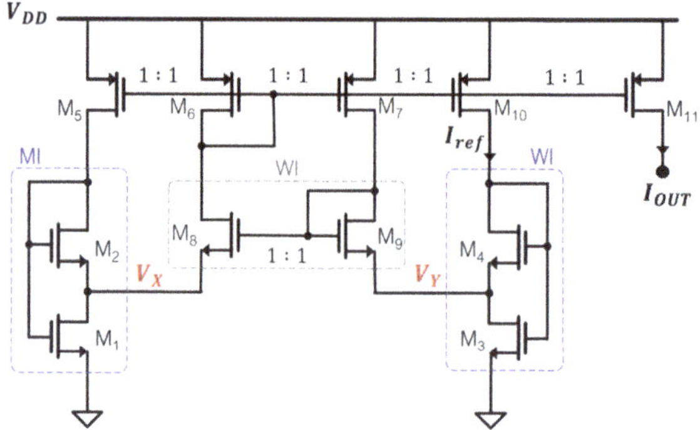

Figure 16. Self-biased current source (SBCS) circuit.

The core of the SBCS in Figure 16 is the self-cascode MOSFET (SCM), composed of transistors M_1 and M_2, which operate in a moderate inversion. Transistors M_3 and M_4 form the second SCM biased in a weak inversion to generate the proportional to absolute temperature (PTAT) voltage V_Y [26,27].

Transistors $M_{2(4)}$ are in a saturation, while $M_{1(3)}$ is in a triode; therefore, $I_{D2} \cong I_{S2}i_{f2}$ and $I_{D1} = I_{S1}(i_{f1} - i_{r1}) = I_{ref}(N+1)$. Since $V_{P1} = V_{P2} = V_P$ and $V_{D1} = V_{S2}$, we have $i_{r1} = i_{f2}$.

The specific current I_S can also be written as $I_S = I_{SH}S$, where I_{SH} is the sheet normalization current and S is the aspect ratio $\frac{W}{L}$, which, combined with (1), yields the relationship (25).

$$\alpha_{12(34)} = \frac{i_{f1(3)}}{i_{f2(4)}} = 1 + \frac{S_2(4)}{S_1(3)}\left(1 + \frac{1}{N}\right) \tag{25}$$

The SCM intermediate voltage $V_{X(Y)}$ relates to the inversion level through the design Equations (26) and (27), which can be directly derived from the ACM using (7) and (25).

$$\frac{V_X}{\phi_t} = \sqrt{1 + \alpha_{12}i_{f2}} - \sqrt{1 + i_{f2}} + \ln\left(\frac{\sqrt{1 + \alpha_{12}i_{f2}} - 1}{\sqrt{1 + i_{f2}} - 1}\right) \tag{26}$$

$$\frac{V_Y}{\phi_t} = \ln \alpha_{34} \tag{27}$$

To simplify the design, we chose $i_{f2} = 15$ and $S_1 = S_2$, which results in $\alpha_{12} = 3$. From this starting point, it is sufficient to extract the sheet normalization current of M_2, as shown in Section 3.1, and to use (1) to determine the aspect ratio. Once V_X is determined,

α and the inversion levels of the other transistors can be calculated, along with their aspect ratios.

Table 6 summarizes the sizes, series/parallel associations and inversion levels of the transistors. Table 7 presents the four parameters extracted for the three transistors used in the SBCS.

Table 6. Sizes and and inversion levels of the transistors of the SBCS.

Transistor	$\frac{W}{L} \times \frac{N_{parallel}}{N_{series}}$	i_f
$M_{1,2}$	$\frac{0.5\ \mu m}{2.0\ \mu m} \times \frac{1}{4}$	15
M_3	$\frac{0.5\ \mu m}{2.0\ \mu m} \times \frac{20}{1}$	0.32
M_4	$\frac{4.0\ \mu m}{2.0\ \mu m} \times \frac{40}{1}$	0.01
$M_{8,9}$	$\frac{0.5\ \mu m}{2.0\ \mu m} \times \frac{35}{1}$	0.1
$M_{5-7,10,11}$	$\frac{0.5\ \mu m}{2.0\ \mu m} \times \frac{1}{1}$	10

Table 7. Extracted parameters of transistors used in the SBCS.

Transistor	NMOS		PMOS
W [μm]	0.5	4.0	0.5
L [μm]	2.0	2.0	2.0
I_S [nA]	29	63	10
V_{T0} [mV]	423	444	-428
n	1.27	1.27	1.31
σ [mV/V]	2.2	2.4	6.5

The DC simulation results in Figure 17 were obtained through the use of either BSIM or the 4PM for a voltage sweep on V_{DD} from 0 to 1.8 V. Both models yielded similar results for I_{OUT}, V_X and V_Y. The SBCS started up for supply voltages above 650 mV. The average values of V_X and V_Y for a V_{DD} higher than 650 mV were approximately 86 mV and 81 mV, respectively, which were very close to the calculated value of 88 mV. The design of the SBCS can be improved and optimized; however, the main goal herein was to compare the results of the 4PM with those of BSIM.

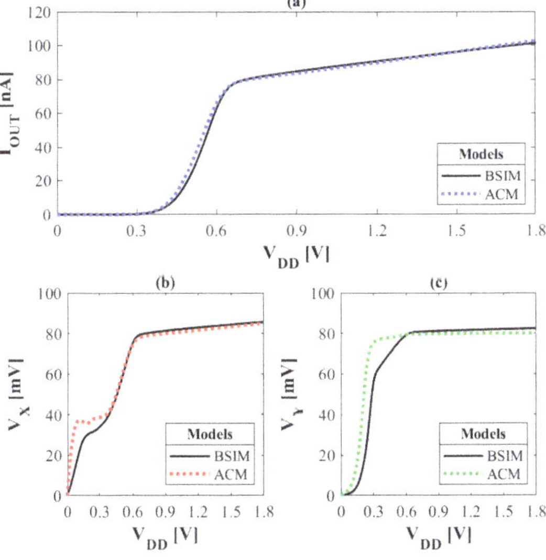

Figure 17. Results of DC analysis for voltage sweep on V_{DD}: (**a**) output current, (**b**) V_X and (**c**) V_Y.

5.4. Common-Source Amplifier

The common-source amplifier in Figure 18 was designed to demonstrate the suitability of the 4PM in the frequency domain in comparison to BSIM. The amplifier was designed for a maximum gain at a frequency of 2 MHz, a bias current of 200 nA and a supply voltage of 1.8 V.

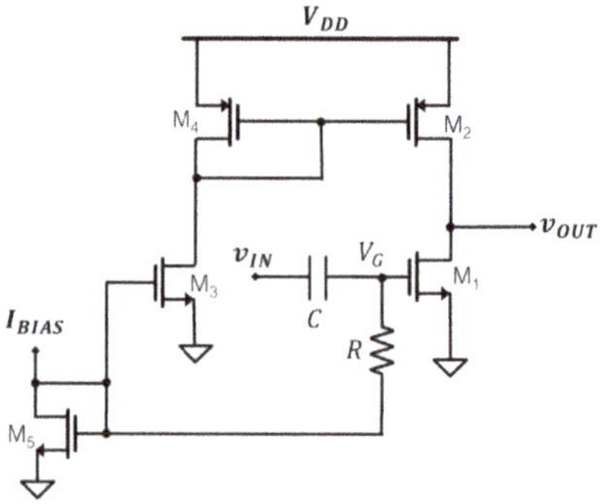

Figure 18. Common-source amplifier.

Table 8 presents the transistors' dimensions and extracted parameters employed in the design. The resistor R of 500 kΩ isolates the node V_G from the bias circuit, while the capacitor C of 150 fF blocks the DC level from the input signal at V_G.

Table 8. Transistor dimensions for the common-source amplifier and extracted parameters.

Transistor	$M_{1,3,5}$	$M_{2,4}$
W [μm]	2.0	0.5
L [μm]	0.18	2.0
I_S [nA]	2000	10
V_{T0} [mV]	518	-428
n	1.36	1.31
σ [mV/V]	21.8	6.5

An AC simulation from 1 kHz to 10 GHz was run for a capacitive load of 10 fF. The results using BSIM and 4PM are depicted in Figure 19, where it is evident that the 4PM managed to follow the BSIM curves in the AC simulation.

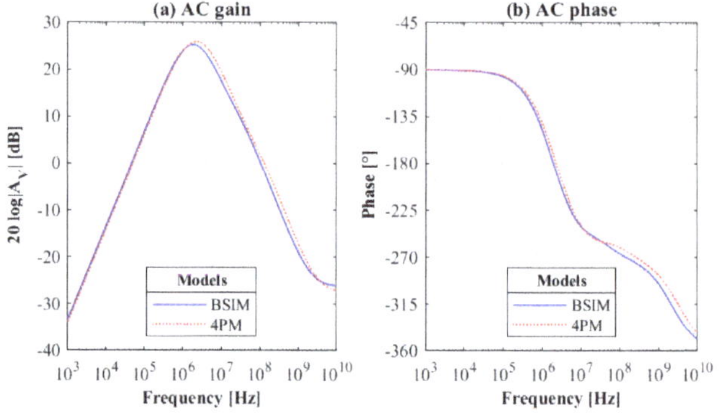

Figure 19. Frequency response of the common-source amplifier using BSIM and 4PM: (**a**) open-loop gain in dB and (**b**) phase.

The center frequency for the 4PM was around 2.14 MHz with a peak gain of 26 dB, while BSIM presented a maximum gain of 25.3 dB at 1.9 MHz. The phase curves presented in Figure 19 show that the 4PM managed to follow BSIM very closely. Two poles were found at 700 kHz and 4.7 MHz for BSIM and at 850 kHz and 5.5 MHz for the 4PM. These differences were expected since the 4PM in Verilog-A does not consider the complete dynamic transistor model.

6. Conclusions

The simulation results of MOS circuits depend on the accuracy of both the MOS model and the extracted transistors' parameters.

The authors of [28] employed an ACM expression of charge density to calculate currents but did not extract the required parameters that should be available in the simulator. Nonetheless, despite using VHDL (the hardware description language VHSIC) to facilitate the widespread use of the model in other simulators, the charge density equations are not familiar to most designers; thus, a gap between hand-design and simulation remains.

This paper introduced a truly compact MOS model composed of only four parameters—enough to describe the DC and small-signal low-frequency characteristics of MOSFET. The 4PM in Verilog-A was used to calculate the current from the UICM, which contains parameters familiar to IC designers. This is significant because a first-order understanding of the MOSFET model along with its associated parameters is indispensable for IC designers since the MOSFET parameters of simulators are numerous and most of them are quite hard to understand.

Besides presenting the 4PM, this paper also introduced the extraction methods employed to obtain accurate parameters, reflected in the consistent results obtained through the simulations of different circuits presented in Section 5.

The four-parameter model is a minimalist model that helps electronic engineers to design MOS circuits and to rapidly find approximate solutions to the circuits' electrical behavior in a way that the troubleshooting can easily be done by directly relating the design parameters to the obtained results before fine tuning through more complex and time-consuming simulations.

The 4PM is particularly useful for the design by hand of low-voltage circuits because fewer parameters are required for accurate results while still maintaining a foundation in physics. Therefore, all things considered, the 4PM helps to bridge the gap between the hand design and simulation of MOS circuits.

Author Contributions: Conceptualization, C.G.-M. and M.C.S.; methodology, C.G.-M., M.C.S., C.M.A. and D.G.A.N.; software, C.M.A. and D.G.A.N.; validation, C.M.A. and D.G.A.N.; writing—original draft preparation, C.M.A. and D.G.A.N.; writing—review and editing, C.G.-M. and M.C.S.; visualization, C.M.A. and D.G.A.N.; supervision, C.G.-M.; funding acquisition, C.G.-M. and M.C.S. All authors have read and agreed to the published version of the manuscript.

Funding: This research was funded by the Brazilian agencies CAPES (finance codes 001 and print #698503P) and CNPq.

Institutional Review Board Statement: Not applicable.

Informed Consent Statement: Not applicable.

Data Availability Statement: The data are contained within the article.

Acknowledgments: The authors would like to thank the Brazilian agencies CAPES and CNPq for supporting this work.

Conflicts of Interest: The authors declare no conflict of interest. The funders had no role in the design of the study; in the collection, analyses, or interpretations of data; in the writing of the manuscript, or in the decision to publish the results.

Abbreviations

The following abbreviations are used in this manuscript:

4PM	Four-parameter model
ACM	Advanced compact MOSFET model
BSIM	Berkeley short-channel IGFET model
CMOS	Complementary metal-oxide semiconductor
CSIG	Common-source intrinsic gain
DC	Direct current
DIBL	Drain-induced barrier lowering
EDA	Electronic design automation
HDLs	Hardware description languages
MI	Moderate inversion
MOSFET	Metal-oxide semiconductor field-effect transistor
PTAT	Proportional to absolute temperature
SBCS	Self-biased current source
SCM	Self-cascode MOSFET
SI	Strong inversion
UCCM	Unified charge-control model
UICM	Unified current-control model
ULV	Ultra-low voltage
VHDL	VHSIC hardware description language
VTC	Voltage transfer characteristic
WI	Weak inversion

Appendix A. Verilog-A Implementation

In Verilog-A, the current flowing from Terminal A to Terminal B is defined using the syntax I(A,B), and the voltage between these two terminals is defined as V(A,B). Therefore, it is very straightforward to set equations and associate voltages and currents.

The sample below contains a definition of the pinch-off voltage in (4), followed by an implementation of Algorithm 443 regarding the source (subscript S) terminal. In the full description, the calculations are performed for both source and drain (subscript D) terminals, which are analogous.

```
analog begin
PhiT = $vt($temperature); // thermal voltage
VP = (V(G,B) - VTH + sigma*V(D,S) + sigma*V(S,B))/n;
// Equation (4), pinch-off voltage

// Condition to calculate WnS
X = exp(((VP - V(S,B))/PhiT)+1);

        if(X < 0.7385) begin
                numeratorS = X + (4/3)*X*X;
                denominatorS = 1 + (7/3)*X+(5/6)*X*X;
                WnS = numeratorS/denominatorS;
        end

        else begin
                numeratorS = log(X)*log(X)+2*log(X)-3;
                denominatorS = 7*log(X)*log(X) + 58*log(X) +127;
                WnS = log(X) - 24*(numeratorS/denominatorS);
        end

// Calculating ZnS
ZnS = log(X) - WnS - log(WnS);

// Calculating EnS
TermC = ZnS/(1 + WnS);

numeratorES = (2*(1+WnS)*(1+WnS+(2/3)*ZnS)-ZnS);
denominatorES = 2*(1+WnS)*(1+WnS+(2/3)*ZnS)-2*ZnS;
```

```verilog
29
30  EnS = TermC*(numeratorES/denominatorES);
31
32  // Finding the qis and ifS
33  qiS = WnS*(1+EnS); // normalized inversion charge at source
34  ifS = (qiS + 1)*(qiS + 1) - 1; // Equation (6), forward inversion level
```

Note that the methodology used to calculate the inversion charges (lines 6–33) is from [22], and we used several variables throughout the description to facilitate the implementation. Afterward, the drain current is calculated from the results of the source, and drain calculations as shown in the sample below. The syntax and guidelines are detailed in [14].

```verilog
1  //Calculating ID
2  I(D,S) <+ = IS*(ifS-irD); // Equation (1),drain-current
```

References

1. Sah, C.-T. A history of MOS transistor compact modeling. In Proceedings of the NSTI Nanotech, Anaheim, CA, USA, 8–12May 2005; pp. 347–390.
2. Liu, W. *MOSFET Models for SPICE Simulation Including BSIM3v3 and BSIM4*; Wiley: Hoboken, NJ, USA, 2001.
3. Singh, Y.; Venugopalan, S.; Karim, M.A.; Khandelwal, S.; Paydavosi, N.; Thakur, P.; Niknejad, A.M.; Hu, C.C. NiknBSIM Industry standard compact MOSFET models. In Proceedings of the European Solid-State Device Research Conference (ESSDERC), Bordeaux, France, 17–21 September 2012; pp. 46–49.
4. Brews, J.R. MOSFET hand analysis using BSIM. *IEEE Circuits Devices Mag.* **2006**, *21*, 28–36. [CrossRef]
5. Jespers, P.G.A.; Murmann, B. *Systematic Design of Analog CMOS Circuits*; Cambridge University Press: Cambridge, UK, 2017.
6. Aunet, S. Ultra low voltage sub-100 mV Vdd CMOS. In Proceedings of the19th IEEE International New Circuits and Systems Conference (NEWCAS), Toulon, France, 13–16 June 2021.
7. Bryant, A.; Brown, J.; Cottrell, P.; Ketchen, M.; Ellis-Monaghan, J.; Nowak, E. Low-power CMOS at Vdd = 4kT/q. In Proceedings of the Device Research Conference. Conference Digest (Cat. No. 01TH8561), Notre Dame, IN, USA, 25–27 June 2001; pp. 22–23.
8. Lotze, N.; Manoli, Y. A 62 mV 0.13 μm CMOS standard-cellbased design technique using Schmitt-trigger logic. *IEEE J. Solid-State Circuits* **2011**, *47*, 47–60. [CrossRef]
9. Choi, J.; Aklimi, E.; Shi, C.; Tsai, D.; Krishnaswamy, H.; Shepard, K.L. Matching the power, voltage, and size of biological systems: A nWscale, 0.023-mm^3 pulsed 33-GHz radio transmitter operating from a 5 kT/q-supply voltage. *IEEE Trans. Circuits Syst. I Regul. Pap.* **2015**, *62*, 1950–1958. [CrossRef]
10. Ballo, A.; Pennisi, S.; Scotti, G.; Venezia, C. A 0.5 V Sub-Threshold CMOS Current-Controlled Ring Oscillator for IoT and Implantable Devices. *J. Low Power Electron. Appl.* **2022**, *12*, 16. [CrossRef]
11. Palumbo, G.; Scotti, G. A Novel Standard-Cell-Based Implementation of the Digital OTA Suitable for Automatic Place and Route. *J. Low Power Electron. Appl.* **2021**, *11*, 42. [CrossRef]
12. Centurelli, F.; Della Sala, R.; Monsurrò, P.; Scotti, G.; Trifiletti, A. A Tree-Based Architecture for High-Performance Ultra-Low-Voltage Amplifiers. *J. Low Power Electron. Appl.* **2022**, *12*, 12. [CrossRef]
13. Galup-Montoro, C.; Schneider, M.C. *MOSFET Modeling for Circuit Analysis and Design*; World Scientific: Singapore, 2007.
14. Technologies, A. Verilog-A Reference Manual. 2004. Available online: https://edadownload.software.keysight.com/eedl/ads/2011_01/pdf/verilogaref.pdf (accessed on 1 August 2021).
15. Tsividis, Y.; McAndrew, C. *Operation and Modeling of the MOS Transistor*; Oxford University Press: Oxford, UK, 2011.
16. Siebel, O.F.; Schneider, M.C.; Galup-Montoro, C. MOSFET threshold voltage definition, extraction and some applications. *Microelectron. J.* **2012**, *43*, 329–336. [CrossRef]
17. Hiblot, G. DIBL–compensated extraction of the channel length modulation coefficient in MOSFETs. *IEEE Trans. Electron Devices* **2018**, *65*, 4015–4018. [CrossRef]
18. Vittoz, E.A. MOS Transistor: Model and modes of operation. In *MEAD Course on "Advanced Analog CMOS IC Design"*; EPF-Lausanne: Lausanne, Switzerland, 2019.
19. Fikry, W.; Ghibaudo, G.; Dutoit, M. Temperature dependence of drain-induced barrier lowering in deep submicrometre MOSFETs. *Electron. Lett.* **1994**, *30*, 911–912. [CrossRef]
20. Chen, Z.; Wong, H.; Han, Y.; Dong, S.; Yang, B.L. Temperature dependences of threshold voltage and drain-induced barrier lowering in 60 nm gate length MOS transistors. *Microelectron. Reliab.* **2014**, *54*, 1109–1114. [CrossRef]
21. Siebel, O.F. *Um Modelo Eficiente do Transistor MOS Para o Projeto de Circuitos VLSI*; Universidade Federal de Santa Catarina: Florianopolis, Brazil, 2007.
22. Fritsch, F.N.; Shafer, R.E.; Crowley, W.P. Algorithm 443: Solution of the transcendental equation wew=x. *Commun. AMC* **1973**, *16*, 123–124.
23. Department of Electrical Engineering and Computer Science, UC Berkeley. *BSIM4v4.5.0 Technical Manual*; Department of Electrical Engineering and Computer Science, UC Berkeley: Berkeley, CA, USA, 2004.

24. Ferreira, J.V.T. *Analysis, Design and Test of Ultra-Low-Voltage CMOS Ring Oscillators*; Universidade Federal de Santa Catarina: Florianopolis, Brazil, 2019.
25. Antúnez-Calistro, G.; Siniscalchi, M.; Silveira, F.; Rossi-Aicardi, C. Variability-Aware Design Method for a Constant Inversion Level Bias Current Generator. *IEEE Trans. Circuits Syst. I Regul. Pap.* **2019**, *66*, 2027–2036. [CrossRef]
26. Heim, P.; Schultz, S.R.; Jabri, M.A. Technology-independent biasing technique for CMOS analogue micropower implementations of neural networks. In Proceedings of the 6th Australian Conference on Neural Networks, Sydney, Australia, 6–8 February 1995; pp. 9–12.
27. Olmos, A.; Boas, A.V.; Soldera, J. A Sub-1V Low Power Temperature Compensated Current Reference. In Proceedings of the 2007 IEEE International Symposium on Circuits and Systems (ISCAS), New Orleans, LA, USA, 27–30 May 2007; pp. 2164–2167.
28. Fonseca, A.K.T.B.; de Souza, F.R. Behavioral modeling of the Advanced Compact MOSFET (ACM) model with VHDL-AMS. In Proceedings of the 2008 Joint 6th International IEEE Northeast Workshop on Circuits and Systems and TAISA Conference, Montreal, QC, Canada, 22–25 June 2008; pp. 169–172.

Journal of
*Low Power Electronics
and Applications*

Article

Hardware Solutions for Low-Power Smart Edge Computing

Lucas Martin Wisniewski [1,†], Jean-Michel Bec [1,†], Guillaume Boguszewski [1] and Abdoulaye Gamatié [2,*,†]

1   CYLEONE S.A.S. Company, 34090 Montpellier, France
2   LIRMM, University Montpellier, CNRS, 34095 Montpellier, France
*   Correspondence: abdoulaye.gamatie@lirmm.fr
†   These authors contributed equally to this work.

Abstract: The edge computing paradigm for Internet-of-Things brings computing closer to data sources, such as environmental sensors and cameras, using connected smart devices. Over the last few years, research in this area has been both interesting and timely. Typical services like analysis, decision, and control, can be realized by edge computing nodes executing full-fledged algorithms. Traditionally, low-power smart edge devices have been realized using resource-constrained systems executing machine learning (ML) algorithms for identifying objects or features, making decisions, etc. Initially, this paper discusses recent advances in embedded systems that are devoted to energy-efficient ML algorithm execution. A survey of the mainstream embedded computing devices for low-power IoT and edge computing is then presented. Finally, CYSmart is introduced as an innovative smart edge computing system. Two operational use cases are presented to illustrate its power efficiency.

Keywords: smart edge computing; energy-efficiency; Internet-of-Things; low-power embedded systems; machine learning; CYSmart

Citation: Martin Wisniewski, L.; Bec, J.-M.; Boguszewski, G.; Gamatié, A. Hardware Solutions for Low-Power Smart Edge Computing. *J. Low Power Electron. Appl.* **2022**, *12*, 61. https://doi.org/10.3390/jlpea12040061

Academic Editor: Orazio Aiello

Received: 28 October 2022
Accepted: 21 November 2022
Published: 25 November 2022

Publisher's Note: MDPI stays neutral with regard to jurisdictional claims in published maps and institutional affiliations.

1. Introduction

The edge computing paradigm [1] is an emerging paradigm for Internet-of-Things systems where computations are distributed across a variety of compact devices in order to bring computing capability closer to data sources, such as environmental sensors and cameras. We can mention the following advantages of edge computing over the traditional centralized computing paradigm found in cloud systems:

- reduced communication bandwidth and power costs as a result of reduced data transfers to centralized cloud servers;
- physical proximity of data and devices facilitates real-time data processing, such as for self-driving cars;
- in-situ processing at the edge devices ensures privacy regarding sensitive data, and prevents their offloading to remote locations;
- as the system is distributed, failure of some nodes can be easily overcome with a minimal impact on the global system and new devices can be added in a modular fashion to increase computing power.

Figure 1 illustrates the hierarchical layers of a typical edge computing infrastructure. Sensors in Layer 0 collect data from the environment first. In subsequent layers, the data are processed with appropriate devices based on the complexity of the processing. To meet the edge computing requirements, devices are placed close to sensors.

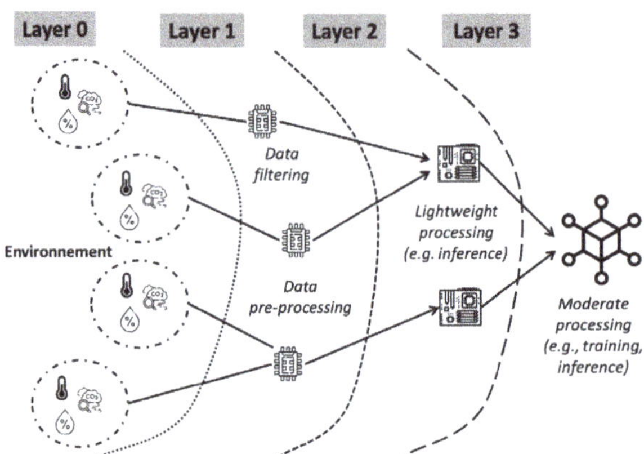

Figure 1. Hierarchical smart edge computing.

1.1. Machine Learning at the Edge

For edge computing nodes, implementing data analytics, particularly machine learning (ML), is a major challenge. Applications that leverage ML techniques at the edge are numerous. In essence, they deal with the inference problem. Real-time video analytics is a prominent application found in systems such as video surveillance, traffic control, and autonomous vehicles. Another notable application is feature extraction from images, for example, detecting areas and objects, identifying handwritten characters, and monitoring healthcare. To ensure people's safety (for instance, in the event of a fire) or to minimize energy consumption by utilizing renewable resources, smart homes and cities incorporate devices that use ML techniques for sensing and controlling the environment. Amazon's Alexa has made automatic speech recognition popular due to its success.

Based on the richness of ML techniques, three main families can be distinguished [2]. In *supervised* learning, classification and regression tasks are often achieved with algorithms such as support vector machines (SVM), artificial neural networks (ANNs), and linear regression. With the use of algorithms like k-means or x-means, *unsupervised* learning can be used for clustering and prediction tasks. *Reinforcement* learning focuses on decision-making through Q-learning. The majority of machine learning algorithms implemented on edge devices currently utilize *inference* (i.e., the process of directly solving ML problems with pre-trained ANNs) rather than *training* (i.e., the process of minimizing error as a function of ANN parameters, given an ML problem). One reason cited in [3] is the high bandwidth and latency costs involved in exchanging network updates across multiple edge devices (centralized model training might be more efficient since updated networks would be transferred directly to devices). Another concern in the design of edge nodes is energy and hardware costs. Structured labeled data is usually used for inference. The case is different for *deep learning* [4], which is used for training tasks in which precision is critical. In this process, complex multi-layer ANNs are applied along with huge amounts of raw data. Edge devices lack the computing power and data storage capacity needed for this.

1.2. Motivation and Contribution of This Study

Smart edge devices rely on embedded systems with limited resources to process sensor data. For ML workloads requiring high computing power, energy-efficient systems are necessary. In recent years, many research efforts have been focused on energy-efficient embedded system designs to solve ML problems [5–9]. These systems are primarily designed to make inferences. However, embedded systems, especially at edge nodes, are likely to require online learning, control, and optimization capabilities. In autonomous cars, for example, using remote cloud-based training could lead to long communication delays. Therefore, embedded training devices are preferred. However, after being trained offline,

ML inference models tend to diverge once in production. The retraining of such models will therefore require online training capabilities.

To understand how low-power embedded computing devices might help fill the aforementioned demand, this paper reviews the main design approaches for energy-efficient ML algorithm execution. In the following sections, it surveys candidates that meet both smart edge computing and Internet of Things requirements for low-power devices. CYSmart is a flexible and low-power smart edge computing system that we present as an example. A few working scenarios are used to evaluate the power efficiency of CYSmart.

1.3. Outline of the Paper

First, we discuss energy-efficient computing systems dedicated to executing machine learning algorithms in Section 2. Section 3 provides a classification of low-power devices for IoT and smart edge computing on the basis of hardware resources and power dissipation constraints. This classification is then used to present a panorama of popular devices. The CYSmart low-power edge computing system is described in Section 4. Moreover, it is briefly compared with selected industrial edge computing technologies. Finally, some concluding remarks are provided in Section 5.

2. A Quick Journey in the Landscape of Energy-Efficient Compute Systems for ML Tasks

Design principle-wise, embedded machine learning can be implemented through a central processing unit (CPU), graphics processing unit (GPU), application-specific integrated circuit (ASIC), and field-programmable gate array (FPGA). There is a wide range of outcomes in terms of power and accuracy of prediction for these implementations [10].

Figure 2 roughly illustrates this tradeoff. With the ASIC implementations [11,12], execution latency is optimized, but ML model accuracy is lower as the model is customized with approximations like quantization of ANN weights (i.e., reducing numerical precision by reducing the number of bits). Contrary to ASIC chip designs, CPUs and GPUs often support high-precision numerical representations, which improve prediction accuracy at the expense of power consumption. ASICs are more energy-efficient than CPUs and GPUs since they use less computing, I/O bandwidth, and memory resources. However, their development can be time-consuming and expensive. FPGAs offer a flexible and cost-effective implementation, allowing better balancing of power consumption, response latency, and prediction accuracy, as evidenced by recent studies [13,14]. Nevertheless, this comes at the expense of programmability, e.g., when compared to CPUs and GPUs.

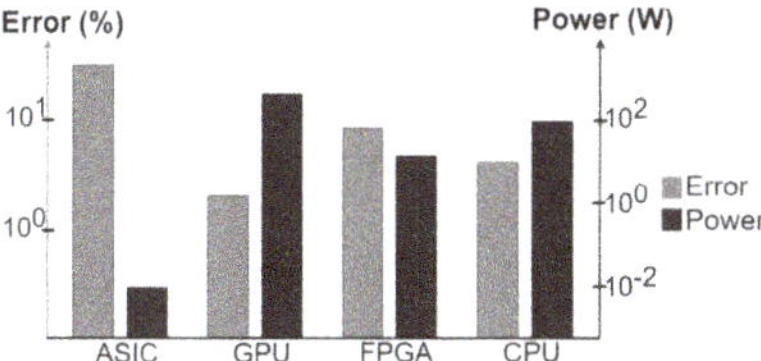

Figure 2. Power and prediction error for four hardware designs (based adapted from [10])—higher is worse.

In Figure 3, another perspective on existing machine learning systems is summarized [7]. Inference is addressed by systems with dissipation less than 100W. Among them are Google's Edge Tensor Processing Unit (TPUEdge) [15] and Intel's MovidiusX processors [16], embedded GPU-based neural engines such as the Apple A12 processor [17] and Huawei Kirin 980 [18], FPGA co-processors like the Zynq-020 [19] and the Stratix-V [20] chips, as well as mobile system-on-chips (SoCs) from Nvidia like Jetson-TX2 [21] and Xavier [22]. Training systems at high performance levels consume more than 100 watts. Typically, they consist of data center chips like the Google TPU3 [23] and the Intel Nervana2 [24] and data center systems like the Nvidia DGX-2 server system [25].

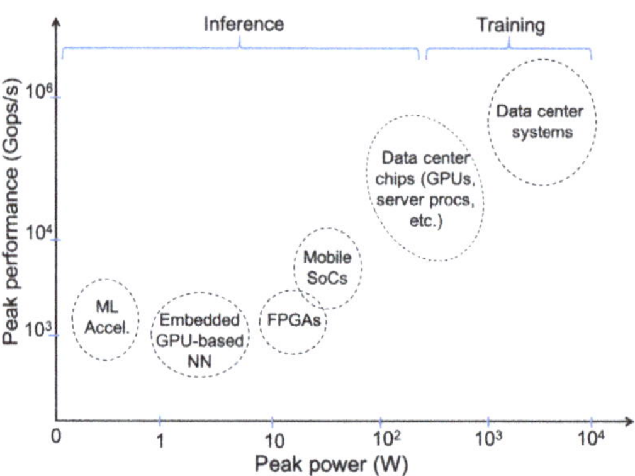

Figure 3. Computing landscape for ML: power vs. performance (adapted from [7]).

2.1. Focus on ML Accelerators, GPUs and FPGAs

Mobile phone SoCs, for example, use ML acceleration to address vector and matrix operations [26]. Various neural processing units and GPUs may be combined to achieve this, as in Qualcomm Snapdragon [27], HiSilicon 600 and 900 series chips [18] and MediaTek Helio P60 [28].

A typical approach toward efficient edge devices is to design hardware accelerators for machine learning models. This is already the case for ANNs for improved energy efficiency and throughput. By minimizing data access costs across the memory hierarchy, these accelerators can enable specialized processing dataflow that better exploits the memory characteristics. In [29], authors highlight several key design specializations tailored to machine learning accelerators: instruction sets that perform linear algebra operations like matrix multiplication and convolutions; on-chip buffers and on-board high-bandwidth memory to efficiently feed data; and high-speed interconnects that enable efficient communication between multiple cores. Additional hardware specializations for inference-only designs include Winograd convolution [30] and non-digital computing [12]. Although accelerators improve the execution performance of individual ML kernels, they may have some negative impact on the overall ML model performance because of costly communications between them and the associated system-on-chip (SoC).

Embedded GPUs and FPGAs are further alternatives for accelerating ML algorithms. As shown in Figure 4, several solutions exist. We only report devices with maximum power consumption of 50W, from the exhaustive list presented in [7]. The selected devices are compared w.r.t. their performance, power consumption and computational precision levels. Accelerators generally offer better precision and power consumption tradeoff, e.g., Google's tensor processing unit for edge computing (TPUEdge) [15] and Eyeriss [31]. On the other hand, GPUs and FPGAs globally provide better performance. The higher their computing precision, the higher their consumption, e.g., Xavier GPU [22] and ZCU102 FPGA [32].

For accelerating ML algorithms, embedded GPUs and FPGAs can also be used. Figure 4 shows several solutions. As part of the exhaustive list presented in [7], we report only devices with a maximum power consumption of 50W. We compare the selected devices in terms of performance, power consumption, and computational precision. There is generally a better tradeoff between precision and power consumption with accelerators, such as Google's tensor processing unit for edge computing (TPUEdge) [15] and Eyeriss [31]. In contrast, GPUs and FPGAs provide better performance globally. In general, the higher the precision of the computation, the more power it consumes, as in Xavier GPU [22] and ZCU102 FPGA [32].

A comprehensive survey on hardware accelerators has been proposed very recently in [33]. The reader can refer to this survey for a full coverage of the state of the art.

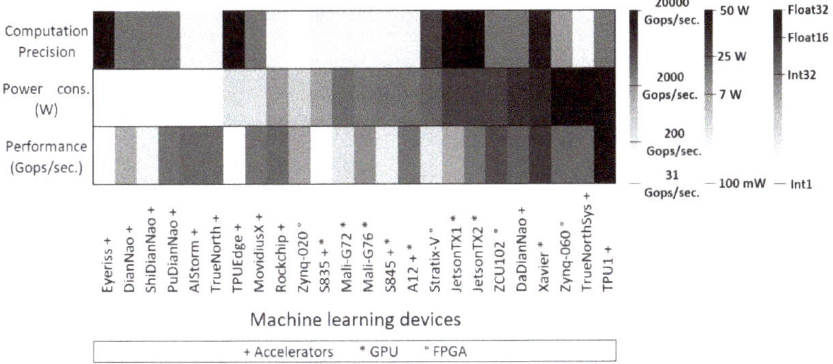

Figure 4. Accelerators, GPUs and FPGAs for embedded ML (adapted from [7])—he darker color, the higher the metric value.

2.2. From Software-Hardware Codesign to Emerging Computing Paradigms

Weight compression [34], parameter pruning, and weight quantization [35] are well-known ML optimization techniques for ANNs. Their goal is to improve energy efficiency by lowering computing complexity, data volume, and hardware resources used during the execution of the networks. Pruning involves gradually suppressing the connections between neurons in an ANN. Quantization involves reducing the number of bits in binary words. It is similar to approximate computing [36], where floating-point representations are converted to fixed-point representations. This reduces the precision of weight values in connections but speeds up execution. Another key aspect is developing compilers and runtime systems [37] that abstract away hardware details. This makes it easier to deploy and train ML models on mobile devices. The extensive software development environment made available to users by Nvidia contributes to the success of Nvidia GPUs for ML.

In the widely adopted von Neumann architectures, ML workloads based on ANNs frequently perform multiply-accumulate operations, which generate multiple data movements between memory and processors. As a result of these exchanges, there is a high execution time and power consumption, and this is known as the "memory wall". Modern ML commodity chips combine CPUs with High-Bandwidth Memory (HBM) via efficient interconnects to address this problem. In parallel, an emerging paradigm, called *near-data processing* [38], has been studied to address the memory wall issue. Computing capability is built into the memory or storage, enabling data stored there to be processed. Mixed-signal circuit design and advanced memory technologies were used to accomplish this. Other near-data processing techniques include in-memory processing [39] and in-storage processing [40]. Integrated 3D technologies and emerging Non-Volatile Memory (NVM) technologies enable such realizations. In comparison to DRAM, NVMs [41,42] like Spin Torque Transfer RAM (STT-RAM) and Resistive RAM (ReRAM) have lower leakage and higher cell density. By using them, edge nodes can mitigate their idle power draw concerns. In Hybrid Memory Cube (HMC) [43], several DRAM dies are stacked above the logic layer using Through-Silicon-Vias (TSV) to address the memory access issue.

3. Classification of Low-Power Devices for IoT and Smart Edge Computing

As IoT and edge computing grow in popularity, multiple sophisticated tiny embedded computing devices have emerged over the last decade. A general and systematic way of assisting designers in choosing low power IoT and smart edge computing devices does not exist. In a recent paper, Neto et al. [5] proposed a classification for IoT devices aimed at smart cities and smart buildings. We revised this classification to better reflect a broader class of edge computing devices encountered beyond smart cities and smart buildings. This includes hardware architectures used by mobile devices such as smartphones. The enhanced classification takes into account the hardware characteristics, including both computing and memory components (which reflect the potential device performance), and the total power dissipated. The resulting classes are accompanied by some typical target algorithms that the corresponding device family can handle.

Our proposed extension of the classification from Neto et al. [5] is represented in Table 1. A total of six device classes are distinguished. The Class 0 devices are based on microcontrollers with limited memory capacity and power consumption. In general, the processed dataset is very small; for example, temperature and humidity measurements. Nevertheless, such devices can perform lightweight inference tasks using simple pretrained models. Hence, all the subsequent device classes can be used for inference. A Class 1 device is one that can store data in addition to collecting and processing data. Such devices generally run on monocore microcontrollers or application cores with larger storage and memory capacities. Typically, such devices process only some basic statistics, such as noise reduction.

Table 1. Device classification for smart and low-power embedded systems (adapted with permission from [5]).

Class	Storage	Memory	Compute Unit Types	Power	Typical Algorithms
0	≤512 MB	≤512 kB	Microcontrollers	≤1 W	Basic computations (lightweight inference)
1	≤4 GB	≤512 MB	Microcontrollers/ Application cores	≤2 W	Basic statistics (inference)
2	≥4 GB	≤2 GB	Application cores	≤4 W	Classification/Regression (inference)
3	≥4 GB	≤8 GB	Application cores	≤16 W	Prediction/Decision-making (inference)
4	≥4 GB	≤16 GB	Application cores	≥16 W	Deep learning, auto-encoders, etc. (inference & training)
5	≥4 GB	≥16 GB	Application cores	≥16 W	

From Class 2, all devices are considered to have one or more application cores. The presence of SD card slots in the majority of these devices makes storage capacity more scalable. Devices of Class 2 are powerful enough to enable CNN inference, such as in image analysis. Their performance is good enough to execute lightweight IoT and edge workloads, as well as more intensive workloads such as training and inference.

In class 3, embedded GPUs make it possible to run lightweight training tasks. It is the first class with sufficient resources to enable real-time video analysis without any special ML accelerators.

In class 4 and class 5, we find devices that can be used in (quasi)autonomous systems such as smartphones or self-driving cars. The devices should be able to withstand environmental changes, while delivering the performance to process large datasets using high-performance accelerators, such as Nvidia GPUs found in server-class systems. A class 4 device is often intended for smartphones and is often more energy-efficient than a class 5 device, which is mostly designed for training and inference purposes.

Quick Survey of Typical Embedded Devices

Following the above classification, we now look at the most popular low-power devices used for IoT and edge computing recently. As shown in Tables 2–4, the devices identified fall into three application families:

- the first application family includes ultra-low-power devices with limited resources suitable for lightweight IoT and edge applications. This applies to all class 0 and class 1 devices;

- the second application family consists of the most popular devices encountered in average edge computing and IoT applications. All devices in class 2 and part of devices in class 3 are included in this class;
- the third application family includes devices with the most powerful hardware resources for performing machine learning and inference tasks. It covers a significant portion of devices in classes 3 and 4 and 5.

We categorize each application family by its device classes, its execution unit (CPU, GPU, accelerator, etc.), the most appropriate ML task (inference vs. training), and some domain application examples. Following the three application families outlined above, we will briefly discuss an application-driven panorama of key IoT and edge devices. We rely on [6] in part for this survey.

Table 2. Ultra low-power devices for IoT and edge computing.

Device	Class	GPU/Accel.	CPU	ML Usage	Application Examples
Arduino Mega	0	-	Microcontroller ATmega 8-bit @16 MHz	inference (ANN)	domotic [44], robotics [45]
Arduino RP2040	0	-	2xARM Cortex-M0+ @133 MHz (RP2040)	inference (ANN)	parking traffic [46], virtual reality [47]
MSP430G2553 LaunchPad	0	-	MSP430 16-Bit RISC Architecture @16 MHz	inference (ANN)	activity recognition [48]
Sony Spresense	0	-	6xARM Cortex-M4F @156 MHz	inference (ANN)	object detection [49]
SparkFun Edge	0	-	32-bit ARM Cortex-M4F @48 MHz	inference (ANN)	speech recognition [50]
STM32F103	0	-	ARM Cortex-M3 @72 MHz	inference (CNN)	image recognition [51]
STM32F765VI	0	-	ARM Cortex-M7 @216 MHz	inference (CNN)	image recognition [52]
Tiny Eats	0	-	ARM Cortex-M0+ @48 MHz	inference (DNN)	audio recognition [53]
Beaglebone Black	1	PowerVR SGX530 GPU	ARM Cortex-A8 single-core @1 GHz	inference (ANN)	robotics [54], camera drones [55]
Hello-Edge	1	-	ARM Cortex-M7 (STM32F746G)	inference (DNN)	keyword spotting [56]
MAX78000	1	Deep CNN Accelerator	ARM Cortex-M4 @100 MHz RISC-V coprocessor @60 MHz	inference (DNN)	object detection [57]
ZedBoard Dev. Board	1	FPGA accel.	2x ARM Cortex-A9 @667 MHz	inference (DNN,CNN)	image recognition [58]

The first application family, shown in Table 2, deals with lightweight data processing. Display devices are usually located close to data sources and are used primarily for domotics. The Arduino board is typically found in smart houses to monitor lightning [44]. Meanwhile, the most powerful devices may come with a compute accelerator integrated into them. Using its CNN accelerator, the MAX78000 device executes a light and optimized object detection algorithm on camera data [57]. Generally, the devices in Table 2 are affordable. They use programming models that are often closer to the hardware, i.e., at a low abstraction level, like assembly code.

Due to its inherent energy-efficiency, ARM technology is often adopted for embedded systems. Cortex-M microcontrollers and Cortex-A application processors are examples. In addition to ARM cores, a few designs use Intel technology, such as the Movidius Myriad 2 vision processing unit (VPU). Using them, deep neural networks (DNNs) can be run in smart cameras, for example. Also worth noting is the emerging GAP8 device, based on the RISC-V open Instruction Set Architecture (ISA) [59]. It paves the way for a new era of processor innovation sustained by a collaborative and dynamic ecosystem.

There are a number of devices that combine ARM CPUs with embedded GPUs (see Tables 3 and 4), except for the Beaglebone Black system (see Table 2). In the powerful Jetson TX series and Tegra X1 systems, Nvidia's Pascal and Maxwell GPUs are combined with ARM Cortex-A57 cores. In both cases, the resulting systems can consume up to 15W of power. Since GPUs are present, they provide higher computing performance at the expense of more power consumption.

GPUs such as those reported in Table 3 are capable of executing the second application family. With Raspberry Pi, this involves image processing to detect tomato disease [60] or image super-resolution [61] or image classification [62] with smartphones. Other applications involve robotics, such as robotic perception [63] implemented using the Robot Operating System (ROS). As a result, the devices used here can support higher abstraction levels of programming.

According to Table 4, all devices in the third application family combine a GPU with at least four application cores, except for the RZ/V2M board. Applications primarily focus on image and video processing. With the Jetson Nano, real-time vehicle object detection is possible [64]. Furthermore, these devices are used in smartphones, such as the Huawei P40 Pro, which is equipped with powerful video super-resolution. Additionally, they can be used with ROS in the robotics field [65] for navigation, perception, and control.

Table 3. Low-power devices at the frontier of IoT and edge computing.

Device	Class	GPU/Accel.	CPU	ML Usage	Application Examples
BeagleBone AI	2	-	2x ARM Cortex-A15 @1,5 GHz 2x ARM Cortex-M4 SoC with 4 EVEs	inference (CNN)	computer vision [66]
Intel Movidius	2	-	Myriad-2 VPU	inference (SVM)	computer vision [67]
Raspberry Pi 3	2	400 MHz VideoCore IV GPU	4xARM A53 @1.2 GHz	inference (SVM, CNN)	video analysis [68] medical data processing [69]
Raspberry Pi Z2 W	2	400 MHz VideoCore IV GPU	4xARM Cortex-A53 @1 GHz	inference (CNN)	object detection [70]
RISC-V GAP8	2	-	8 RISC-V 32-bit @250 MHz + HW ConvolutionEngine	inference (CNN)	image, audio processing [71]
Samsung Galaxy S3 (Exynos 4412 SoC)	2	Mali-400 MP GPU	4xARM Cortex-A9 quad-core @1.4 GHz	inference (CNN)	image classif. [72]
Khadas VIM 3	2,3	4xARM Mali-G52 @800 MHz	4xARM Cortex-A73 @2.2 GHz 2xARM Cortex-A53 @1.8 GHz	inference (CNN)	robotics [63]
Raspberry Pi 4	2,3	500 MHz VideoCore VI GPU	4xARM Cortex-A72 @1.5 GHz	inference (SVM, CNN)	image analysis [60]
Motorola Z2 Force (Snapdragon 835 SoC)	2, 3	Qualcomm Adreno 540 GPU	4x Kryo 280 @ 2.45 GHz 4x Kryo 280 @ 1.9 GHz	inference (CNN)	image classif. [62] recognition [73]
Xiaomi Redmi 4X (Snapdragon 435 SoC)	2, 3	Qualcomm Adreno 505 GPU	8xARM Cortex-A53 @1.4 GHz	inference (CNN)	image super resolution [61]

Table 4. Powerful embedded devices for ML at the edge.

Device	Class	GPU / Accel.	CPU	ML Usage	Application Examples
Coral Development Board	3	GC7000 Lite GPU + TPUEdge accel.	NXP i.MX 8M SoC (4x ARM Cortex-A53 + Cortex-M4F)	inference (CNN)	image processing [74]
Google Pixel C (Tegra X1 SoC)	3	256-core Maxwell GPU	4x ARM Cortex-A57 + 4x ARM Cortex-A53	inference (SVM)	pedestrian recognition [75]
Jetson Nano	3	128-core Maxwell GPU	4x ARM Cortex-A57	inference (CNN)	video image recognition [64]
Jetson TX1	3	256-core Maxwell GPU	4x ARM Cortex-A57 2x MB L2	inference (CNN)	video, image analysis [76], robotics [77]
Odroid-XU4 (Exynos 5422 SoC)	3	ARM Mali-T628 MP6 GPU	4x ARM Cortex-A15 + 4x ARM Cortex-A7	inference/ training (ANN)	urban flooding, automobile traffic [78]
RZ/V2M Evaluation Board	3	DRP-AI	1x ARM Cortex-A53 @996 MHz	inference (CNN)	image processing [79]
Samsung Galaxy S8 (Exynos 8895 SoC)	3	ARM Mali-G71 GPU	4x ARM Cortex-A53 @ 1.7 GHz 4x Exynos M2 @2.5 GHz	inference (CNN)	image recognition [73]
Odroid-M1	3,4	4xARM Mali-G52 @ 650 MHz	4xARM Cortex-A55 @2 GHz	inference (CNN)	video image recognition [80]
Huawei P40 PRO (Kirin 990)	4	16xARM Mali-G76 @600 MHz	2x ARM Cortex-A76 @2.86 GHz 2x ARM Cortex-A76 @2.09 GHz 4x ARM Cortex-A55 @1.86 GHz	inference (CNN)	video super resolution [81]
Jetson TX2	4	256-core Pascal GPU	2x Denver2, 2 MB L2 + 4x ARM Cortex-A57, 2 MB L2	inference (CNN, DNN, SVM)	video, image analysis [76], robotics [82]
One Plus 9 Pro (Snapdragon 888)	4	Adreno 660 GPU	1x ARM Cortex-X1 @ 2.84 GHz 3x ARM Cortex-A78 @2.42 GHz 4x ARM Cortex-A55 @1.80 GHz	inference (CNN)	image classification [83]
Jetson AGX Orin	5	2048xCUDA cores 64xTensor cores @1.3 GHz	12xARM Cortex-A78 @2.2 GHz	inference (CNN)	robotics [65]
Jetson AGX Xavier	5	512xVolta GPU	8xNVIDIA Carmel	inference (CNN)	real-time object detection [84]

A few devices, however, combine CPUs with specific ML accelerators. ZedBoard and Coral Dev Board both integrate an FPGA-based accelerator and Google's TPUEdge [15]. Based on the computing complexity of algorithm execution, this diversity of cores enables maximizing the efficiency of the combined processing elements.

Homogeneous multicore devices appear mostly in devices for the first application family (Table 2), and in a few cases for the second application family (Table 3). These devices dissipate only a few milliwatts or a few Watts, such as SparkFun Edge and Hello-Edge. They, however, deliver less performance than heterogeneous devices.

Lastly, it is worth noting that most of the devices reported in Tables 2 and 3 are used for inference tasks rather than training (which is more expensive) due to their limited computing resources. Only some devices listed in Table 4 has been considered for lightweight training tasks, e.g., Odroid-XU4 board.

4. Low-Power Smart Edge Computing with CYSmart Solution

CYSmart is an edge computing system that gathers, processes, and displays locally measured data with minimal power consumption. It is capable of providing real-time feedback to domain experts. There are a number of low-power devices in this system, called

CYComs, which collect data from sensors at points of interest, pre-process it, and transmit it to the CYEdge via LoRa networks, as illustrated in Figure 5.

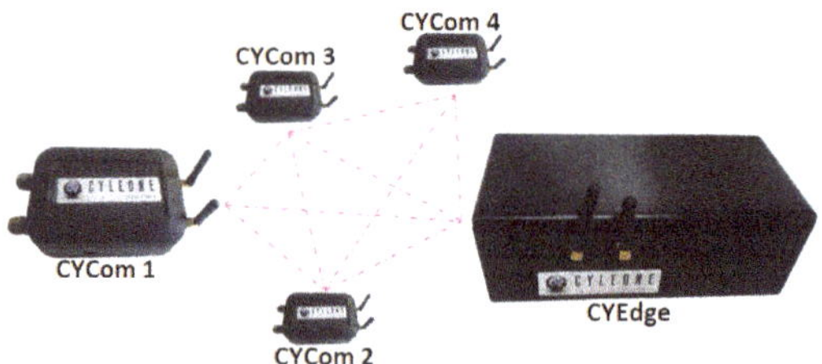

Figure 5. Overview of the CYSmart solution.

A CYCom implements the services provided by Layers 0 and 1 in Figure 1. As a result, CYSmart is able to perform some preliminary lightweight analyses on the collected data. This analysis can be performed to filter it before sending the result to the other components of the system. The outputs CYComs can then be processed and displayed by the CYEdge component, which typically implements Layers 2 and 3 of Figure 1. Data processing algorithms can determine which device class is appropriate for implementation of a CYEdge. For energy-efficient and secure computations, the latter is deployed close to CYComs.

- Measurement identifier and name of the point of interest
- Type of measurement performed
- Unit of measurement used
- Range of measurement desired
- Operating mode of the module (continuous measurement, on demand, sleep. . .)
- Time range of system activity
- Battery level of CYComs
- Limit ranges of expected values
- Alert generation
- Transmission signal strength

Every measure is stored in the CYEdge internal memory and remains accessible at any time through:

- a visualization tool that displays the measurement curves versus time;
- a download of all the information stored on a USB flash drive, computer, or server. Depending on the needs of the customer or the third party software used, the file type and format are adapted.

Figure 6 presents a more detailed technical presentation of CYSmart. The CYCom and CYEdge components are detailed in the next two sections, followed by some use case scenarios.

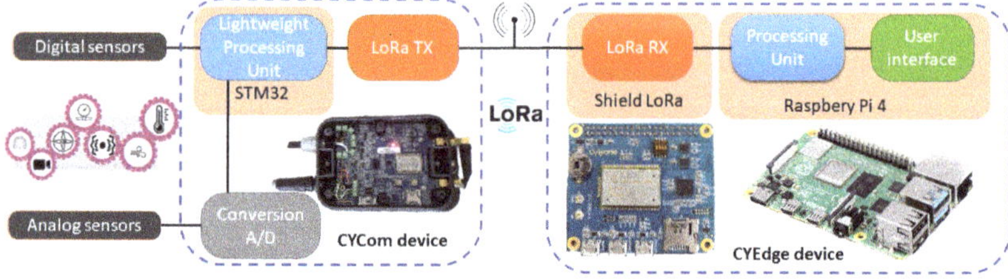

Figure 6. Detailed view of CYSmart solution.

4.1. Data Acquisition Device: CYCom

A CYCom is a device used to acquire data in the CYSmart system. Physical data can be recovered using this low-power technology running on an external battery. Data can come from digital sensors with serial communication (SPI, UART, I2C) or from analog sensors with a 16-bit Analog to Digital Converter (0–10V, 4–20 mA, or thermal resistance input). An STM32 microcontroller allows the CYCom to pre-process data (threshold detection, filtering, conversion...) before sending it by LoRa to the CYEdge. Each device can be physically configured to communicate with the CYEdge unit using one of two frequency bands (433 MHz or 866 MHz), each with four transmission channels. In the event the receiving device is not reachable, the sent frame can be stored on the receiving device's internal flash memory (8 MB) or SD card and sent back when the connection is restored. The device can also make use of other modules, such as a micro-USB port (currently used for CYCom updates), Bluetooth module, 3-axis inertial measurement unit (IMU) module (acceleration, angular, magnetic), or GPS module directly integrated within the device.

4.2. Centralized Early Data Processing: CYEdge

Data centralisation and setup of the sensor network is achieved with the CYEdge technology. It is a box that can be connected to an external battery or to the socket. The technology consists of two parts. The first one is a Raspberry Pi 4 (Processing Unit) and the second one is a proprietary shield developed by CYleone that enables LoRa communication with CYComs. This shield is a LoRa gateway that allows the processing of AT commands sent by the Raspberry Pi. These commands are sent to setup the CYComs but also to retrieve the data frames from the CYComs. The Raspberry Pi acts mainly as a data processing and graphic display unit. It reads, processes, saves and displays data from the shield on the graphical user interface. The interface allows the configuration of the sensor network and the retrieval of all the data and configurations of each CyCom. It can be accessed by connecting via WiFi or Ethernet to the local network created by the board itself, or to an existing network. The Raspberry can also be used for other parallel tasks such as GPS measurements, digital data retrieval and synchronization of this data with that received via LoRa.

4.3. Use Case Evaluation of CYSmart

One use case application of CYSmart is to collect data from different points of interest every interval of ten minutes, in a critical environment. As shown in Figure 6, the CYComs collect data from sensors where it is difficult to take and transmit measurements, such as aeraulic measurements in basements or bunkers. Data can be stored in the CYComs and sent to the CYEdge after lightweight processing, such as threshold detection or filtering. By sending only useful and ready-to-use data, this application minimizes LoRa communication. On the CYEdge, complementary data processing can be performed. A diagram of the CYCom's operations is shown in Figure 7.

A second use case application relies on LoRa to wirelessly transmit raw data from a CYCom directly to the CYEdge. The latter performs all data processing and displays the evolution of the values. This scenario is used to track the evolution of a process or a physical value in time, such as a refrigerator temperature in catering. Here, in most cases, the interval between two measures is around a few seconds, e.g., four seconds in our case. In addition, a CYEdge can only be paired with one to five CYComs. Figure 8 shows the corresponding flow diagram.

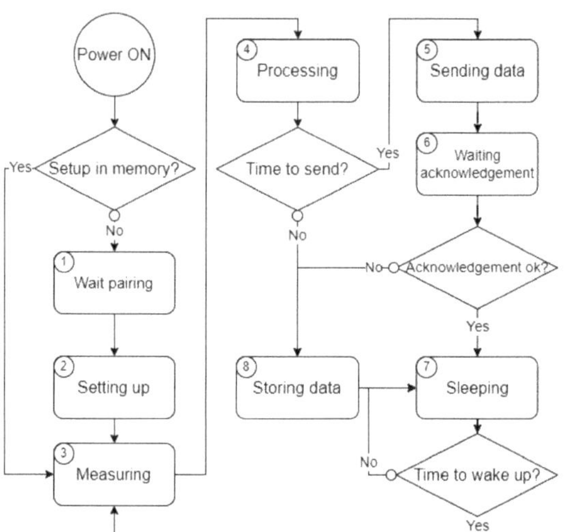

Figure 7. Operational diagram of the first application use case (Case I).

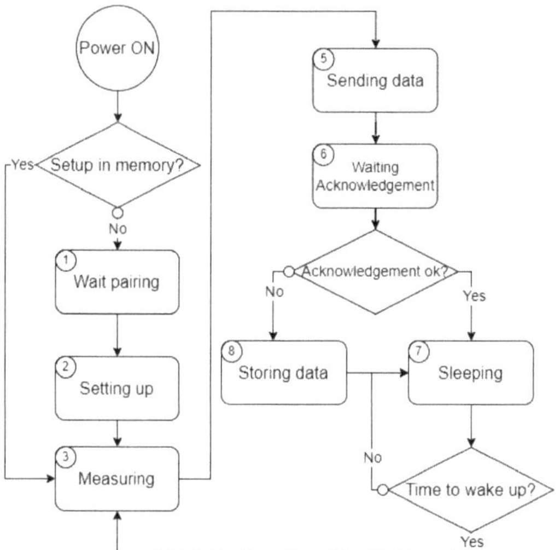

Figure 8. Operational diagram of the second application use case (Case II).

CYSmart devices are now ready to be evaluated according to the above use cases. As a processing unit, CYEdge utilizes a Raspberry PI 4 device with a LoRa shield. The CYCom uses a homogeneous architecture based on a STM32L496 microcontroller and ARM Cortex-M4 CPU (class 0 device hardware). The CYCom has a storage capacity of 1 MB from the STM32 and 8 MB from an external flash memory (class 0 device storage), as well as an SD card slot. There is also 320 KB memory in the STM32, corresponding to a class 0 memory.

According to the different steps of the two operational diagrams, the evaluation of CYCom is presented in Table 5. During each step, the power is measured using the generator voltage (constant 15V) and the CYCom current draw. A CYCom is connected to a generator through an ampere-meter to measure the current draw with the greatest precision between the generator value and the ampere-meter display.

Table 5. Power consumption and duration of each steps in the use case scenarios.

Step Labels	Detailed of the Step	Power Consumption	Time Duration
1	Wait pairing and synchronizing from the CYEdge	412.5 mW	20 s
2	Setting up measurement parameters from the CYEdge	468 mW	7 s
3	Measuring digital and analog data from sensor	357 mW	10 s (Case I) 200 ms (Case II)
4	Data processing (filtering, conversion)	387 mW	50 ms
5	Sending stored and measured data to the CYEdge (LoRa)	377.5 mW	2–10 s
6	Waiting acknowledgement from the CYEdge	252 mW	1–25 s
7	Sleep until next measuring	177 mW	1 s–1 min
8	Storing not sent data in RAM memory	256.5 mW	50 ms

From Figures 7 and 8, we note two parts in the operational diagram. The first part, with steps 1 and 2, concerns the setup of a CYCom. Both steps are performed only once during the setup of the entire system and the CYComs. The latter retain their configuration in memory until they are reinitialized by the use of a hardware reset (push button inside). The second part relates to the execution routine of the device. In this routine, the device is woken up, a measurement is taken, data is processed and sent (depending on the use case), before returning to sleep until the next measurement. Steps 3 to 8 also belong to the routine and are executed in an infinite loop.

Figure 9 shows the duration distribution for the steps in one routine iteration, considering a worst-case scenario. In Case I (see Figure 9a), the worst case scenario occurs when there is 1 min between each data measure in the CYCom and there are frequent connection issues between the CYCom and CYEdge (i.e., low communication quality). In this worst-case scenario, (i) the CYCom transmits data to the CYEdge during 2 s and waits for an acknowledgement during 5 s; (ii) if no acknowledgment is received, the CYCom repeats phase (i) up to five times, otherwise it moves on to step 7. After five attempts (about 35 s), if no acknowledgement is received, the CYCom stores the data in its local memory, and proceeds to step 7.

Figure 9. Time distribution between the different steps in the worst-case scenario, for two use cases. (**a**) Use case I; (**b**) Use case II.

For Case II in Figure 9b, we assume that the CYEdge and CYCom are close to each other. This minimizes the loss of communication between both components during data exchanges. It enables the CYCom to send data to the CYEdge only once during 2 s and wait for the corresponding acknowledgement during 1 s. If no acknowledgement is received, the data is stored in the local memory of the CYCom. The worst-case scenario requires 3 s to reach step 7.

Despite their high power consumption, steps 1 and 2 only consist of system setup functions executed during installation. For this reason, as shown in Figures 10a,b, various components are activated during these steps to set them up.

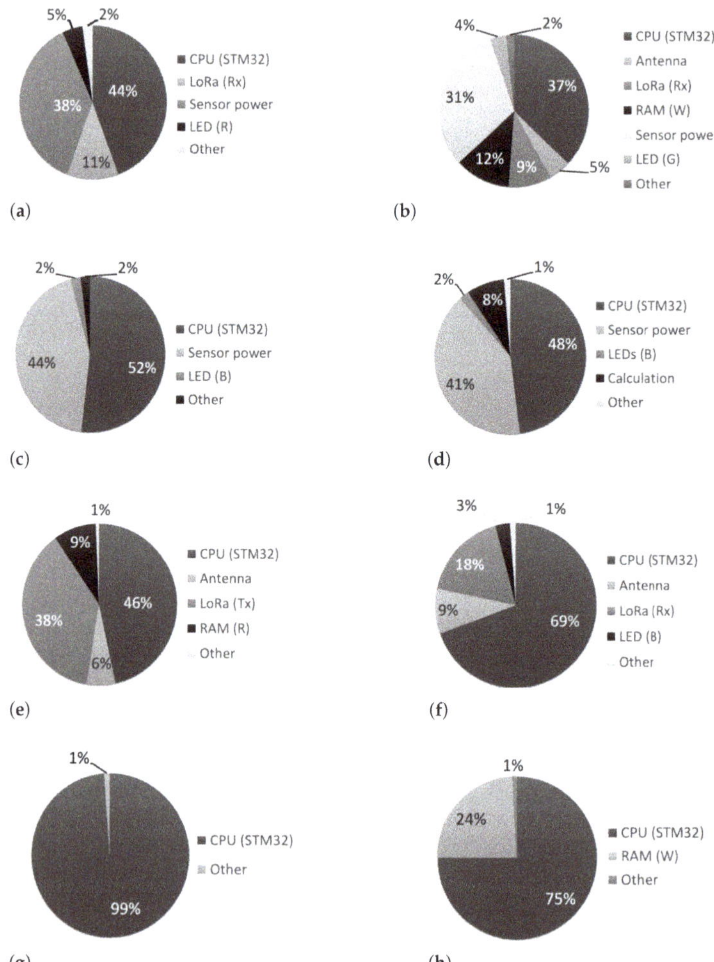

Figure 10. Power consumption breakdown for the different steps occurring in Figures 7 and 8. (**a**) Step 1; (**b**) Step 2; (**c**) Step 3; (**d**) Step 4; (**e**) Step 5; (**f**) Step 6; (**g**) Step 7; (**h**) Step 8.

During the execution of the routine, the sensor power management is activated only during steps 3 and 4. Figures 10c,d show that it is the second most power consuming function after the CPU. It is activated for 10 s, as in Case I. This represents 9% of the routine execution duration in Figure 9a and only 5% in Figure 9b. This allows for stable data collection from analog sensors. To provide the most precise data, it is necessary to have so much time. However, digital sensors can take less time to acquire data than analog sensors in Case II. This explains the shorter duration in Case II. The complexity of the data processing affects the processing time, but not the power consumption.

The CYCom sends data to CYEdge in steps 5 and 6. Here, the duration depends on the communication load between both devices. The CYCom will attempt to send data to the CYEdge five times before storing it (step 8). During steps 5 and 6, Figure 10e,f shows that the LoRa communication is activated and represents up to 38% of the power consumption during the data transmission, i.e., the third most power consumer function.

Lastly, the process performs the sleep function, i.e., step 7. In Figure 10g, it is the step with the least activated functions: only the CPU is activated, resulting in the lowest power consumption. According to the use case, this function can be more or less time consuming. Figure 9a shows that in the worst-case scenario in Case I, CYCom is in this state 57% of the execution routine. Additionally, depending on the frequency of the data measurement, this step can be repeated and reach more than 95% of the process duration. In some cases, this duration can be shortened depending on the measuring frequency and the total number of CYComs deployed in the network to improve the bandwidth. With 1 s sleep duration in Case II, the routine spends 25% of its time in the worst-case scenario.

4.4. Gained Insights and Discussion

As a representative ultra-low-power device of the CYSmart system, the CYComs were the primary focus of the above use cases. There is also another component, the CYEdge, which embeds a Raspberry Pi 4 and a LoRa communication shield. Based on the power measurements of the CYEdge under normal operating conditions, it can be classified as a class 3 device, as described in Table 1. Below are some insights regarding CYSmart's current implementation and potential improvements.

Gained insights. The CYCom component of the CYSmart system utilizes a commercial-off-the-shelf (COTS) microcontroller manufactured by STMicroelectronics. Choosing this approach reduces the cost of the component as well as the development time. A CYCom's CPU is the primary energy consumer in the aforementioned use cases, followed by the LoRa module and the sensor power supply. Each of these three modules can be improved.

STM32 boards are based on von Neumann microarchitecture, leading to costly data movement between different hardware units. As a result, future improvements could include designing a customized solution that meets the requirements of the domain applications. This is consistent with the notion of domain-specific hardware accelerators as described in [85]. There is a lot of power consumed during one routine iteration in the second use case from the previous section without any data processing being performed. This unnecessary power consumption must be eliminated in order to improve the energy efficiency of the system. This problem may be solved by means of power gating, for example. A customized solution that incorporates such a mechanism is therefore desirable. Suitable design approaches should be considered for design space exploration by selecting high-level methodologies, e.g., [86–88], covering different abstraction levels: high-level analytical modeling [89–92], transaction-level modeling [93–95], cycle-accurate design [96–98], or register transfer level [99]. As surveyed in this paper, it is possible to implement the architecture using FPGA or ASIC designs at the expense of higher costly implementation efforts. As for CYComs, the CYEdge power consumption can be reduced by applying the same design methods.

Sensor power consumption is difficult to reduce since it is heavily dependent on the type of sensor being used. A wide range of digital and analog sensors can be interfaced with the CYCom. External 24 V lithium-ion batteries are currently used to power the integrated sensors. Instead of analog sensors, digital sensors with internal 3.3 V batteries could be considered here to reduce power consumption. Depending on the measurement environment, LoRa modules consume varying amounts of power. In both Cases I and II, the system can communicate across a reinforced concrete wall 90 cm thick with the initial parameters. In the case of a 15 dBm data transmission capacity and a spreading factor of 12, the maximum transmission delay is 2 s. Therefore, its maximum power consumption is 166 mW. These parameters can be adjusted according to the operational environment in order to reduce the LoRa module's power consumption.

Comparison of CYSmart w.r.t. selected industrial solutions. As a mature low-power edge computing solution, the CYSmart system can be compared with a number of industrial technologies. For this purpose, we consider some relevant criteria, described as follows:

- *Device classes*: supported device classes as defined in Table 1. This criterion implicitly suggests a range of power consumption;
- *Sensor diversity*: diversity of sensor types supported by a technology, such as digital versus analog sensors, as well as sensor voltage ranges. The criterion is qualitative in nature and can be rated on three levels: high, average, and low.
- *Transmission speed*: the speed of data transmission between the sensors at the edge frontier and the gateway or centralized system that is responsible for pre-processing the data. Generally, it is measured in terms of the number of samples per second (S/sec) or bits per second (b/sec);
- *Communication distance*: the distance over which a technology communicates wirelessly. It is essential in critical environments, such as basements, bunkers, and nuclear power plants;
- *Number of edge layers*: the number of layers considered in the hierarchical edge computing implementation, as shown in Figure 1;
- *Measurement points*: the number of data measurement points (i.e., sensors) managed by a single gateway or centralized system;
- *Dimension of measurement device*: the form factor of a device that incorporates sensors to collect data during the deployment of a technology;
- *Dimension of gateway/central device*: the form factor of a gateway or centralized system that manages sensor data;
- *Easy deployment*: the effort required for an easy deployment of a technology. This is a qualitative criterion;
- *Application diversity*: it refers to the variety of applications that can be leveraged by a technology, such as smart-home, smart-industry, and smart-city. The criterion is also qualitative in nature.

In light of the aforementioned criteria, Table 6 provides a comparison of CYSmart w.r.t. the industrial edge computing technologies summarized in the sequel. The *TMI-Orion* company proposes a solution for the design and manufacture of high level technologies that target harsh environments. A key component of its edge computing technology is a network of smart sensors such as NanoVACQ Fullradio [100], which communicate via a 2.4 GHz radio protocol with a Radio Transceiver [101]. Using a serial protocol, the latter transmits data to a host computer that manages and displays data from a sensor network. The *Gravio* company develops an IoT platform that is capable of connecting several sensors. Using the ZigBee wireless protocol, these sensors communicate with the Gravio Hub [102]. Data can be viewed and managed by users.

The *moneo appliance* is an edge solution manufactured by IFM company [103]. It consists of a dedicated software toolbox that allows for the management of sensor parameters as well as data display. Templates are provided in the toolbox for defining network configurations. Sensors are connected to the moneo appliance via an IO-Link Master, which serves as an interface between the appliance and the gateway computer.

The *Advantech* company developed an IoT solution that relies on data measurement devices named WISE (e.g., WISE-4060 [104]) and an intelligent edge server (e.g., EIS-D150 [105]). By using the WiFi protocol, the WISE devices send data from the sensors to the edge server. Users are provided with a real-time dashboard for managing WISE devices. The *InHand Networks* company has defined a specific gateway [106], which provides data optimization in the IoT infrastructure and provides real-time response times. The gateway device can be connected to a local network. It is compatible with real-time Ethernet protocols and supports the Docker software system.

The MCM200 series components (e.g., MCM-204 [107]) are edge computing solutions designed by the *Adlink* company. They are standalone data acquisition devices (i.e., no host computer is required) that can monitor, analyze, and execute real-time actions. WiFi or Ethernet ports are available for communication.

Finally, the *Analog Devices* company offers the *SmartMesh Wireless HART* technology which consists of a small network manager (LTP5903-WHR [108]) that communicates with a number of sensor nodes called "motes" (e.g., LTP5900-WHR [109]). The network manager and motes must be programmed by the user. The network manager is responsible for centralizing data and communicating it to the host computer. Using analog data from sensors, the motes transmit data to the network manager.

Table 6 globally illustrates that CYSmart and Advantech technologies offer several advantages over other solutions. There are many similarities between these two technologies; however, CYSmart is capable of supporting a larger wireless communication distance than Advantech's solution. Because of this, CYSmart is well suited to critical environments, such as nuclear power plants.

Table 6. Comparison of CYSmart with similar edge computing technologies.

	CYSmart	TMI Orion [100,101]	Gravio [102]	Moneo Appliance [103]	Advantech [104,105]	InHand Networks [106]	ADLINK [107]	Smartmesh WirelessHART [108,109]
Device classes	0 and 3	0 and 2	2	4	0 and 3	2	2	0 and 2
Sensor diversity	high	low	low	average	high	low	high	average
Transmission speed	0.3 S/sec	10 S/sec	-	2500 S/sec	-	1000 Mb/sec	256 KS/sec	250 Kb/sec
Communication distance (in meters)	800	30	100	wired	110	wired	wired	200
Number of edge layers	2	2	2	2	3	2	2	2
Measurement points	20	4	64	16	-	6	20	500
Dimension of measurement device (in millimeters)	$170 \times 90 \times 65$	$31 \times 129 \times 79$	$36 \times 36 \times 9$	-	$80 \times 98 \times 25$	-	-	$39 \times 24 \times 8$
Dimension of gateway/central device (in millimeters)	$245 \times 110 \times 85$	$127 \times 8 \times 46$	$97 \times 97 \times 29$	$35 \times 105 \times 150$	$260 \times 140 \times 54$	$180 \times 115 \times 45$	$110 \times 40 \times 126$	$103 \times 56 \times 20$
Easy deployment	average	average	high	high	average	low	average	low
Application diversity	high	low	average	low	high	average	average	high

5. Summary

Embedded architectures for future edge devices likely will need to support training, control, and optimization capabilities, according to the current trends in edge computing. In this paper, we discuss recent efforts regarding energy-efficient hardware solutions for machine learning at the edge. We reviewed current design approaches and devices targeted at implementing IoT and smart edge computing with limited computing and power capabilities. Candidate low-power devices that could meet IoT and smart edge computing requirements have been surveyed. CYSmart, a flexible low-power edge computing system, was demonstrated as an interesting solution. To evaluate its power efficiency, a few working scenarios have been considered. Finally, a brief comparison of CYSmart with selected industrial edge computing technologies was presented.

Author Contributions: Conceptualization, L.M.W., J.-M.B., G.B. and A.G.; methodology, L.M.W., J.-M.B. and A.G.; software, L.M.W. and J.-M.B.; validation, L.M.W., J.-M.B. and A.G.; writing—original draft preparation, L.M.W., J.-M.B., G.B. and A.G.; writing—review and editing, L.M.W. and A.G.; supervision, J.-M.B., G.B. and A.G. All authors have read and agreed to the published version of the manuscript.

Funding: This research received no external funding.

Institutional Review Board Statement: Not applicable.

Informed Consent Statement: Not applicable.

Acknowledgments: The authors would like to thank Guillaume Devic and Gilles Sassatelli for their feedback in early discussions on part of the current work.

Conflicts of Interest: The authors declare no conflict of interest.

Abbreviations

The following abbreviations are used in this manuscript:

AI	Artificial Intelligence
ANN	Artificial Neural Network
ASIC	Application-Specific Integrated Circuit
CNN	Convolution Neural Network
COTS	Commercial-Of-The-Shelf
CPU	Central Processing Unit
DRAM	Dynamic Random Access Memory
FPGA	Field-Programmable Gate Array
GPU	Graphics Processing Unit
HBM	High-Bandwidth Memory
HMC	Hybrid Memory Cube
I/O	Input/Output
IMU	Inertial Measurement Unit
IoT	Internet of Thing
ISA	Instruction Set Architecture
LoRa	Long Range
ML	Machine Learning
NVM	Non-Volatile Memory
RAM	Random Access Memory
ReRAM	Resistive RAM
ROS	Robot Operating System
SoC	System-on-Chip
STT-RAM	Spin Transfer Torque RAM
SVM	Support Vector Machines
TPU	Tensor Processing Unit
TSV	Through-Silicon-Vias

References

1. Satyanarayanan, M. The Emergence of Edge Computing. *Computer* **2017**, *50*, 30–39. [CrossRef]
2. Qiu, J.; Wu, Q.; Ding, G.; Xu, Y.; Feng, S. A survey of machine learning for big data processing. *EURASIP J. Adv. Signal Process.* **2016**, *2016*, 67. [CrossRef]
3. Kukreja, N.; Shilova, A.; Beaumont, O.; Huckelheim, J.; Ferrier, N.; Hovland, P.; Gorman, G. Training on the Edge: The why and the how. In Proceedings of the IEEE IPDPS Workshops,Rio de Janeiro, Brazil, 20–24 May 2019; pp. 899–903.
4. LeCun, Y.; Bengio, Y.; Hinton, G.E. Deep learning. *Nature* **2015**, *521*, 436–444. [CrossRef] [PubMed]
5. Neto, A.R.; Soares, B.; Barbalho, F.; Santos, L.; Batista, T.; Delicato, F.C.; Pires, P.F. Classifying Smart IoT Devices for Running Machine Learning Algorithms. In Proceedings of the XLV Integrated SW and HW Seminar, Natal, Brazil, 14–19 July 2018.
6. Murshed, M.G.S.; Murphy, C.; Hou, D.; Khan, N.; Ananthanarayanan, G.; Hussain, F. Machine Learning at the Network Edge: A Survey. *ACM Comput. Surv.* **2022**, *54*, 1–37. [CrossRef]
7. Reuther, A.; Michaleas, P.; Jones, M.; Gadepally, V.; Samsi, S.; Kepner, J. Survey and Benchmarking of Machine Learning Accelerators. In Proceedings of the 2019 IEEE High Performance Extreme Computing Conference (HPEC), Waltham, MA, USA, 24–26 September 2019.
8. Andrade, L.; Prost-Boucle, A.; Pétrot, F. Overview of the state of the art in embedded machine learning. In Proceedings of the DATE Conference, Dresden, Germany, 19–23 March 2018; pp. 1033–1038.
9. Gamatié, A.; Devic, G.; Sassatelli, G.; Bernabovi, S.; Naudin, P.; Chapman, M. Towards Energy-Efficient Heterogeneous Multicore Architectures for Edge Computing. *IEEE Access* **2019**, *7*, 49474–49491. [CrossRef]
10. Deng, Y. Deep Learning on Mobile Devices—A Review. *Proc. SPIE* **2019**, 109930A. [CrossRef]
11. Chen, T.; Du, Z.; Sun, N.; Wang, J.; Wu, C.; Chen, Y.; Temam, O. DianNao: A Small-footprint High-throughput Accelerator for Ubiquitous Machine-learning. In Proceedings of the 19th International Conference on Architectural Support for Programming Languages and Operating Systems (ASPLOS '14), Salt Lake City, UT, USA, 1–5 March 2014; pp. 269–284. [CrossRef]
12. Shafiee, A.; Nag, A.; Muralimanohar, N.; Balasubramonian, R.; Strachan, J.P.; Hu, M.; Williams, R.S.; Srikumar, V. ISAAC: A Convolutional Neural Network Accelerator with In-Situ Analog Arithmetic in Crossbars. In Proceedings of the 2016 ACM/IEEE 43rd Annual International Symposium on Computer Architecture (ISCA), Seoul, Republic of Korea, 18–22 June 2016; pp. 14–26. [CrossRef]
13. Nurvitadhi, E.; Venkatesh, G.; Sim, J.; Marr, D.; Huang, R.; Ong Gee Hock, J.; Liew, Y.T.; Srivatsan, K.; Moss, D.; Subhaschandra, S.; et al. Can FPGAs Beat GPUs in Accelerating Next-Generation Deep Neural Networks? In Proceedings of the 2017 ACM/SIGDA International Symposium on Field-Programmable Gate Arrays (FPGA'17), Monterey, CA, USA, 22–24 February 2017; pp. 5–14. [CrossRef]
14. Lacey, G.; Taylor, G.W.; Areibi, S. Deep Learning on FPGAs: Past, Present, and Future. *arXiv* **2016**, arXiv:1602.04283.
15. Google. Edge TPU. Available online: https://coral.ai/products (accessed on 27 October 2022).
16. Marantos, C.; Karavalakis, N.; Leon, V.; Tsoutsouras, V.; Pekmestzi, K.; Soudris, D. Efficient support vector machines implementation on Intel/Movidius Myriad 2. In Proceedings of the International Conference on Modern Circuits and Systems Technologies (MOCAST), Thessaloniki, Greece, 7–9 May 2018; pp. 1–4.
17. Peng, T. AI Chip Duel: Apple A12 Bionic vs Huawei Kirin 980. Available online: https://syncedreview.com/2018/09/13/ai-chip-duel-apple-a12-bionic-vs-huawei-kirin-980 (accessed on 27 October 2022).
18. HiSilicon. Kirin. 2019. Available online: https://www.hisilicon.com/en/SearchResult?keywords=Kirin (accessed on 27 October 2022).
19. Guo, K.; Sui, L.; Qiu, J.; Yu, J.; Wang, J.; Yao, S.; Han, S.; Wang, Y.; Yang, H. Angel-Eye: A Complete Design Flow for Mapping CNN Onto Embedded FPGA. *IEEE Trans. Comput.-Aided Des. Integr. Circuits Syst.* **2018**, *37*, 35–47. [CrossRef]
20. Podili, A.; Zhang, C.; Prasanna, V. Fast and efficient implementation of Convolutional Neural Networks on FPGA. In Proceedings of the IEEE 28th International Conference on Application-specific Systems, Architectures and Processors (ASAP), Seattle, WA, USA, 10–12 July 2017; pp. 11–18. [CrossRef]
21. NVIDIA. Jetson TX2. 2019. Available online: https://www.nvidia.com/fr-fr/autonomous-machines/embedded-systems/jetson-tx2 (accessed on 27 October 2022).
22. Hruska, J. Nvidia's Jetson Xavier Stuffs Volta Performance Into Tiny Form Factor. 2018. Available online: https://www.extremetech.com/computing/270681-nvidias-jetson-xavier-stuffs-volta-performance-into-tiny-form-factor (accessed on 27 October 2022).
23. Teich, P. Tearing Apart Google's TPU 3.0 AI Coprocessor. 2018. Available online: https://www.nextplatform.com/2018/05/10/tearing-apart-googles-tpu-3-0-ai-coprocessor/ (accessed on 27 October 2022).
24. Rao, N. Beyond the CPU or GPU: Why Enterprise-Scale Artificial Intelligence Requires a More Holistic Approach. 2018. Available online: https://newsroom.intel.com/editorials/artificial-intelligence-requires-holistic-approach/ (accessed on 27 October 2022).
25. Cutress, I. NVIDIA's DGX-2: Sixteen Tesla V100s, 30TB of NVMe, Only \$400K. 2018. Available online: https://www.anandtech.com/show/12587/nvidias-dgx2-sixteen-v100-gpus-30-tb-of-nvme-only-400k (accessed on 27 October 2022).
26. Ignatov, A.; Timofte, R.; Chou, W.; Wang, K.; Wu, M.; Hartley, T.; Gool, L.V. AI Benchmark: Running Deep Neural Networks on Android Smartphones. In Proceedings of the European Conference on Computer Vision (ECCV) Workshops, Munich, Germany, 8–14 September 2018.

27. Qualcomm. Neural Processing SDK for AI. 2019. Available online: https://developer.qualcomm.com/software/qualcomm-neural-processing-sdk (accessed on 27 October 2022).

28. MediaTek. Helio P60. 2019. Available online: https://www.mediatek.com/products/smartphones/mediatek-helio-p60 (accessed on 27 October 2022).

29. Ananthanarayanan, R.; Brandt, P.; Joshi, M.; Sathiamoorthy, M. Opportunities and Challenges Of Machine Learning Accelerators In Production. In Proceedings of the USENIX Conference on Operational Machine Learning, Santa Clara, CA, USA, 20 May 2019; pp. 1–3.

30. Lavin, A.; Gray, S. Fast Algorithms for Convolutional Neural Networks. In Proceedings of the IEEE Conference on Computer Vision and Pattern Recognition (CVPR), Las Vegas, NV, USA, 27–30 June 2016. [CrossRef]

31. Chen, Y.H.; Yang, T.J.; Emer, J.; Sze, V. Eyeriss v2: A Flexible Accelerator for Emerging Deep Neural Networks on Mobile Devices. *IEEE J. Emerg. Sel. Top. Circuits Syst.* **2018**, *9*, 292–308. [CrossRef]

32. Xilinx. Tearing Apart Google's TPU 3.0 AI Coprocessor. 2019. Available online: https://www.xilinx.com/products/boards-and-kits/ek-u1-zcu102-g.html (accessed on 27 October 2022)

33. Peccerillo, B.; Mannino, M.; Mondelli, A.; Bartolini, S. A survey on hardware accelerators: Taxonomy, trends, challenges, and perspectives. *J. Syst. Archit.* **2022**, *129*, 102561. [CrossRef]

34. Han, S.; Mao, H.; Dally, W.J. Deep Compression: Compressing Deep Neural Network with Pruning, Trained Quantization and Huffman Coding. In Proceedings of the 4th International Conference on Learning Representations, ICLR, San Juan, Puerto Rico, 2–4 May 2016.

35. Marculescu, D.; Stamoulis, D.; Cai, E. Hardware-aware Machine Learning: Modeling and Optimization. In Proceedings of the International Conference on Computer-Aided Design (ICCAD '18), San Diego, CA, USA, 5–8 November 2018; pp. 137:1–137:8.

36. Gupta, S.; Agrawal, A.; Gopalakrishnan, K.; Narayanan, P. Deep Learning with Limited Numerical Precision. In Proceedings of the 32nd International Conference on Machine Learning, Lille, France, 6–11 July 2015. [CrossRef]

37. C4ML organizers. Compilers for ML. 2019. Available online: https://www.c4ml.org/ (accessed on 27 October 2022).

38. Balasubramonian, R.; Chang, J.; Manning, T.; Moreno, J.H.; Murphy, R.; Nair, R.; Swanson, S. Near-Data Processing: Insights from a MICRO-46 Workshop. *IEEE Micro* **2014**, *34*, 36–42. [CrossRef]

39. Liu, J.; Zhao, H.; Ogleari, M.A.; Li, D.; Zhao, J. Processing-in-Memory for Energy-Efficient Neural Network Training: A Heterogeneous Approach. In Proceedings of the IEEE/ACM MICRO Symposium, Fukuoka, Japan, 20–24 October 2018; pp. 655–668.

40. Choe, H.; Lee, S.; Park, S.; Kim, S.J.; Chung, E.; Yoon, S. Near-Data Processing for Machine Learning. 2017. Available online: https://openreview.net/pdf?id=H1_EDpogx (accessed on 27 October 2022).

41. Endoh, T.; Koike, H.; Ikeda, S.; Hanyu, T.; Ohno, H. An Overview of Nonvolatile Emerging Memories—Spintronics for Working Memories. *IEEE JETCAS* **2016**, *6*, 109–119. [CrossRef]

42. Senni, S.; Torres, L.; Sassatelli, G.; Gamatié, A.; Mussard, B. Exploring MRAM Technologies for Energy Efficient Systems-On-Chip. *IEEE J. Emerg. Sel. Top. Circuits Syst.* **2016**, *6*, 279–292. [CrossRef]

43. Pawlowski, J.T. Hybrid memory cube (HMC). In Proceedings of the IEEE Hot Chips Symposium (HCS), Stanford, CA, USA, 17–19 August 2011; pp. 1–24.

44. Kusriyanto, M.; Putra, B.D. Smart home using local area network (LAN) based arduino mega 2560. In Proceedings of the 2nd International Conference on Wireless and Telematics (ICWT), Yogyakarta, Indonesia, 1–2 August 2016; pp. 127–131.

45. Drgoňa, J.; Picard, D.; Kvasnica, M.; Helsen, L. Approximate model predictive building control via machine learning. *Appl. Energy* **2018**, *218*, 199–216. [CrossRef]

46. Sousa, R.d.S. Remote Monitoring and Control of a Reservation-Based Public Parking. Ph.D. Thesis, Universidade de Coimbra, Coimbra, Portugal, 2021.

47. Brun, D.; Jordan, P.; Hakkila, J. Demonstrating a Memory Orb—Cylindrical Device Inspired by Science Fiction. In Proceedings of the 20th International Conference on Mobile and Ubiquitous Multimedia, Leuven, Belgium, 5–8 December 2021; pp. 239–241.

48. Stolovas, I.; Suárez, S.; Pereyra, D.; De Izaguirre, F.; Cabrera, V. Human activity recognition using machine learning techniques in a low-resource embedded system. In Proceedings of the 2021 IEEE URUCON, Montevideo, Uruguay, 24–26 November 2021; pp. 263–267.

49. Edge Impulse. Detect objects with centroids (Sony's Spresense). Available online: https://docs.edgeimpulse.com/docs/tutorials/detect-objects-using-fomo (accessed on 27 October 2022).

50. SparkFun Electronics. Edge Hookup Guide. 2019. Available online: https://learn.sparkfun.com/tutorials/sparkfun-edge-hookup-guide/all (accessed on 27 October 2022).

51. Jin, G.; Bai, K.; Zhang, Y.; He, H. A Smart Water Metering System Based on Image Recognition and Narrowband Internet of Things. *Rev. D'Intelligence Artif.* **2019**, *33*, 293–298. [CrossRef]

52. Alasdair Allan. Deep Learning at the Edge on an Arm Cortex-Powered Camera Board. 2018. Available online: https://aallan.medium.com/deep-learning-at-the-edge-on-an-arm-cortex-powered-camera-board-3ca16eb60ef7 (accessed on 27 October 2022).

53. Nyamukuru, M.T.; Odame, K.M. Tiny eats: Eating detection on a microcontroller. In Proceedings of the 2020 IEEE Second Workshop on Machine Learning on Edge in Sensor Systems (SenSys-ML), Sydney, Australia, 21 April 2020; pp. 19–23.

54. Sharad, S.; Sivakumar, P.B.; Narayanan, V.A. The smart bus for a smart city—A real-time implementation. In Proceedings of the IEEE International Conference on Advanced Networks and Telecommunications Systems (ANTS), Bangalore, India, 6–9 November 2016; pp. 1–6.

55. Nayyar, A.; Puri, V. A Review of Beaglebone Smart Board's-A Linux/Android Powered Low Cost Development Platform Based on ARM Technology. In Proceedings of the 9th International Conference on Future Generation Communication and Networking (FGCN), Jeju Island, South Korea, 25–28 November 2015; pp. 55–63.

56. Zhang, Y.; Suda, N.; Lai, L.; Chandra, V. Hello Edge: Keyword Spotting on Microcontrollers. *arXiv* **2017**, arXiv:1711.07128.

57. Wang, G.; Bhat, Z.P.; Jiang, Z.; Chen, Y.W.; Zha, D.; Reyes, A.C.; Niktash, A.; Ulkar, G.; Okman, E.; Hu, X. BED: A Real-Time Object Detection System for Edge Devices. *arXiv* **2022**, arXiv:2202.07503.

58. Wang, C.; Yu, Q.; Gong, L.; Li, X.; Xie, Y.; Zhou, X. DLAU: A Scalable Deep Learning Accelerator Unit on FPGA. *IEEE Trans. Comput.-Aided Des. Integr. Circuits Syst.* **2016**, *36*, 513–517. [CrossRef]

59. RISC-V Foundation. RISC-V: The Free and Open RISC ISA. 2019. Available online: https://riscv.org/ (accessed on 27 October 2022).

60. Gonzalez-Huitron, V.; León-Borges, J.A.; Rodriguez-Mata, A.; Amabilis-Sosa, L.E.; Ramírez-Pereda, B.; Rodriguez, H. Disease detection in tomato leaves via CNN with lightweight architectures implemented in Raspberry Pi 4. *Comput. Electron. Agric.* **2021**, *181*, 105951. [CrossRef]

61. Ledig, C.; Theis, L.; Huszar, F.; Caballero, J.; Cunningham, A.; Acosta, A.; Aitken, A.; Tejani, A.; Totz, J.; Wang, Z.; et al. Photo-Realistic Single Image Super-Resolution Using a Generative Adversarial Network. In Proceedings of the The IEEE Conference on Computer Vision and Pattern Recognition (CVPR), Honolulu, HI, USA, 21–26 July 2017.

62. Jacob, B.; Kligys, S.; Chen, B.; Zhu, M.; Tang, M.; Howard, A.; Adam, H.; Kalenichenko, D. Quantization and Training of Neural Networks for Efficient Integer-Arithmetic-Only Inference. In Proceedings of the The IEEE Conference on Computer Vision and Pattern Recognition (CVPR), Salt Lake City, UT, USA, 18–22 June 2018.

63. Rodríguez-Gómez, J.P.; Tapia, R.; Paneque, J.L.; Grau, P.; Eguíluz, A.G.; Martínez-de Dios, J.R.; Ollero, A. The GRIFFIN perception dataset: Bridging the gap between flapping-wing flight and robotic perception. *IEEE Robot. Autom. Lett.* **2021**, *6*, 1066–1073. [CrossRef]

64. Valladares, S.; Toscano, M.; Tufiño, R.; Morillo, P.; Vallejo-Huanga, D. Performance Evaluation of the Nvidia Jetson Nano Through a Real-Time Machine Learning Application. In *Proceedings of the Intelligent Human Systems Integration 2021*; Russo, D., Ahram, T., Karwowski, W., Di Bucchianico, G., Taiar, R., Eds.; Springer International Publishing: Cham, Switzerland, 2021; pp. 343–349.

65. Chemel, T.; Duncan, J.; Fisher, S.; Jain, R.; Morgan, R.; Nikiforova, K.; Reich, M.; Schaub, S.; Scherlis, T. Tartan Autonomous Underwater Vehicle Design and Implementation of TAUV-22: Kingfisher. 2020. Available online: https://robonation.org/app/uploads/sites/5/2022/06/RS2022_Carnegie_Mellon_University_TartanAUV_TDR.pdf (accessed on 27 October 2022).

66. Long, C. BeagleBone AI Makes a Sneak Preview. 2019. Available online: https://beagleboard.org/blog/2019-05-16-beaglebone-ai-preview (accessed on 27 October 2022).

67. Hochstetler, J.; Padidela, R.; Chen, Q.; Yang, Q.; Fu, S. Embedded Deep Learning for Vehicular Edge Computing. In Proceedings of the IEEE/ACM Symposium on Edge Computing (SEC), Bellevue, WA, USA, 25–27 October 2018; pp. 341–343.

68. Xu, R.; Nikouei, S.Y.; Chen, Y.; Polunchenko, A.; Song, S.; Deng, C.; Faughnan, T. Real-Time Human Objects Tracking for Smart Surveillance at the Edge. In Proceedings of the International Conference on Communications (ICC), Kansas City, MO, USA, 20–24 May 2018; pp. 1–6.

69. Triwiyanto, T.; Caesarendra, W.; Purnomo, M.H.; Sułowicz, M.; Wisana, I.D.G.H.; Titisari, D.; Lamidi, L.; Rismayani, R. Embedded Machine Learning Using a Multi-Thread Algorithm on a Raspberry Pi Platform to Improve Prosthetic Hand Performance. *Micromachines* **2022**, *13*, 191. [CrossRef] [PubMed]

70. Willems, L. Detect People on a Device that Fits in the Palm of Your Hands. Bachelor's Thesis, University of Twente, Enschede, The Netherlands, 2020.

71. Flamand, E.; Rossi, D.; Conti, F.; Loi, I.; Pullini, A.; Rotenberg, F.; Benini, L. GAP-8: A RISC-V SoC for AI at the Edge of the IoT. In Proceedings of the International Conference on Application-Specific Systems, Architectures and Processors (ASAP), Milan, Italy, 10–12 July 2018; pp. 1–4.

72. Howard, A.G.; Zhu, M.; Chen, B.; Kalenichenko, D.; Wang, W.; Weyand, T.; Andreetto, M.; Adam, H. MobileNets: Efficient Convolutional Neural Networks for Mobile Vision Applications. *arXiv* **2017**, arXiv:1704.04861.

73. Kang, D.; Kang, D.; Kang, J.; Yoo, S.; Ha, S. Joint optimization of speed, accuracy, and energy for embedded image recognition systems. In Proceedings of the 2018 Design, Automation Test in Europe Conference Exhibition (DATE), Dresden, Germany, 19–23 March 2018; pp. 715–720. [CrossRef]

74. Cass, S. Taking AI to the edge: Google's TPU now comes in a maker-friendly package. *IEEE Spectr.* **2019**, *56*, 16–17. [CrossRef]

75. Campmany, V.; Silva, S.; Espinosa, A.; Moure, J.; Vázquez, D.; López, A. GPU-based Pedestrian Detection for Autonomous Driving. *Procedia Comput. Sci.* **2016**, *80*, 2377–2381. [CrossRef]

76. Liu, Q.; Huang, S.; Han, T. Fast and Accurate Object Analysis at the Edge for Mobile Augmented Reality: Demo. In Proceedings of the 2nd ACM/IEEE Symposium on Edge Computing, SEC'17, San Jose/Fremont, CA, USA, 12–14 October 2017; pp. 33:1–33:2.

77. Ezra Tsur, E.; Madar, E.; Danan, N. Code Generation of Graph-Based Vision Processing for Multiple CUDA Cores SoC Jetson TX. In Proceedings of the International Symposium on Embedded Multicore/Many-core SoC (MCSoC), Hanoi, Vietnam, 12–14 September 2018; pp. 1–7.

78. Beckman, P.; Sankaran, R.; Catlett, C.; Ferrier, N.; Jacob, R.; Papka, M. Waggle: An open sensor platform for edge computing. In Proceedings of the 2016 IEEE SENSORS, Orlando, FL, USA, 30 October–3 November 2016; pp. 1–3.

79. Morishita, F.; Kato, N.; Okubo, S.; Toi, T.; Hiraki, M.; Otani, S.; Abe, H.; Shinohara, Y.; Kondo, H. A CMOS Image Sensor and an AI Accelerator for Realizing Edge-Computing-Based Surveillance Camera Systems. In Proceedings of the 2021 Symposium on VLSI Circuits, Kyoto, Japan, 13–19 June 2021; pp. 1–2. [CrossRef]

80. Hardkernel. Odroid-M1. 2022. Available online: https://www.hardkernel.com/2022/03/ (accessed on 27 October 2022).¶

81. Liu, S.; Zheng, C.; Lu, K.; Gao, S.; Wang, N.; Wang, B.; Zhang, D.; Zhang, X.; Xu, T. Evsrnet: Efficient video super-resolution with neural architecture search. In Proceedings of the Proceedings of the IEEE/CVF Conference on Computer Vision and Pattern Recognition, Nashville, TN, USA, 19–25 June 2021; pp. 2480–2485.

82. Chinchali, S.; Sharma, A.; Harrison, J.; Elhafsi, A.; Kang, D.; Pergament, E.; Cidon, E.; Katti, S.; Pavone, M. Network Offloading Policies for Cloud Robotics: A Learning-based Approach. *Auton. Robot.* **2021**, *45*, 997–1012. [CrossRef]

83. Pouget, A.; Ramesh, S.; Giang, M.; Chandrapalan, R.; Tanner, T.; Prussing, M.; Timofte, R.; Ignatov, A. Fast and accurate camera scene detection on smartphones. In Proceedings of the IEEE/CVF Conference on Computer Vision and Pattern Recognition, Virtual, 19–25 June 2021; pp. 2569–2580.

84. Dextre, M.; Rosas, O.; Lazo, J.; Gutiérrez, J.C. Gun Detection in Real-Time, using YOLOv5 on Jetson AGX Xavier. In Proceedings of the 2021 XLVII Latin American Computing Conference (CLEI), Cartago, Costa Rica, 25–29 October 2021; pp. 1–7. [CrossRef]

85. Dally, W.J.; Turakhia, Y.; Han, S. Domain-Specific Hardware Accelerators. *Commun. ACM* **2020**, *63*, 48–57. [CrossRef]

86. Apvrille, L.; Bécoulet, A. Prototyping an Embedded Automotive System from its UML/SysML Models. In Proceedings of the Embedded Real Time Software and Systems (ERTS2012), Toulouse, France, 29 January–1 February 2012.

87. Dekeyser, J.L.; Gamatié, A.; Etien, A.; Ben Atitallah, R.; Boulet, P. Using the UML Profile for MARTE to MPSoC Co-Design. Available online: https://www.researchgate.net/profile/Pierre-Boulet/publication/47363143_Using_the_UML_Profile_for_MARTE_to_MPSoC_Co-Design/links/09e415083fb08c939b000000/Using-the-UML-Profile-for-MARTE-to-MPSoC-Co-Design.pdf (accessed on 27 October 2022).

88. Yu, H.; Gamatié, A.; Rutten, É.; Dekeyser, J. Safe design of high-performance embedded systems in an MDE framework. *Innov. Syst. Softw. Eng.* **2008**, *4*, 215–222. [CrossRef]

89. Parashar, A.; Raina, P.; Shao, Y.S.; Chen, Y.H.; Ying, V.A.; Mukkara, A.; Venkatesan, R.; Khailany, B.; Keckler, S.W.; Emer, J. Timeloop: A Systematic Approach to DNN Accelerator Evaluation. In Proceedings of the IEEE International Symposium on Performance Analysis of Systems and Software (ISPASS), Madison, WI, USA, 24–26 March 2019; pp. 304–315. [CrossRef]

90. An, X.; Boumedien, S.; Gamatié, A.; Rutten, E. CLASSY: A Clock Analysis System for Rapid Prototyping of Embedded Applications on MPSoCs. In *Proceedings of the 15th International Workshop on Software and Compilers for Embedded Systems, SCOPES'12*; Association for Computing Machinery: New York, NY, USA, 2012; pp. 3–12. [CrossRef]

91. Caliri, G.V. Introduction to analytical modeling. In Proceedings of the 26th International Computer Measurement Group Conference, Orlando, FL, USA, 10–15 December 2000; pp. 31–36.

92. Corvino, R.; Gamatié, A.; Geilen, M.; Józwiak, L. Design space exploration in application-specific hardware synthesis for multiple communicating nested loops. In Proceedings of the 2012 International Conference on Embedded Computer Systems: Architectures, Modeling, and Simulation, SAMOS XII, Samos, Greece, 16–19 July 2012; pp. 128–135. [CrossRef]

93. Ghenassia, F. *Transaction-Level Modeling with SystemC: TLM Concepts and Applications for Embedded Systems*; Springer: New York, NY, USA, 2006.

94. Mello, A.; Maia, I.; Greiner, A.; Pecheux, F. Parallel simulation of systemC TLM 2.0 compliant MPSoC on SMP workstations. In Proceedings of the Design, Automation Test in Europe Conference Exhibition (DATE'10), Dresden, Germany, 8–12 March 2010; pp. 606–609. [CrossRef]

95. Schirner, G.; Dömer, R. Quantitative Analysis of the Speed/Accuracy Trade-off in Transaction Level Modeling. *ACM Trans. Embed. Comput. Syst.* **2009**, *8*, 1–29. [CrossRef]

96. Binkert, N.; Beckmann, B.; Black, G.; Reinhardt, S.K.; Saidi, A.; Basu, A.; Hestness, J.; Hower, D.R.; Krishna, T.; Sardashti, S.; et al. The Gem5 Simulator. *SIGARCH Comput. Archit. News* **2011**, *39*, 1–7. [CrossRef]

97. Butko, A.; Gamatié, A.; Sassatelli, G.; Torres, L.; Robert, M. Design Exploration for next Generation High-Performance Manycore On-chip Systems: Application to big.LITTLE Architectures. In Proceedings of the ISVLSI: International Symposium on Very Large Scale Integration; Montpellier, France, 8–10 July 2015; pp. 551–556. [CrossRef]

98. Butko, A.; Garibotti, R.; Ost, L.; Lapotre, V.; Gamatié, A.; Sassatelli, G.; Adeniyi-Jones, C. A trace-driven approach for fast and accurate simulation of manycore architectures. In Proceedings of the 20th Asia and South Pacific Design Automation Conference, Chiba, Japan, 19–22 January 2015; pp. 707–712. [CrossRef]

99. Breuer, M.; Friedman, A.; Iosupovicz, A. A Survey of the State of the Art of Design Automation. *Computer* **1981**, *14*, 58–75. [CrossRef]

100. TMI Orion nano Vacq FUll Radio. Available online: https://www.tmi-orion.com/medias/pdf/en/NanoVACQ-PT-FullRadio-EN.pdf (accessed on 27 October 2022).

101. TMI Orion Transceiver. Available online: https://www.tmi-orion.com/medias/pdf/en/Radio-transceiver-en.pdf (accessed on 27 October 2022).

102. Gravio Hub. Available online: https://doc.gravio.com/manuals/gravio4/1/en/topic/gravio-hub (accessed on 27 October 2022).

103. Moneo Appliance. Available online: https://www.ifm.com/us/en/us/moneo-us/moneo-appliance (accessed on 27 October 2022).
104. Advantech WISE-4060. Available online: https://advdownload.advantech.com/productfile/PIS/WISE-4060/file/WISE-4060-B_DS(122121)20221020155553.pdf (accessed on 27 October 2022).
105. Advantech EIS-D150. Available online: https://advdownload.advantech.com/productfile/PIS/EIS-D150/file/EIS-D150_DS(050922)20220509111551.pdf (accessed on 27 October 2022).
106. inHand Networks Edge Gateway. Available online: https://inhandnetworks.com/upload/attachment/202210/19/InHand%20Networks_InGateway902%20Edge%20Gateway_Prdt%20Spec_V4.1.pdf (accessed on 27 October 2022).
107. Adlink MCM Edge DAQ. Available online: https://www.adlinktech.com/Products/Download.ashx?type=MDownload&isDatasheet=yes&file=1938%5cMCM-210_Series_datasheet_20210412.pdf (accessed on 27 October 2022).
108. SmartMesh WirelessHART Network Manager. Available online: https://www.analog.com/media/en/technical-documentation/data-sheets/5903whrf.pdf (accessed on 27 October 2022).
109. SmartMesh WirelessHART 5900. Available online: https://www.analog.com/media/en/technical-documentation/data-sheets/5900whmfa.pdf (accessed on 27 October 2022).

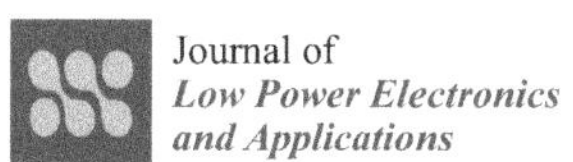

Journal of
Low Power Electronics and Applications

Article

0.6-V 1.65-µW Second-Order G_m-C Bandpass Filter for Multi-Frequency Bioimpedance Analysis Based on a Bootstrapped Bulk-Driven Voltage Buffer

Juan M. Carrillo [1,*] and **Carlos A. de la Cruz-Blas** [2]

[1]   Department of Electrical, Electronic and Automation Engineering, University of Extremadura, Avenida de Elvas s/n, 06006 Badajoz, Spain
[2]   Institute of Smart Cities, IEEC Department, Public University of Navarre, 31006 Pamplona, Spain
*   Correspondence: jmcarcal@unex.es

Abstract: A bootstrapping technique used to increase the intrinsic voltage gain of a bulk-driven MOS transistor is described in this paper. The proposed circuit incorporates a capacitor and a cutoff transistor to be connected to the gate terminal of a bulk-driven MOS device, thus achieving a quasi-floating-gate structure. As a result, the contribution of the gate transconductance is cancelled out and the voltage gain of the device is correspondingly increased. The technique allows for implementing a voltage follower with a voltage gain much closer to unity as compared to the conventional bulk-driven case. This voltage buffer, along with a pseudo-resistor, is used to design a linearized transconductor. The proposed transconductance cell includes an economic continuous tuning mechanism that permits programming the effective transconductance in a range sufficiently wide to counteract the typical variations that process parameters suffer during fabrication. The transconductor has been used to implement a second-order G_m-C bandpass filter with a relatively high selectivity factor, suited for multi-frequency bioimpedance analysis in a very low-voltage environment. All the circuits have been designed in 180 nm CMOS technology to operate with a 0.6-V single-supply voltage. Simulated results show that the proposed technique allows for increasing the linearity and reducing the input-referred noise of the bootstrapped bulk-driven MOS transistor, which results in an improvement of the overall performance of the transconductor. The center frequency of the bandpass filter designed can be programmed in the frequency range from 6.5 kHz to 37.5 kHz with a power consumption ranging between 1.34 µW and 2.19 µW. The circuit presents an in-band integrated noise of 190.5 μV_{rms} and is able to process signals of 110 mV_{pp} with a THD below -40 dB, thus leading to a dynamic range of 47.4 dB.

Keywords: bandpass filter; bootstrapping; bulk-driven; linearized transconductor; quasi-floating gate; voltage follower

Citation: Carrillo, J.M.; de la Cruz-Blas, C.A. 0.6-V 1.65-µW Second-Order G_m-C Bandpass Filter for Multi-Frequency Bioimpedance Analysis Based on a Bootstrapped Bulk-Driven Voltage Buffer. *J. Low Power Electron. Appl.* **2022**, *12*, 62. https://doi.org/10.3390/jlpea12040062

Academic Editor: Orazio Aiello

Received: 31 October 2022
Accepted: 28 November 2022
Published: 30 November 2022

Publisher's Note: MDPI stays neutral with regard to jurisdictional claims in published maps and institutional affiliations.

1. Introduction

The electrical bioimpedance technique allows for characterizing indirectly the properties of a biological media in a noninvasive way [1]. An AC excitation signal is applied to the impedance under test, Z_{BIO}, and the corresponding response is acquired by means of an instrumentation amplifier [2], conditioned and processed. This technique is being widely used nowadays to assist in the diagnosis of different diseases extended among the population as well as for monitoring physiological variables [3,4]. Frequently, the response of the sample is required to be repeated at different frequencies in order to obtain a more complete information, which is known as bioimpedance spectroscopy. The typical frequency range, known as dispersion range, varies from several hundreds of Hz to a few MHz. The frequency analysis can be carried out sequentially, by modifying the frequency of the excitation signal. Nevertheless, when the bioimpedance of the media varies rapidly, a multi-frequency analysis is required in order to obtain all the responses at the same time. In this case, as illustrated in Figure 1, different AC excitation signals are generated and simultaneously applied to the impedance, being subsequently separated with the help of bandpass filter (BPF) sections, being the G_m-C a flexible and suitable approach for monolithic integration [5–15]. The resulting solution is susceptible of being incorporated in an Internet of Things (IoT) platform [16]. Nevertheless, different specifications must be met for this purpose, which can be especially stringent in terms of total power consumption when the overall application is intended to be incorporated into a wearable device.

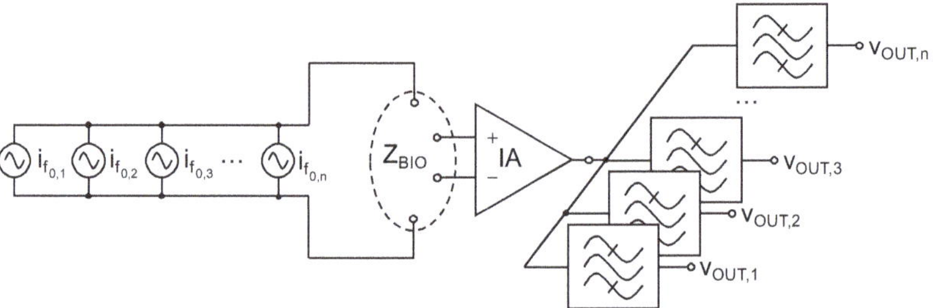

Figure 1. Block diagram of a multi-frequency bioimpedance system.

The bulk-driven technique is well-suited for low-voltage CMOS analog design, as it allows for operation with very low supply voltages and overcomes the non-zero threshold voltage constraint [10,17–25]. Indeed, in a bulk-driven transistor, the DC voltage required to switch the device on and the signal to be processed are decoupled and applied, respectively, to the gate and bulk terminal, which allows for providing and extending the input voltage range with respect to the conventional gate-driven device. Nevertheless, one of the main drawbacks of such technique is the reduction of the effective transconductance, due to the lower value of the bulk transconductance, g_{mb}, as compared to the gate transconductance, g_m. As a consequence, an increase of input-referenced magnitudes, such as the offset voltage or the noise, takes place. Different techniques have been proposed to electronically enhance the effective transconductance of a bulk-driven transistor, consequently increasing area and power consumption [26,27].

In this contribution, the application of a bootstrapping effect to a bulk-driven MOS transistor to increase its intrinsic voltage gain is proposed. The technique has been used to design a low-voltage voltage buffer, in which the noise contribution is reduced and the linearity is increased. The voltage buffer has been incorporated in the implementation of a linearized transconductor, which, in turn, is the basic building block of a second-order G_m-C BPF aimed to signal separation in a multi-frequency bioimpedance measurement system. All the circuits have been designed in 180 nm CMOS technology to operate with a 0.6-V single supply. The rest of the manuscript has been organized as follows: In Section 2, the voltage buffer is described and analyzed, whereas simulated results are used to confirm its principle of operation. The design of the linearized transconductor is detailed in Section 3 and the implementation of the filter is presented in Section 4. Simulated results are provided in Section 5 and conclusions are drawn in Section 6.

2. Boostrapped Bulk-Driven Voltage Follower

2.1. Bulk Driven Buffer: Simulation and Analytical Results

Figure 2a illustrates a conventional bulk-driven flipped voltage follower, where the input voltage is applied to the bulk of transistor MD, a bias voltage V_{BIAS} is applied to its gate, and the output voltage V_{OUT} is obtained at the source. A negative feedback loop is established around transistors MF and MD, which forces the current I_B via the constant voltage V_{BN} to flow through the drain of device MD, and ensures a very low output resistance.

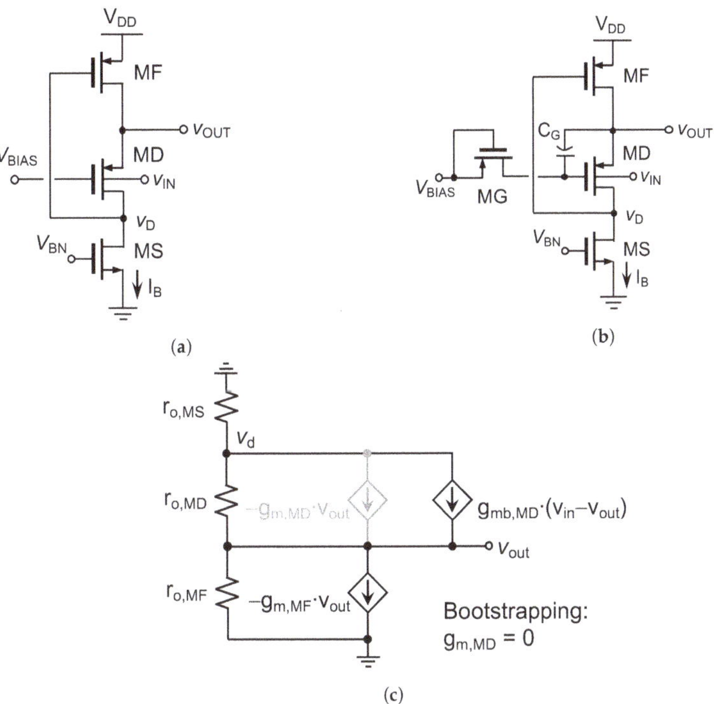

Figure 2. Bulk-driven FVF cell: (**a**) conventional approach; (**b**) proposed bootstrapped version; and (**c**) small–signal circuit ($g_{m,MD} = 0$ for the bootstrapped case).

The proposed circuit is implemented by adding a capacitor C_G between the gate and source terminals of MD and a cutoff transistor MG acting as a pseudo-resistor between V_{BIAS} and the gate of MD, as shown in Figure 2b, in a similar way as in the quasi-floating gate transistor technique [28]. It is worth noting that these elements are the ones usually employed to design a bootstrapping circuit [29,30], but they are used here to cancel out the gate transconductance of transistor MD, i.e., $g_{m,MD} = 0$, thus enhancing the voltage gain of the cell.

Figure 2c depicts the equivalent small-signal circuit of Figure 2a and the main parameters of the cell are summarized in the second column of Table 1, where $g_{m,Mi}$, $g_{mb,Mi}$, and $r_{o,Mi}$ are the gate transconductance, the bulk transconductance, and the output resistance of transistor Mi, respectively. In addition, $R_{D,MD}$ and $R_{S,MD}$ are the equivalent resistances seen from the drain and source terminals of MD, also respectively. The small-signal equivalent circuit of the buffer in Figure 2b is very similar to the one illustrated in Figure 2c, but due to the bootstrapping effect $g_{m,MD} = 0$. As a result, the corresponding small-signal expressions are modified accordingly for the proposed approach, as shown in the third column of Table 1. Note that, for the case of the voltage gain, the proposed circuit avoids the signal attenuation inherent in the bulk-driven technique. In return, the values of R_{out} and $R_{S,MD}$ are incremented due to the cancellation of $g_{m,MD}$. On the other hand, the open loop gain is the same for both circuits, i.e., $g_{mh,MD} \cdot r_{o,MD}$, whereas the loop gain can be expressed as $(g_{m,MD} + g_{mb,MD}) \cdot r_{o,MD}$ and $g_{mb,MD} \cdot r_{o,MD}$ for the conventional and the bootstrapped version, respectively [31].

Table 1. Small-signal parameter comparison of the conventional and bootstrapped buffers.

	Conventional	Bootstrapped
Gain	$\dfrac{g_{mb,MD}}{g_{m,MD}+g_{mb,MD}}$	≈ 1
R_{out}	$\dfrac{1}{g_{m,MF}\cdot(g_{m,MD}+g_{mb,MD})\cdot(r_{o,MD}\|r_{o,MS})}$	$\dfrac{1}{g_{m,MF}\cdot g_{mb,MD}\cdot(r_{o,MD}\|r_{o,MS})}$
$R_{D,MD}$	$\dfrac{1}{g_{m,MF}}$	$\dfrac{1}{g_{m,MF}}$
$R_{S,MD}$	$\dfrac{1}{g_{m,MD}+g_{mb,MD}}$	$\dfrac{1}{g_{mb,MD}}$
Open loop gain	$g_{mb,MD}\cdot r_{o,MD}$	$g_{mb,MD}\cdot r_{o,MD}$
Loop gain	$(g_{m,MD}+g_{mb,MD})\cdot r_{o,MD}$	$g_{mb,MD}\cdot r_{o,MD}$

2.2. Analytical and Simulated Results

In this subsection, analytical expressions and simulation results of the conventional and proposed buffer are provided. The simulations have been obtained using a standard 180 nm CMOS technology with the following aspect ratios for the common transistors $W_{MD}/L_{MD} = 20$ µm/1 µm, $W_{MF}/L_{MF} = 1$ µm/1 µm, $I_B = 100$ nA, set by a simple current mirror with $W_{MS}/L_{MS} = 4$ µm/1 µm. For the bootstrapped implementation, $C_G = 0.25$ pF and transistor MG ($W_{MG}/L_{MG} = 240$ nm/340 nm) is connected as a pseudo-resistor, implemented by a thick oxide device to obtain a larger value of resistance when it is compared to standard transistors. As a consequence, a lower operating cutoff frequency can be achieved. The supply voltage was set equal to 0.6 V; both cells were loaded with an output capacitor of 50 fF, and V_{BIAS} was fixed to 0.1 V.

Gain, area, and power consumption: Figure 3 shows a comparison of the AC small-signal response of the conventional and the bootstrapped buffers. The technique operates properly for frequencies higher than 3 Hz, obtaining a gain of 0.21 V/V (-13.4 dB) and 0.92 V/V (-0.7 dB) for the conventional and the proposed cell, respectively. For obtaining operation at lower frequencies, capacitor C_G should be made larger or the configuration of the pseudo-resistor could be modified to increase its value. In the case of the high cutoff frequency, the value for the proposed cell is lower as compared to the conventional solution, since the output resistance of the proposed cell has been increased. The small overdamping observed in the magnitude response of the proposed circuit at frequencies slightly higher than 1 MHz can be easily cancelled by connecting a very small capacitor at the drain terminal of the driver transistors MD in Figure 2b. In any case, it does not affect the stability of the feedback loop implicit in the buffer. The power consumption is the same in both designs, 60 nW (not including the bias circuits), whereas in terms of silicon area, the proposed cell is twice as large as the conventional technique due to the presence of capacitor C_G. However, larger capacities (in the order of tenths of pF) will be used in the final application, thus making this increase in area not very significant. In addition, it is worth mentioning that, in the used technology, metal–insulator–metal capacitors can be placed on top of the active devices, which allows for reducing the total area occupation of the voltage buffer.

Figure 4 shows the voltage gain of the conventional and the bootstrapped buffers as a function of the input differential-mode (DM) voltage in a range from -200 mV to 200 mV with respect to a common-mode (CM) voltage of 300 mV. Note that the gain of the proposed cell is more than four times higher than that of the conventional cell in the voltage range between -150 mV and 150 mV, and it is much closer to unity. In addition, the proposed cell has a more constant response than the conventional cell, leading to a more linear behavior, as it will be demonstrated next.

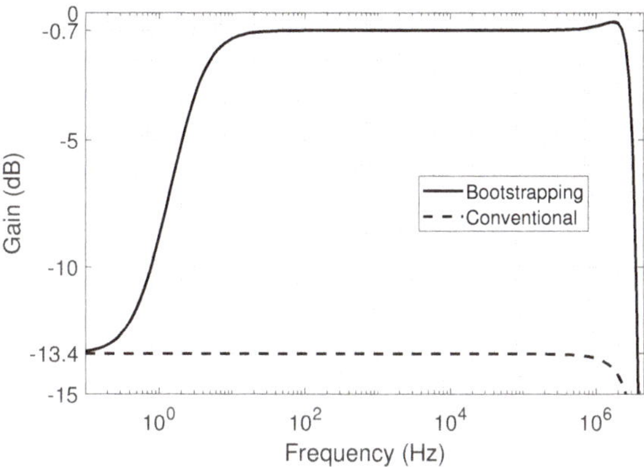

Figure 3. Frequency response comparison of the conventional and bootstrapped buffers.

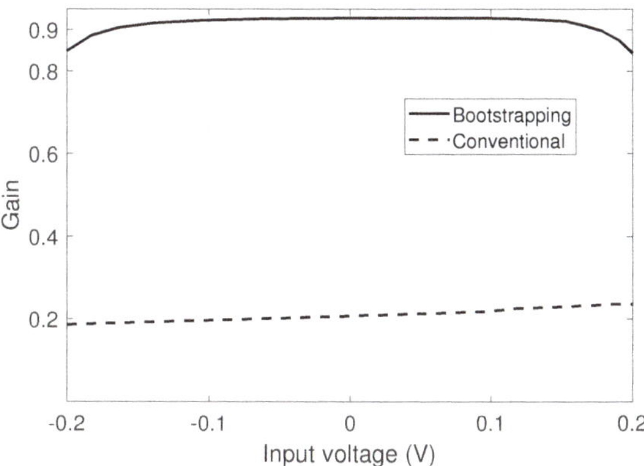

Figure 4. Gain versus input DM voltage of the two voltage buffers.

THD analysis: Considering that the PMOS transistors in Figure 2 operate saturated in the weak inversion region, and neglecting the channel length modulation effect, their drain current can be defined as [22]

$$i_D = I_T \left(\frac{W}{L} \right) exp \left(\frac{V_{SG} + V_{th}}{nV_T} \right) \left[1 - exp \left(\frac{V_{SD}}{V_T} \right) \right] \tag{1}$$

where I_T, V_{th}, n, and V_T are the technology current, the threshold voltage, the subthreshold slope, and the thermal potential, respectively. In a bulk-driven transistor, the signal is implicit in the threshold voltage, which can be expressed as

$$V_{th} = V_{th0} - \gamma_P \left(\sqrt{2\phi + V_{BS}} - \sqrt{2\phi} \right) \tag{2}$$

where V_{th0} is the threshold voltage when $V_{BS} = 0$ and ϕ and γ_P are fabrication process constants. It is worth pointing out that, for a PMOS transistor, the values of V_{th}, V_{th0}, and γ_P are negative. Using these expressions, it is possible to find a closed-form relationship between v_{OUT} and v_{IN} for the circuits in Figure 2. Indeed, the large-signal input/output voltage expression for the conventional bulk-driven FVF cell is the solution of a quadratic function that can be written as follows:

$$v_{OUT} = \frac{-(2A + \gamma_P^2) \pm \sqrt{\gamma_P^4 + \gamma_P^2(4A + 8\phi) + 4\gamma_P^2 v_{IN}}}{2} \tag{3}$$

with $A = -V_{BIAS} + V_{th0} + \gamma_P \sqrt{2\phi} - nV_T \ln\left(\frac{I_T}{I_S(W/L)}\right)$. An evident nonlinear behavior can be observed in the input/output transfer characteristic of the conventional voltage follower. On the other hand, the $v_{OUT} - v_{IN}$ transfer characteristic of the proposed buffer is inherently linear and given by:

$$v_{OUT} = 2\phi - \frac{A^2}{\gamma_P^2} + v_{IN} \tag{4}$$

As inferred from (4), the linearity of the proposed cell is improved since the AC signal at the source terminal of transistor MD is copied to its gate, allowing the input/output voltage relationship to become linear. As a consequence, the THD performance is better for the proposed bootstrapped buffer as compared to the conventional structure.

Figure 5 shows the simulated THD comparison for a sinusoidal input signal of 1 kHz with an amplitude swept from 10 mV to 250 mV. The dominant distortion contribution in both cases is due to the second-order harmonic. Note that the proposed cell has a THD lower than 1% (-40 dB) for input signals up to 180 mV, with a corresponding output voltage of 166 mV, whereas, for the conventional cell, an input signal of only 50 mV, corresponding to an output voltage of 10 mV, is allowed to achieve the same distortion level. This represents an increase of almost 5 and 20 times of the maximum input and output signal levels, respectively, that can be processed.

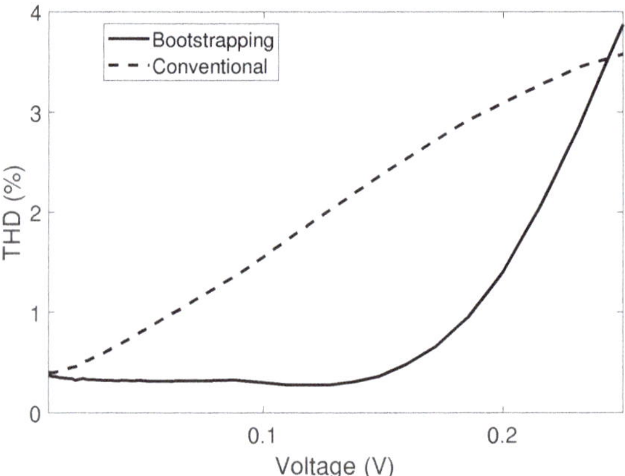

Figure 5. THD comparison.

Noise response: A straightforward analysis of the noise equivalent circuit of the conventional buffer reveals that the power spectral density of the input-referred noise is:

$$\frac{\overline{n_{iC}^2}}{\Delta f} = \frac{\overline{i_{n,MF}^2}}{\Delta f} \frac{1}{g_{m,MF}^2 g_{mb,MD}^2 (r_{o,MD} \parallel r_{o,MS})^2} + \frac{\overline{i_{n,MD}^2}}{\Delta f} \frac{1}{g_{mb,MD}^2} + \frac{\overline{i_{nb,MF}^2}}{\Delta f} \frac{(g_{m,MD} + g_{mb,MD})^2}{g_{m,MF}^2 g_{mb,MD}^2} \tag{5}$$

where the subscripts of the noise current sources are related to the names of the transistors in Figure 2. On the other hand, for the bootstrapped version of the voltage buffer, we have:

$$\frac{\overline{n_{iB}^2}}{\Delta f} = \frac{\overline{i_{n,MF}^2}}{\Delta f} \frac{1}{g_{m,MF}^2 g_{mb,MD}^2 (r_{o,MD} \parallel r_{o,MS})^2} + \frac{\overline{i_{n,MD}^2}}{\Delta f} \frac{1}{g_{mb,MD}^2} + \frac{\overline{i_{nb,MF}^2}}{\Delta f} \frac{1}{g_{m,MF}^2} \tag{6}$$

As it can be seen in (5) and (6), the first two noise contributions are equal because the ratio of R_{out} to gain and $R_{S,MD}$ to gain are the same in both circuits. The difference relies on the last term, related to the ratio of $R_{D,MD}$ to gain, which is different in both implementations. Subtracting both equations and defining $g_{mb,MD} = \eta g_{m,MD}$ and $g_{mb,MD} = \lambda g_{m,MF}$, the extra noise for the conventional buffer is:

$$\frac{\overline{n_{iC}^2}}{\Delta f} - \frac{\overline{n_{iB}^2}}{\Delta f} = \frac{\overline{i_{nb,MF}^2}}{\Delta f} \cdot \frac{\frac{2\lambda^2}{\eta} + \frac{\lambda^2}{\eta^2}}{g_{mb,MD}^2} \tag{7}$$

In Figure 6, it is evidenced by simulations that the noise corresponding to the bootstrapped buffer is lower than in the case of the conventional solution, according also to the prediction in (7).

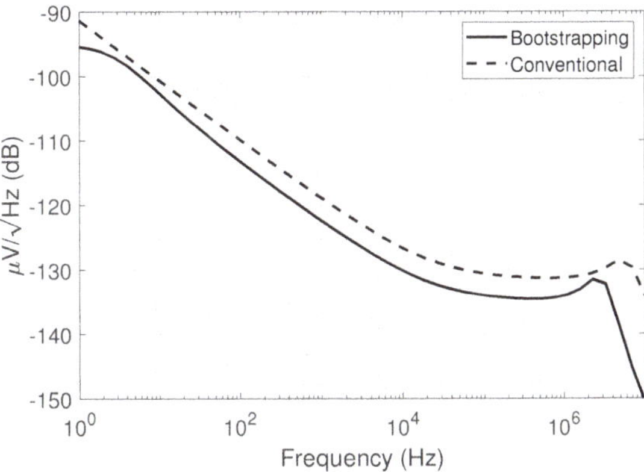

Figure 6. Noise comparison. The input power spectral density is represented in dB on the y-axis to illustrate more clearly the tendencies.

3. Proposed Linearized Transconductor

The circuit schematic of the proposed transconductor, consisting of a linearization resistor and two voltage followers, is illustrated in Figure 7. The input signals, v_{IN}^+ and v_{IN}^-, are applied to the bulk terminal of the driver transistors MD1 and MD2, producing a buffered replica of these voltages, $v_{IN,B}^+$ and $v_{IN,B}^-$, at their source terminal. The bootstrapping action applied to the bulk-driven transistors leads to a gain close to unity for the voltage followers, as detailed in the previous section. The corresponding DM signal, $v_{IN,B}^+ - v_{IN,B}^-$, is applied to a pseudo-resistor, implemented by transistors MR1 and MR2, where voltage-to-current (V-to-I) conversion takes place.

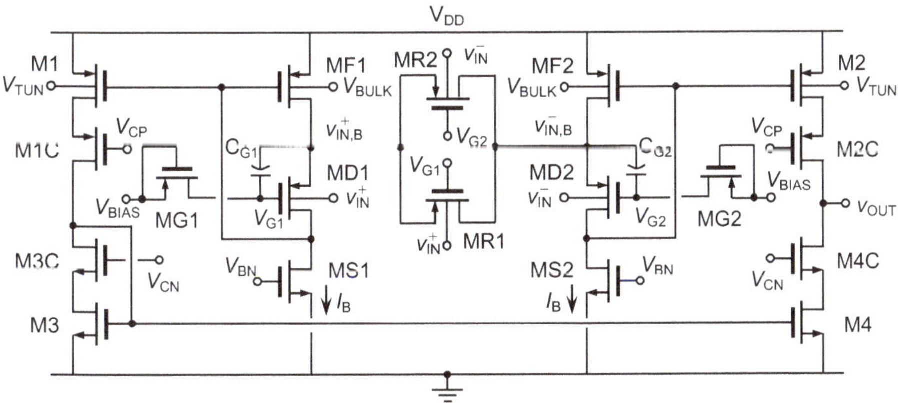

Figure 7. Proposed linearized transconductor.

Assuming that the parallel connection of transistors MR1 and MR2 leads to a resistor with an approximately constant value R_{LIN} for small values of their source-to-drain voltage, the effective transconductance of the *V*-to-*I* converter has been determined by means of a hand analysis, and can be expressed as:

$$G_{m,eff} = \frac{2}{R_{LIN}} \cdot \alpha_{BD} \cdot \frac{1}{1 + \frac{2}{R_{LIN}} \cdot \frac{1}{g_{mb,MD} + g_{m,MD}} \cdot \frac{g_{o,MD} + g_{o,MS}}{g_{m,MF}}} \approx \frac{2}{R_{LIN}} \tag{8}$$

where $g_{mb,Mi}$, $g_{m,Mi}$, and $g_{o,Mi}$ are the bulk transconductance, gate transconductance, and output conductance of transistor M*i*, respectively, and α_{BD} is the intrinsic gain of the bulk-driven follower. In the case of a conventional bulk-driven FVF, $\alpha_{BD} = g_{mb,MD} / (g_{mb,MD} + g_{m,MD})$, causing a noticeable signal attenuation that leads to a transconductance degeneration. The signal attenuation can result adequate in a low-voltage environment, as it reduces the signal swing at the intermediate nodes of the transconductor. Nevertheless, this decrease of the effective input transconductance leads to an increase of input-referred magnitudes, such as the noise or the offset voltage. Alternatively, when the proposed bootstrapped bulk-driven FVF is used, it happens that $\alpha_{BD} \approx 1$ and, hence, there is an enhancement of the transconductance of the cell.

The response of the transconductor is linearized by connecting the bulk terminals of the transistors in the active resistor, MR1 and MR2, to the input terminals of the transconductor, v_I^+ and v_I^-, whereas the gate terminals are connected to the bootstrapping network in order to also benefit from this effect. This solution, first proposed in [32] and adapted to operate with bulk-driven transistors in [22], is modified here to also take advantage of the bootstrapping effect. Indeed, the common connection of the gate, source, and bulk terminals of transistors MD1-MR1 and MD2-MR2 in the core of the transconductor leads to equal V_{SG} and V_{SB} voltages for each pair of devices and, hence, to a linearized response that is also insensitive to variations in the input CM voltage [22]. The general expression of the drain current of a MOS transistor operated in the subthreshold region, given by (1), can be approximated by means of the Taylor series when the transistor operates in triode, i.e., when v_{DS} is very small. In particular, the Taylor series can be truncated at the linear term, thus obtaining

$$i_{D,triode} = \frac{I_T}{V_T} \left(\frac{W}{L} \right) exp \left(\frac{V_{SG} + V_{th}}{nV_T} \right) v_{SD} \tag{9}$$

Similarly, the expression of the threshold voltage can be linearized as [23]

$$V_{th} = V_{th0} - (n - 1)v_{BS} \tag{10}$$

Considering the expressions in (9) and (10), the output conductance of a MOS transistor biased in the subthreshold region and operated in triode can be written as:

$$g_o \equiv \frac{di_D}{dv_{DS}} \approx \frac{I_T}{V_T} \left(\frac{W}{L} \right) exp \left(\frac{V_{SG} + V_{th0} - (n - 1)v_{BS}}{nV_T} \right) \tag{11}$$

As transistors MR1 and MR2 in Figure 7 are connected in parallel, the effective conductance of the composite structure, $g_{LIN} = R_{LIN}^{-1}$, is the sum of the individual conductances of both devices. Assuming that the signal v_{BS} applied at the bulk terminals of devices MR1 and MR2 has a CM DC component, V_{BS}, and a purely DM signal contribution, v_i and $-v_i$, respectively, the value of the linearization resistor can be approximated as:

$$R_{LIN} = \frac{1}{g_{LIN}} = \frac{1}{g_{o,MR1} + g_{o,MR2}} =$$

$$= \left[\frac{I_T}{V_T} \left(\frac{W}{L} \right) exp \left(\frac{V_{SG} + V_{th0} - (n - 1)V_{BS}}{nV_T} \right) \cdot 2 \left(1 + \left(\frac{(n - 1)v_i}{nV_T} \right)^2 + \left(\frac{(n - 1)v_i}{nV_T} \right)^4 + ... \right) \right]^{-1} \tag{12}$$

The odd-power terms of the signal cancel out each other, whereas the even-power terms are summed. Taking into account only the linear term of v_i signal, the expression of the linearization resistor can be further approximated as

$$R_{LIN} = \left[2\frac{I_T}{V_T}\left(\frac{W}{L}\right) exp\left(\frac{V_{SG} + V_{th0} - (n-1)V_{BS}}{nV_T}\right) \right]^{-1}. \tag{13}$$

The circuit section used to bias the transconductor is shown in Figure 8. In particular, voltages V_{BN} and V_{BP} are used to generate the different replicas of the biasing current I_B required in the *V*-to-*I* converter. Furthermore, voltages V_{CN} and V_{CP} allow for biasing NMOS and PMOS cascode devices. An ultra-low-voltage environment connecting the gate of NMOS and PMOS cascode transistors to V_{DD} and ground, respectively, seems to be a straightforward biasing solution leading to a reduction of the total current consumption. Nevertheless, appropriate bias conditions would be only ensured in typical mean conditions and at the nominal value of the supply voltage and the temperature. The use of the simple and well-known structure in Figure 8 allows for tracking PVT variations and translate them to the bias voltage of the cascode transistors. A similar situation arises in the biasing of the gates of the bulk-driven MOS transistors through the bootstrapping network, the reason why the DC signal V_{BIAS} is also generated.

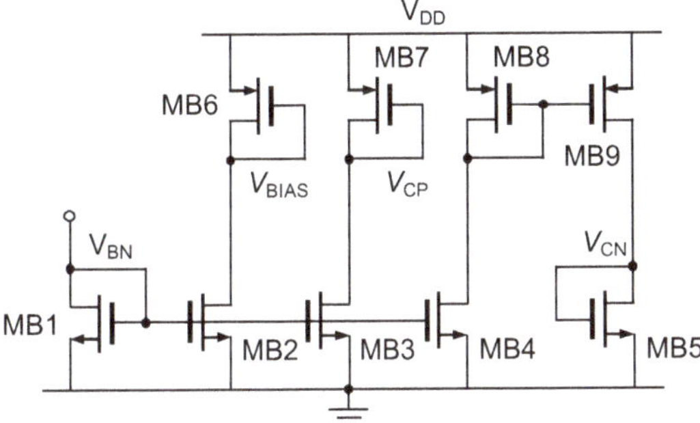

Figure 8. Circuit section used to generate biasing voltages and currents.

Conventionally, the transconductance of the *V*-to-*I* converter illustrated in Figure 7 is tuned by modifying the value of the tail current of the FVF cells. As current I_B changes, the V_{SG} of the driver transistors also does, modifying the effective value of R_{LIN} and, hence, of $G_{m,eff}$. Here, a different tuning mechanism, based on controlling the gain of the PMOS current mirrors formed by transistors MF1-M1 and MF2-M2, is proposed. The bulk terminal of the input transistors of the current mirror, MF1 and MF2, is connected to a fixed DC voltage V_{BULK}, whereas a variable voltage V_{TUN} is applied to the bulk terminal of the output transistors, M1 and M2. When $V_{TUN} > V_{BULK}$, the effective threshold voltage of the output transistors is higher and the current flowing though the output branch is lower, thus having a current attenuation. Conversely, for $V_{TUN} < V_{BULK}$, the effective value of V_{th} of the output transistors of the current mirror becomes lower than that of the input transistors, obtaining a higher output current and, hence, a signal amplification. The voltage V_{TUN} finds its upper bound in the supply voltage V_{DD} and, theoretically, can be decreased until the ground level is reached. Nevertheless, considering that the source and the bulk of these transistors form a *pn* junction, deep forward biasing of this parasitic diode must be avoided. To this end, the exponential behavior of the current flowing through the bulk terminal of a PMOS transistor when the bulk voltage is changed has been considered in order to determine a practical lower bound for the tuning range of voltage V_{TUN}. In particular, in Figure 9, the bulk current of transistors M1 and M2 in Figure 7, I_{BULK}, is represented as a function of the tuning variable V_{TUN}. A current level equal to 1% of the biasing current, i.e., $0.01I_B$, has been selected as a reasonable limit in order to avoid deep forward operation of the source-bulk *pn* junction of transistors M1 and M2. As a result, a value of 200 mV for V_{TUN} is selected as the lower bound of the tuning variable.

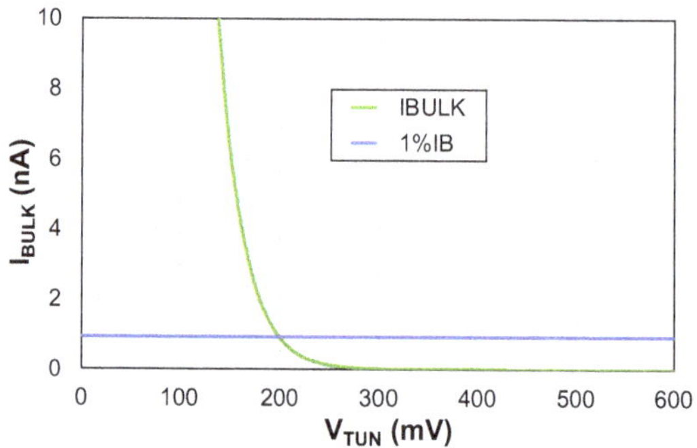

Figure 9. Bulk current over the tuning variable V_{TUN}.

4. Second-Order G_m-C Bandpass Filter

The second-order G_m-C BPF illustrated in Figure 10 has been implemented by using the linearized transconductor described in the previous section and depicted in Figure 7, which is based in turn on the bootstrapped bulk-driven voltage buffer shown in Figure 2b. The filter structure incorporates four transconductors in order to be able to set independently the center frequency, ω_0, the gain at the center frequency, $|H(\omega_0)|$, and the quality factor, Q. In our application, only ω_0 is intended to be swept, whereas $|H(\omega_0)|$ and Q will have fixed values. Nevertheless, the configuration selected allows for keeping constant a given quality factor while the center frequency is swept. In addition, there is an additional degree of freedom in the structure that allows for maximizing the dynamic range of the BPF. Indeed, the other node in the filter, $v_{OUT,LP}$ in Figure 10, provides a lowpass response. The lowpass response presents an overdamping at the frequency of the poles that is a function of the quality factor selected for the BPF. As a consequence, a noticeable peak appears at that node at ω_0, thus limiting the dynamic response of the overall biquad. This fact can be avoided with the structure illustrated in Figure 10, as the value of Q can be set through the ratios of the active (transconductance) or the passive (capacitor) elements, which allows for decreasing the overall gain of the lowpass response, thus decreasing the maximum signal amplitude achieved at $v_{OUT,LP}$ at the center frequency of the BPF.

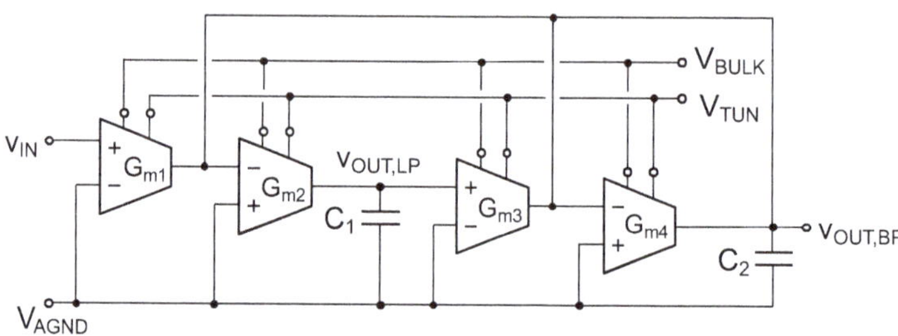

Figure 10. Second-order G_m-C bandpass filter.

The transfer function of the selected BPF can be written as:

$$H(s)_{BP} = \frac{\frac{G_{m1}}{C_2}s}{s^2 + \frac{G_{m4}}{C_2}s + \frac{G_{m2}G_{m3}}{C_1 C_2}} \tag{14}$$

where G_{mi}, with $i = 1$ to 4, represents the effective transconductance of the i-th transconductor and C_1 and C_2 are integrated capacitors. The gain at the center frequency, $|H(\omega_0)|$, the center frequency, ω_0, and the quality factor, Q, can be obtained from (14) in a straightforward manner and expressed as:

$$|H(\omega_0)| = \frac{G_{m1}}{G_{m4}} \tag{15a}$$

$$\omega_0 = \sqrt{\frac{G_{m2}G_{m3}}{C_1 C_2}} \tag{15b}$$

$$Q = \sqrt{\frac{C_2}{C_1} \cdot \frac{G_{m2}G_{m3}}{G_{m4}^2}} \tag{15c}$$

The intended application of the BPF is the separation of signals with different frequencies in a multi-frequency bioimpedance measurement system. Thus, the selectivity of the filter must be relatively high, which requires a moderately high value of the quality factor. A hand-analysis of the response at node $v_{OUT,LP}$ of the filter reveals that an optimal choice in order not to limit the dynamic range of the BPF response is obtained when $C_1 = C_2 = C$. Thus, the following equality has been established for the transconductances $G_{m2} = G_{m3} = k \cdot G_{m4} = k \cdot G_m$ so that the factor Q is equal to parameter k. In addition, transconductors G_{m1} and G_{m4} have been sized to be equal, $G_{m1} = G_{m4} = G_m$, in order to have a gain at the center frequency equal to unity. Therefore, the expressions in (15a–15c) can be rewritten as:

$$|H(\omega_0)| = 1 \tag{16a}$$

$$\omega_0 = k \cdot \frac{G_m}{C} \tag{16b}$$

$$Q = k \tag{16c}$$

The factor k has been achieved by properly sizing the pseudo-resistor in each transconductor, whereas the rest of the V-to-I converter has been kept equal. The response of the BPF, in particular the center frequency, can be programmed by fixing voltage V_{BULK} to an appropriate value and by tuning the value of the control voltage V_{TUN} around it. For $V_{TUN} = V_{DD}$, the transconductors achieve their minimum transconductance value, thus leading to the lowest value of ω_0. Conversely, when V_{TUN} reaches the minimum reliable value, the G_m is maximized and also is the value of the center frequency.

5. Simulated Results

The bootstrapped bulk-driven voltage buffer in Figure 2b, the linearized transconductor in Figure 7, and the second-order G_m-C BPF in Figure 10 have been designed in 180 nm CMOS technology to operate with a single-supply of 0.6 V. The simulated results corresponding to the voltage buffer have already been provided in Section 2 in order to demonstrate its principle of operation and, hence, the metrics corresponding to the other two blocks are described here.

The sizes of the main transistors involved in the implementation of the linearized transconductor are reported in Table 2, whereas the value of capacitors C_{G1} and C_{G2} was set equal to 0.25 pF. The circuit was biased with a current $I_B = 100$ nA and the value of the voltages V_{BULK} and V_{TUN} was nominally set equal to 400 mV. In addition, a load capacitor of 1 pF was connected to the output terminal. The transconductor was first characterized at low frequency, as the bootstrapped structure is not DC coupled. The effective transconductance, $G_{m,eff}$, was simulated and is represented in Figure 11 as a function of the input DM voltage when the value of the tuning variable V_{TUN} is swept from 200 mV to 600 mV. As observed, the transconductance can be programmed in a range of approximately 5×, showing a linearized behavior, even though some dependence on the level of the input signal can also be noticed, as predicted by (12). The open-loop frequency response of the transconductor is illustrated in Figure 12, where the magnitude and the phase of the voltage gain are represented. The low frequency corner due to the bootstrapping network is located at around 2.5 Hz, whereas the voltage gain in the low frequency band is 54.2 dB with a unity gain frequency is equal to 94.2 kHz and a phase margin of 85.6°. The low frequency corner achieved is compatible with the frequency range of interest in the intended application. If, for any reason, a lower cutoff frequency is required, a larger value for the gate capacitor C_G or the pseudo-resistor MG in the bootstrapping network has to be implemented, as already indicated in Section 2. The stability of the transconductor is easily ensured with the value of the load capacitor selected, as the phase margin ranged between 83.5° and 87.6° when V_{TUN} was swept in the range [200 mV, 600 mV]. The transient behavior to a square wave of the G_m cell connected in unity-gain non-inverting configuration allowed for confirming its stability.

Table 2. Aspect ratios (μm/μm) for the main transistors of the transconductor in Figure 7.

Device	W/L	Device	W/L
MD1, MD2	20/1	M1, M2, M3, M4	1/1
MF1, MF2	1/1	M1C, M2C	30/0.5
MS1, MS2	4/1	M3C, M4C	10/0.5
MG1, MG2	0.24/0.34	MR1, MR2	1/0.5

The robustness of the proposed transconductor has been checked by considering in the simulations mismatches as well as process, voltage, and temperature (PVT) variations. In particular, a 1000-run Monte Carlo analysis with process and mismatch variations in a 3-σ range has been carried out. Under these stringent mismatch conditions, the values of the open-loop voltage gain, unity-gain frequency, and phase margin were found to be 45.0 ± 12.0 dB, 131.9 ± 17.9 kHz, and $83.7 \pm 25.2°$. In addition, the closed-loop BW of the transconductor was 110.0 ± 24.1 kHz. In all of these results, the data are represented as the mean value plus/minus the standard deviation. Corner analyses were also run in order to determine the impact of PVT variations on the performance of the transconductor. For the active devices' typical mean (*tt*), fast-fast (*ff*), slow-slow (*ss*) fast-slow (*fs*), and slow-fast (*sf*) conditions were considered, whereas the values of the passive components were varied between the minimum and maximum ranges indicated by the foundry. Additionally, the supply voltage was varied $\pm 10\%$ and the temperature, with nominal value equal to 27 °C, was moved in the range between -20 °C and 80 °C. Considering a total of 45 corners, the open-loop gain, unity-gain frequency, and phase margin varied in the ranges [41.8, 55.6] dB, [84.8, 101.1] kHz, and [84.8, 86.4]°, the closed-loop BW being constrained between 61.4 kHz and 125.4 kHz.

The overall performance of the transconductor is summarized in Table 3, where is it also compared to other similar solutions previously reported. The following figure-of-merit (FoM) has been used for a fair comparison of the transconductors:

$$FoM_T = 100 \cdot \frac{BW \cdot C_L}{P} \tag{17}$$

where BW is the bandwidth of the transconductor connected in non-inverting unity-gain configuration, C_L is the load capacitor, and P the power consumption. As observed in Table 3, the proposed low-voltage linearized transconductor is competitive in terms of the FoM_T, whereas it presents a high open-loop gain at low frequency and provides the largest BW in the comparative.

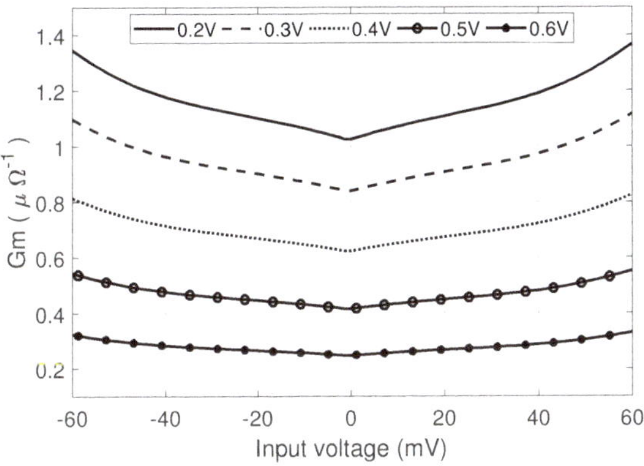

Figure 11. Effective transconductance of the linearized transconductor vs. $v_{I,DM}$.

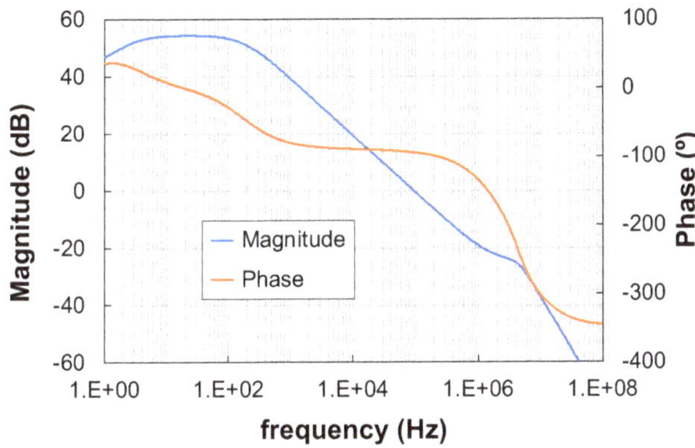

Figure 12. Frequency response of the transconductor (left axis: magnitude, right axis: phase).

Table 3. Simulated performance of the linearized transconductor and comparison with other similar solutions previously reported.

Parameter	[17] ALOG'12	[18] ALOG'14	[22] Access'21	[24] TCAS-II'22	This Work
Technology (µm)	0.35	0.13	0.18	0.13	0.18
Results	Measured	Measured	Simulated	Measured	Simulated
V_{DD} (V)	0.8	0.25	0.5	0.3	0.6
Power (nW)	40	10	0.278–535	708	361.2
G_m (nA/V)	66	22	0.34–383	4070	248.3–1024.9
Open-loop gain (dB)	61	NA	31.2	15	54.2
BW (kHz)	0.195	NA	2.67×10^{-3}	6	99.5
SR^+/SR^- (V/ms)	0.12	94600	NA	NA	3.15/1.56
THD (dB)	−48.2 @ 600 mV$_{pp}$	−45.5 @ 100 mV$_{pp}$	−46.0 @ 480 mV$_{pp}$	−54.4 @ 100 mV$_{pp}$	−52.6 @ 200 mV$_{pp}$
FoM_T (kHz·pF/nW)	12.2	NA	19.2-11.5	84.7	27.5

The BPF was implemented by using four transconductors exactly equal excluding the linearization active resistor. Indeed, blocks G_{m1} and G_{m4} have a nominal transconductance nominally equal to G_m and, thus, the sizes of devices MR1 and MR2 correspond to those indicated in Table 2, that is, $1/0.5$ µm/µm. Nevertheless, as circuit sections G_{m2} and G_{m3} were sized with a transconductance equal to $4G_m$, transistors MR1 and MR2 in these cases were provided with aspect ratios equal to $3.8/0.5$ µm/µm. The biasing current for all the transconductors was set again equal to 100 nA, leading to a total DC power consumption of 2.74 µA. The capacitors in the BPF were implemented as metal–insulator–metal devices, with equal values $C_1 = C_2 = 25$ pF. With these transconductance and capacitor ratios, the quality factor of the BPF was nominally set equal to 4. The reason for selecting relatively high capacitor values is to separate the filter center frequency from the secondary poles of the transconductors, thus avoiding as much as possible any overdamping in the frequency response.

The magnitude response of the BPF over the frequency is depicted in Figure 13 for different values of the tuning variable V_{TUN}. As observed, the filter center frequency ranges between 6.5 kHz and 37.5 kHz, which demonstrates that the tuning mechanism results are suitable to avoid the parameter variations due to the fabrication process with a very economical implementation. When $V_{TUN} = V_{BULK} = 400$ mV, the center frequency is equal to 19.1 kHz. The gain of the BPF at the center frequency, nominally set equal to 0 dB as already indicated in (16a), increases slightly as the value of V_{TUN} is decreased, due to the slight overdamping caused by the approaching of f_0 to the position of the secondary poles in a system with a relatively high quality factor. The noise of the BPF has been integrated in the -3-dB band for the same tuning conditions previously indicated, obtaining a value of 190.5 µV$_{rms}$. Furthermore, the -40-dB THD criterion has been used to determine the maximum input signal amplitude that can be processed with a given linearity, obtaining a maximum amplitude of 55 mV. At this point, it is interesting to mention that the large value of the time constant associated with capacitor C_G and pseudo-resistor MG in the bootstrapping network leads to a transient response in the BPF output signal of around 1 s before the steady-state regime is achieved. Additionally, the compression curve of the BPF output signal and the third-order intermodulation distortion are represented in Figures 14 and 15, respectively. The IMD3 has been obtained by applying two input tones separated ±100 Hz with respect to the BPF center frequency. In addition, from Figure 14, the input-referred 1-dB compression point has been determined to be -19.13 dBm.

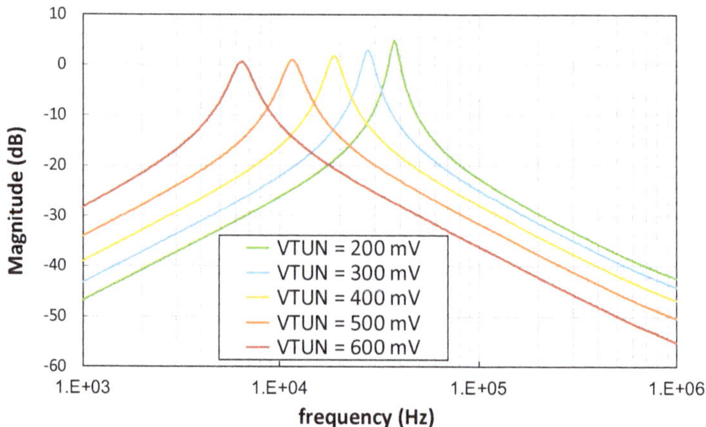

Figure 13. Magnitude response vs. frequency of the BPF for different values of V_{TUN}.

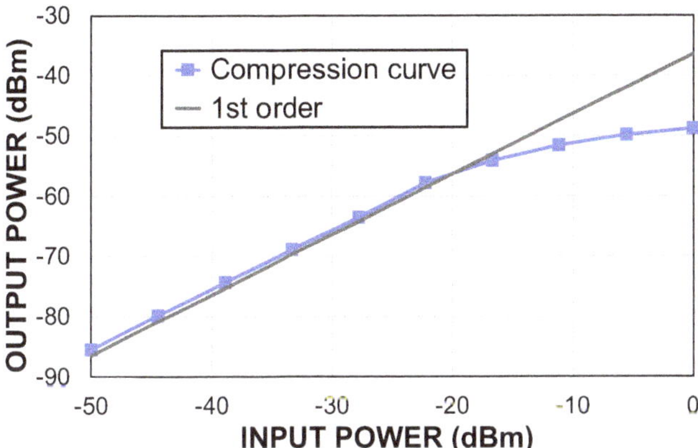

Figure 14. Compression curve of the BPF.

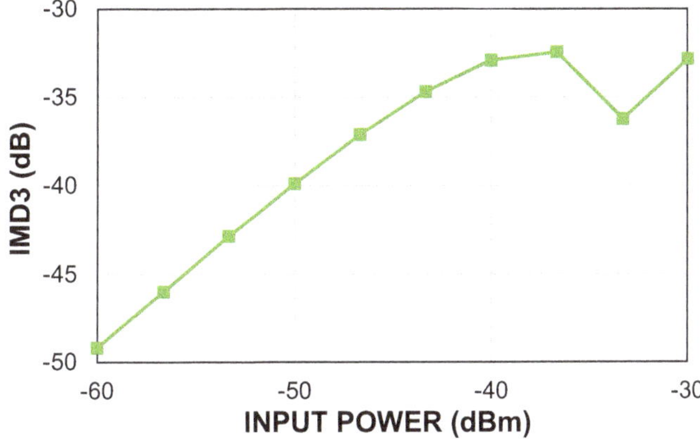

Figure 15. IMD3 vs the input signal.

The impact of mismatches and PVT variations on the response of the proposed BPF has been estimated by means of Monte Carlo and corner analyses in the same conditions as described in the case of the linearized transconductor. Regarding Monte Carlo simulations, the center frequency demonstrated itself to be very stable, with a value of 19.4 ± 1.3 kHz, showing worst-case responses equal to 16.1 kHz and 20.7 kHz in the corners.

The performance of the proposed BPF is reported in Table 4, where it is compared to other similar solutions previously reported. In order to establish an objective comparison between the different BPF structures, the following FoM has been used [7]

$$FoM_{BPF} = \frac{P \cdot V_{DD}}{n \cdot f_0 \cdot DR} \tag{18}$$

where P is the power consumption, V_{DD} the supply voltage, n the filter order, f_0 the center frequency, and DR the dynamic range. It is worth pointing out that the DR has been calculated as the ratio of the input signal leading to a THD of -40 dB and the in-band input-referred integrated noise. As observed, the proposed approach features a reduced power consumption in a low supply voltage, which results in being very suitable for bioimpedance-based IoT applications. In addition, the FoM is competitive as compared to the other solutions, with an acceptable DR taking into account the stringent operating conditions at the used supply voltage.

Table 4. Simulated performance of the proposed G_m-C filter and comparison with similar BPF solutions.

Parameter	[7] TBCAS'07	[9] TCAS-II'12	[10] * MEJ'15	[14] ICECS'20	[15] * ICECS'21	[23]* Access'21	This Work *
Technology (μm)	0.35	0.35	0.05	0.13	0.18	0.18	0.18
V_{DD} (V)	1	3.3	0.4	1.2	0.8	0.5	0.6
Power (μW)	44.3	75.4	31.8	256.0	24.0	0.06	1.65
Filter order	6	2	2	8	2	3	2
f_0 (kHz)	0.67	20	10	100	72.7	0.25	19.1
$f_0^{min} - f_0^{max}$ (Hz)	~100–20 k	20–20 k	1–30 k	2–100 k	72.7 k	250	6.5–37.5 k
Q	N.A	3	1	4.8/5.2	5	N.A.	5.9
$v_{IN,max}$ (mV_{pp})	40	245 ‡	178 †	140 ‡	800	N.A.	110 †
In-band noise (μV)	70.8	58.7	53.0	100	266.6	240.0	190.5
DR (dB)	49.0	63.5	68.4	49.0	60.5	60.4	47.4
$FoM_{BPF} \times 10^{-13}$ (SI)	3.4	979.6	93.1	64.0	21.9	0.377	5.5

* Simulated, † @ −40 dB THD, ‡ @ 1-dB compression point.

6. Conclusions

The bootstrapping effect has been applied to a bulk-driven MOS transistor in order to enhance its voltage gain up to a value close to unity. As a result, a voltage follower with improved noise and linearity responses and able to operate in extremely low voltage conditions can be obtained. This voltage buffer has been used, along with a low-voltage pseudo-resistor, to implement a linearized transconductor, which is the basic building block of a second-order G_m-C BPF aimed at multi-frequency bioimpedance analysis. These circuits have been designed in a 180 nm CMOS process to operate with a supply voltage as low as 0.6 V. The performance of the filter is compatible with the requirements of IoT applications, especially in terms of power consumption, and is comparable to other state-of-the-art solutions previously reported.

Author Contributions: Conceptualization, J.M.C. and C.A.d.l.C.-B.; methodology, J.M.C. and C.A.d.l.C.-B.; software, J.M.C. and C.A.d.l.C.-B.; formal analysis, J.M.C. and C.A.d.l.C.-B.; investigation, J.M.C. and C.A.d.l.C.-B.; resources, J.M.C. and C.A.d.l.C.-B.; data curation, J.M.C. and C.A.d.l.C.-B.; writing—original draft preparation, J.M.C. and C.A.d.l.C.-B.; writing—review and editing, J.M.C. and C.A.d.l.C.-B.; visualization, J.M.C. and C.A.d.l.C.-B.; supervision, J.M.C. and C.A.d.l.C.-B.; project administration, J.M.C. and C.A.d.l.C.-B.; funding acquisition, J.M.C. All authors have read and agreed to the published version of the manuscript.

Funding: Work funded by projects RTI2018-095994-B-I00, from MCIN/AEI/10.13039/501100011033, and IB18079, from *Junta de Extremadura* R&D Plan, and by Fondo Europeo de Desarrollo Regional (FEDER) Una manera de hacer Europa.

Data Availability Statement: Not applicable.

Conflicts of Interest: The authors declare no conflict of interests.

References

1. Grimnes, S.; Martinsen, O.G. *Bioimpedance and Bioelectricity Basics*, 3rd ed.; Academic Press: Cambridge, MA, USA, 2015.
2. Corbacho, I.; Carrillo, J.M.; Ausín, J.L.; Domínguez, M.A.; Pérez-Aloe, R.; Duque-Carrillo, J.F. Wide-bandwidth electronically programmable CMOS instrumentation amplifier for bioimpedance spectroscopy. *IEEE Access* **2022**, *10*, 95604–95612. [CrossRef]
3. Kusche, R.; Klimach, P.; Ryschka, M. A multichannel real-time bioimpedance measurement device for pulse wave analysis. *IEEE Trans. Biomed. Circuits Syst.* **2018**, *12*, 614–622. [CrossRef] [PubMed]
4. Anand, G.; Yu, Y.; Lowe, A.; Kalra, A. Bioimpedance analysis as a tool for hemodynamic monitoring: Overview, methods and challenges. *Physiol. Meas.* **2021**, *42*, 03TR01.

5. Rodriguez-Villegas, E.; Yufera, A.; Rueda, A. A 1.25-V micropower G_m-C Filter Based FGMOS Transistors Oper. Weak Inversion. *IEEE J. Solid-State Circuits* **2004**, *39*, 100–111. [CrossRef]
6. Graham, D.W.; Hasler, P.E.; Chawla, R.; Smith, P.D. A low-power programmable bandpass filter section for higher order filter applications. *IEEE Trans. Circuits Syst. I Regul. Pap.* **2007**, *54*, 1165–1176. [CrossRef]
7. Corbishley, P.; Rodriguez-Villegas, E. A nanopower bandpass filter for detection of an acoustic signal in a wearable breathing detector. *IEEE Trans. Biomed. Circuits Syst.* **2007**, *1*, 163–171. [CrossRef]
8. Lo, T.Y.; Hung, C.C. A wide tuning range G_m-C continuous-time analog filter. *IEEE Trans. Circuits Syst. I Regul. Pap.* **2007**, *54*, 713–722. [CrossRef]
9. Rumberg, B.; Graham, D.W. A low-power and high-precision programmable analog filter bank. *IEEE Trans. Circuits Syst. II Express Briefs* **2012**, *59*, 234–238. [CrossRef]
10. Kulej, T.; Khateb, F. 0.4-V bulk-driven differential-difference amplifier. *Microelectron. J.* **2015**, *46*, 362–369. [CrossRef]
11. Veerendranath, P.; Vasantha, M.; Kumar, Y.N.; Bonizzoni, E. A novel low power G_m-C continuous-time analog filter with wide tuning range. In Proceedings of the 2018 31st International Conference on VLSI Design and 2018 17th International Conference on Embedded Systems (VLSID), Pune, India, 6–10 January 2018; pp. 214–219.
12. Della Sala, R.; Monsurrò, P.; Scotti, G.; Trifiletti, A. Area-Efficient Low-Power Bandpass Gm-C Filter for Epileptic Seizure Detection in 130 nm CMOS. In Proceedings of the 2019 26th IEEE International Conference on Electronics, Circuits and Systems (ICECS), Genoa, Italy, 27–29 November 2019; pp. 298–301.
13. Ballo, A.; Grasso, A.D.; Pennisi, S.; Venezia, C. High-Frequency Low-Current Second-Order Bandpass Active Filter Topology and Its Design in 28-nm FD-SOI CMOS. *J. Low Power Electron. Appl.* **2020**, *10*, 27. [CrossRef]
14. Long, G.B.; Ericson, M.N.; Britton, C.L.; Roehrs, B.D.; Farquhar, E.D.; Frank, S.S.; Yen, A.; Blalock, B.J. A sub-threshold low-power integrated bandpass filter for highly-integrated spectrum analyzers. In Proceedings of the 2020 27th IEEE International Conference on Electronics, Circuits and Systems (ICECS), Glasgow, UK , 23–25 November 2020; pp. 1–4.
15. Corbacho, I.; Carrillo, J.M.; Ausín, J.L.; Domínguez, M.A.; Duque-Carrillo, J.F. 0.8-V CMOS G_m-C bandpass filter for electrical bioimpedance spectroscopy. In Proceedings of the 2021 28th IEEE International Conference on Electronics, Circuits and Systems (ICECS), Dubai, United Arab Emirates, 28 November–1 December 2021; pp. 1–4.
16. Vela, L.M.; Kwon, H.; Rutkove, S.B.; Sanchez, B. Standalone IoT bioimpedance device supporting real-time online data access. *IEEE Internet Things J.* **2019**, *6*, 9545–9554. [CrossRef]
17. Cotrim, E.D.C.; Ferreira, L.H.d.C. An ultra-low-power CMOS symmetrical OTA for low-frequency G_m-C applications. *Analog Integr. Circuits Signal Process.* **2012**, *71*, 275–282. [CrossRef]
18. Colletta, G.D.; Ferreira, L.H.C.; Pimenta, T.C. A 0.25-V 22-nS symmetrical bulk-driven OTA for low-frequency G_m-C applications in 130-nm digital CMOS process. *Analog Integr. Circuits Signal Process.* **2014**, *81*, 377–383. [CrossRef]
19. Khateb, F.; Kulej, T.; Akbari, M.; Steffan, P. 0.3-V Bulk-Driven Nanopower OTA-C Integrator in 0.18 µm CMOS. *Circuits Syst. Signal Process.* **2019**, *38*, 1333–1341. [CrossRef]
20. Ballo, A.; Grasso, A.D.; Pennisi, S. Active load with cross-coupled bulk for high-gain high-CMRR nanometer CMOS differential stages. *Int. J. Circuit Theory Appl.* **2019**, *47*, 1700–1704. [CrossRef]
21. Kulej, T.; Khateb, F.; Kumngern, M. 0.3-V Nanopower Biopotential Low-Pass Filter. *IEEE Access* **2020**, *8*, 119586–119593. [CrossRef]
22. Khateb, F.; Kulej, T.; Akbari, M.; Kumngern, M. 0.5-V high linear and wide tunable OTA for biomedical applications. *IEEE Access* **2021**, *9*, 103784–103794. [CrossRef]
23. Khateb, F.; Prommee, P.; Kulej, T. MIOTA-based filters for noise and motion artifact reductions in biosignal acquisition. *IEEE Access* **2022**, *10*, 14325–14338. [CrossRef]
24. Kulej, T.; Khateb, F.; Arbet, D.; Stopjakova, V. A 0.3-V High Linear Rail-to-Rail Bulk-Driven OTA in 0.13 µm CMOS. *IEEE Trans. Circuits Syst. II Express Briefs* **2022**, *69*, 2046–2050. [CrossRef]
25. Khateb, F.; Kulej, T.; Akbari, M.; Tang, K.T. A 0.5-V Multiple-Input Bulk-Driven OTA in 0.18-µm CMOS. *IEEE Trans. Very Large Scale Integr. VLSI Syst.* **2022**, *30*, 1739–1747. [CrossRef]
26. Zhao, X.; Fang, H.; Ling, T.; Xu, J Transconductance improvement technique for bulk-driven OTA in nanometre CMOS process. *Electron. Lett.* **2015**, *51*, 1758–1759. [CrossRef]
27. Wang, Y.; Zhao, X.; Zhang, Q.; Lv, X. Adjustably transconductance enhanced bulk-driven OTA with the CMOS technologies scaling. *Electron. Lett.* **2018**, *44*, 917–918. [CrossRef]
28. Lopez-Martin, A.; Garde, M.P.; Algueta-Miguel, J.M.; Beloso-Legarra, J.; Carvajal, R.G.; Ramirez-Angulo, J. Energy Efficient Amplifiers Based on Quasi Floating Gate Techniques. *Appl. Sci.* **2021**, *7*, 3271. [CrossRef]
29. Horowitz, P.; Hill, W. *The Art of Electronics*, 3rd ed.; Cambridge University Press: Cambridge, UK, 2015.
30. Cinco-Izquierdo, O.J.; de la Cruz-Blas, C.A.; Sanz-Pascual, M.T. High-linearity tunable low-G_m transconductor based on bootstrapping. *IEEE Trans. Circuits Syst. II Express Briefs* **2022**, *69*, 259–263.
31. Ochoa, A. *Feedback in Analog Circuits*, 1st ed.; Springer: Berlin/Heidelberg, Germany, 2016.
32. Krummenacher, F.; Joehl, N. A 4-MHz CMOS continuous-time filter with on-chip automatic tuning. *IEEE J. Solid-State Circuits* **1988**, *23*, 750–758. [CrossRef]

Journal of
*Low Power Electronics
and Applications*

Article

All-Standard-Cell-Based Analog-to-Digital Architectures Well-Suited for Internet of Things Applications

Ana Correia [1,2], Vítor Grade Tavares [3], Pedro Barquinha [2] and João Goes [1,*]

[1] CTS/UNINOVA, Departamento de Engenharia Electrotécnica e de Computadores (DEEC), NOVA School of Science and Technology (FCT NOVA), Universidade NOVA de Lisboa, 2829-516 Caparica, Portugal
[2] CENIMAT/I3N, Departamento de Ciência dos Materiais (DCM), and CEMOP/UNINOVA, NOVA School of Science and Technology (FCT NOVA), Universidade NOVA de Lisboa, 2829-516 Caparica, Portugal
[3] INESC-TEC and Faculdade de Engenharia da Universidade do Porto (FEUP), Rua Dr. Roberto Frias, 4200-465 Porto, Portugal
* Correspondence: goes@fct.unl.pt

Abstract: In this paper, the most suited analog-to-digital (A/D) converters (ADCs) for Internet of Things (IoT) applications are compared in terms of complexity, dynamic performance, and energy efficiency. Among them, an innovative hybrid topology, a digital–delta (Δ) modulator (ΔM) ADC employing noise shaping (NS), is proposed. To implement the active building blocks, several standard-cell-based synthesizable comparators and amplifiers are examined and compared in terms of their key performance parameters. The simulation results of a fully synthesizable Digital-ΔM with NS using passive and standard-cell-based circuitry show a peak of 72.5 dB in the signal-to-noise and distortion ratio (SNDR) for a 113 kHz input signal and 1 MHz bandwidth (BW). The estimated FoM$_{\text{Walden}}$ is close to 16.2 fJ/conv.-step.

Keywords: analog-to-digital converters; high resolution; digital–delta modulator ADC; noise shaping; all-standard-cell-based; Internet of Things

Citation: Correia, A.; Tavares, V.G.; Barquinha, P.; Goes, J. All-Standard-Cell-Based Analog-to-Digital Architectures Well-Suited for Internet of Things Applications. *J. Low Power Electron. Appl.* **2022**, *12*, 64. https://doi.org/10.3390/jlpea12040064

Academic Editor: Andrea Acquaviva

Received: 24 October 2022
Accepted: 3 December 2022
Published: 7 December 2022

Publisher's Note: MDPI stays neutral with regard to jurisdictional claims in published maps and institutional affiliations.

1. Introduction

The Internet of things (IoT) is heavily driven by significant semiconductor and nanotechnology breakthroughs. Low-cost, reliable, and highly integrated circuits and systems have been designed, allowing for the introduction of important features such as remote access control and the operation of large amounts of data [1].

High-resolution analog-to-digital (A/D) converters (ADCs) are relevant building blocks in different IoT systems. Applications such as high-precision sensor networks, communications, imaging, and signal processing require outstanding ADC performance, including high-accuracy, low-power consumption, and, in some cases, wide-bandwidth (BW) specifications [2].

To accomplish a high resolution, delta-sigma ($\Delta\Sigma$) modulators ($\Delta\Sigma$M) and successive approximation register (SAR) ADCs (SAR-ADCs) are frequently utilized. While in $\Delta\Sigma$M, larger sampling frequencies (F_S) are used to achieve higher resolutions, in conventional SAR-ADCs, energy efficiency is often sacrificed to reach the target resolution. Furthermore, old-fashioned architectures and techniques were revisited, and hybrid structures are currently a reality, mixing these schemes with popular structures. Employing noise shaping (NS) in an SAR-ADC and using a delta modulation in a $\Delta\Sigma$M, resulting in a delta-delta-sigma ($\Delta\Delta\Sigma$) modulator ($\Delta\Delta\Sigma$M), are examples of this new era of hybrid ADC architectures pursuing the most outstanding and efficient ADC [3–5].

Taking into consideration that the ADC design is a complex and time-consuming task, it is desirable to reduce this effort, especially when porting between different nodes or technologies is required. Moreover, the lower nodes' technology constraints (low intrinsic gain, reduced supply voltage, leakage current, etc.) bring other challenges in porting tasks, sometimes requiring a complete redesign or the implementation of different circuit schemes.

Circuits based on digital logic can be quickly realized and modified to accomplish specifications and technological changes. Therefore, the use of digital circuits is becoming popular in analog or mixed-signal circuit design, such as in the case of ADCs. Consequently, synthesizable solutions using standard cells are used to further reduce redesign time and effort [6–8].

In this work, the most-suited ADC architectures for IoT applications are described. A hybrid ADC solution, a digital–delta (Δ) modulator (ΔM) with NS, is proposed that can be implemented using only passive and digital circuitry based on standard cells. Circuit details and some simulation results are also provided.

This paper is organized as follows. Section 2 presents the most popular ADC architectures for IoT applications, where complexity, dynamic performance, and energy efficiency are compared. In Section 3, some standard-cell-based active building blocks, comparators, and integrators, are presented with reference to their advantages and the main challenges during their integration in complex systems. Section 4 provides schematic details regarding the proposed standard-cell-based digital-ΔM employing NS and some simulation results. Lastly, the main conclusions are drawn in Section 5.

2. Most-Suited Analog-to-Digital Converter (ADC) Architectures for Internet of Things (IoT)

The most suited ADC architectures for IoT applications are described in this section. In spite of being an old-fashioned topology and not directly implemented in IoT applications, ΔM is mentioned because it is the basis of $\Delta\Sigma$M and has inspired some other hybrid architectures, such as SAR-ADC with NS, $\Delta\Delta\Sigma$M or the proposed digital-ΔM with NS ADCs.

Lastly, a qualitative comparison between them is provided considering complexity, dynamic performance, and energy efficiency.

2.1. Delta Modulator (ΔM) ADC

In a patent from 1946 submitted by Deloraine et al., delta modulation was referred to for the first time as a method to transmit analog data by means of a one-bit code [9].

As shown in Figure 1a, the basic ΔM transforms an analog input signal, V_{in}, into a synchronous digital output, D_{out}. It employs a 1-bit quantizer, a digital-to-analog (D/A) converter (DAC), and an integrator in the feedback path as an attempt to anticipate the input signal. Thus, this integrator acts as a predictor [10].

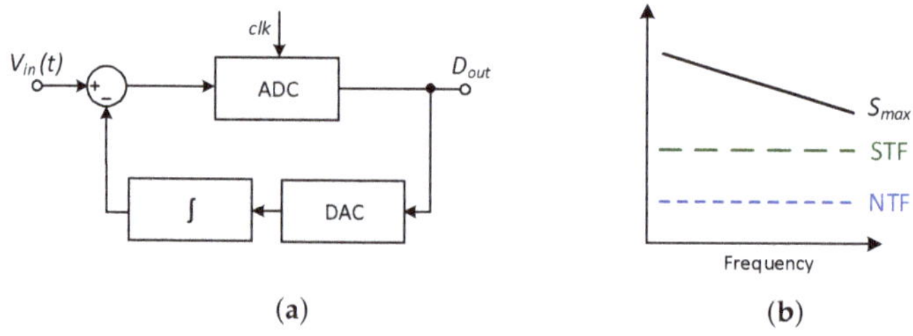

Figure 1. ΔM ADC: (**a**) block diagram and (**b**) illustrative magnitude of STF, NTF and S_{max}.

Noise can negatively impact ΔM performance in two different ways: through granular noise or slope overload. While the former results from the quantization of a continuous signal (the signal is forced to assume a discrete value), the latter is dominant when the step size of the integrator is too small, resulting in the incorrect tracking of the input signal [10,11].

Despite the good robustness to transmission errors, simple filtering requirements, and low associated complexity, the nonidealities associated with the integrator in the feedback path can limit linearity, noise performance, and system accuracy. Furthermore, the amplitude of V_{in} and ADC performance are inversely proportional to the input signal frequency, F_{in}. Therefore, as Figure 1b illustrates, while the signal transfer function (STF) and the noise transfer function (NTF) are constant, the maximal signal, S_{max}, decreases with the frequency [5].

Both noise and nonidealities can be problematic and severely restrict the maximal dynamic performance of the converter [4].

2.2. SAR-ADC with Noise Shaping (NS)

SAR-ADCs are currently one of the most popular topologies to realize A/D conversion due to their energy efficiency, low die area, and low circuit complexity [12]. However, higher resolutions are difficult to achieve without sacrificing energy efficiency.

In conventional topologies, the circuit relies essentially on a 1-bit comparator, an N-bit DAC, and a sample-and-hold (S/H) block. A binary-search algorithm is used to reduce the analog residue to less than one least significant bit (LSB) [13].

Like in other architectures, the most critical building block is the DAC because its nonidealities, the associated noise, and its settling time dominated by the reference settling directly affect the ADC performance. This aspect is even more crucial for high-resolution converters.

In the last decade, the introduction of oversampling and NS in the conventional SAR-ADC allowed for better higher dynamic performance beyond 14 bits of resolution (i.e., 12.5 bits of effective number of bits (ENOB)) [3]. The main idea is to use the analog residue that still remains after the SAR operation, the residue voltage (V_{res}), and integrate it to perform a NS, spreading the noise through a higher BW than the band of interest. The block diagram of a SAR-ADC employing NS is shown in Figure 2a, in which the ADC can simply be a single comparator. In Figure 2b, the STF, NTF, and S_{max} magnitudes as a function of frequency are illustrated, with the NTF slope characteristic from systems employing NS being notable.

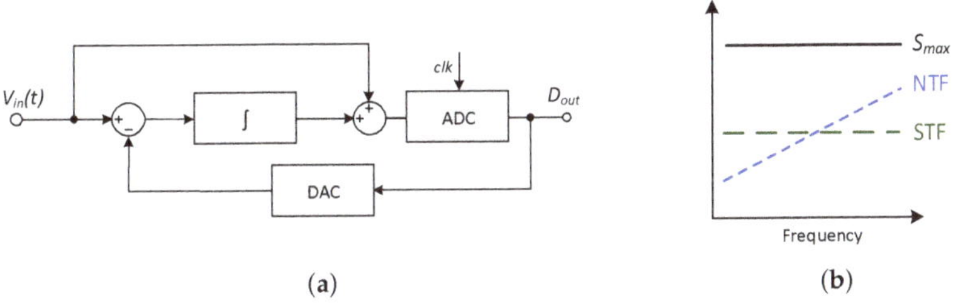

Figure 2. SAR-ADC employing NS: (**a**) block diagram and (**b**) illustrative magnitude of STF, NTF and S_{max}.

Given the absence of amplifiers in the pure topology (besides the comparator), in these hybrid structures, the same strategy has been pursued, maintaining the circuit simplicity, and relaxing the specifications of the comparator and DAC [14]. Thus, different works have been proposed using passive NS structures [15–17].

2.3. First-Order Delta–Sigma (ΔΣ) Modulator (ΔΣM) ADC

The ΔΣM topology emerged to avoid the shortcomings of the ΔM by moving the integrator from the feedback to the forward path. As illustrated in Figure 3a, in which the local quantizer (ADC) can again simply be a single comparator, the integrator operates over the error difference instead of the signal estimation, as in the ΔM case.

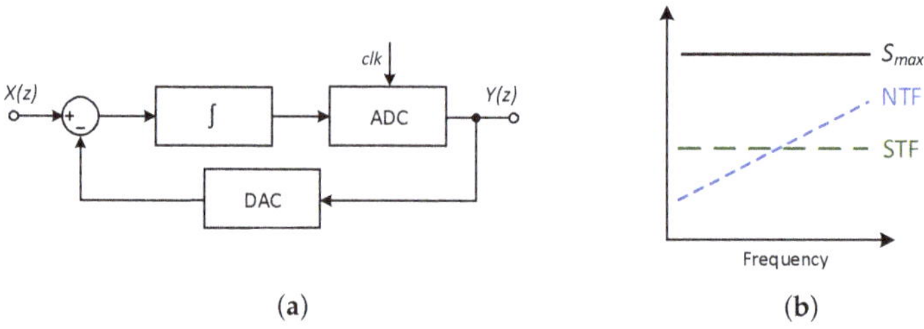

(a)   (b)

Figure 3. First-order ΔΣM ADC: (**a**) block diagram and (**b**) illustrative magnitude of STF, NTF and S_{max}.

This architecture relies essentially on an analog filter (integrator), a quantizer, and a DAC [4,11]. Using the linear additive white-noise model for the quantizer, this system can be represented in the Z domain, resulting in the following STF and NTF:

$$STF(z) = \frac{H(z)}{1 + H(z)} \tag{1}$$

$$NTF(z) = \frac{1}{1 + H(z)} \tag{2}$$

where $H(z)$ is the integrator transfer function. The NS effect is more effective for higher $H(z)$ filter orders, promoting higher-resolution converters; however, extra complexity is added to the system. These functions are represented in Figure 3b.

2.4. Delta–Delta–Sigma (ΔΔΣ) Modulator (ΔΔΣM) ADC

ΔΔΣM is another example of a hybrid architecture. Combining delta modulation with a 1st-order ΔΣM, ΔΔΣM was proposed in [5]. As depicted in Figure 4a, two integrators are used in this topology, one in the feedforward and another in the feedback paths.

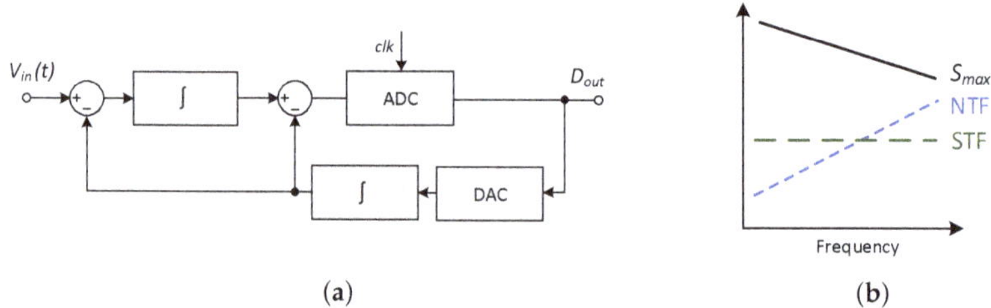

(a)   (b)

Figure 4. ΔΔΣM ADC [5]: (**a**) block diagram and (**b**) illustrative magnitude of STF, NTF and S_{max}.

Since two integrators are involved, the complexity of the architecture is higher. Furthermore, the nonidealities of the DAC (placed in the feedback path) impact the ADC linearity. Thus, a high dynamic resolution is difficult to achieve, especially with low energy efficiency. Despite the inherent complexity, small modifications can be performed to the architecture, such as changing the relative position of the DAC and integrator in the feedback path. The integrator becomes a digital accumulator, reducing the complexity and rendering the architecture more suitable for IoT.

As represented in Figure 4b, while the NS imposes an inclination to the NTF curve, shaping the noise for higher frequencies, delta modulation impacts the S_{max} (similarly to the ΔM topology).

2.5. Proposed Hybrid ADC: Digital–Delta (Δ) Modulator (ΔM) with Noise Shaping (NS)

A digital-ΔM employing NS was initially proposed by the authors of this paper in [18]. It utilizes oversampling and NS to improve the overall performance by minimizing the impact of thermal and quantization noise. As depicted in Figure 5, it comprises a 1-bit comparator, an accumulator, an N-bit DAC, an S/H circuit, and an integrator in the NS section.

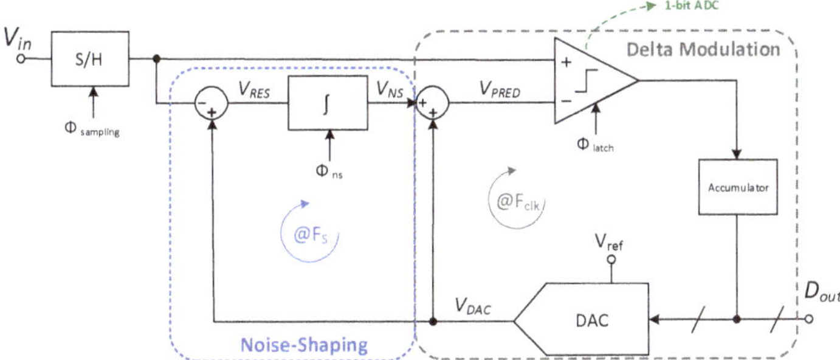

Figure 5. Diagram of the proposed digital-ΔM ADC employing NS.

This architecture was initially inspired by an SAR-ADC. However, instead of the typical SAR logic, this topology uses an accumulator in the digital domain. Therefore, the search algorithm is based on the prediction of the next V_{in} working as a ΔM.

Comparing the block diagram of the proposed ADC architecture, depicted in Figure 6a, with other topologies shows that the comparator also connects to the sampled V_{in} to perform a direct comparison with the estimation. Additionally, this architecture employs an accumulator placed between the comparator and the DAC, which is a relevant advantage to achieve a fully synthesizable ADC. However, since this topology utilizes delta modulation, S_{max} is dependent on the frequency (Figure 6b).

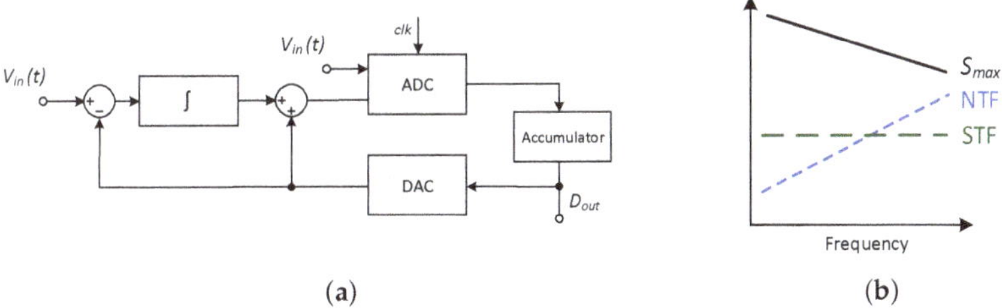

(**a**)   (**b**)

Figure 6. Proposed digital-ΔM employing NS: (**a**) block diagram and (**b**) illustrative magnitude of STF, NTF and S_{max}.

2.6. Comparison among the Most-Suited Architectures

All the described architectures are qualitatively compared in Table 1.

In all the described topologies, the most critical building blocks are the comparator, integrator, and the DAC, since their nonidealities impact ADC performance. However, depending on the ADC architecture and the circuit location, their effect can be distinct.

Generally, SAR-ADC and digital-ΔM, both employing NS, present higher complexity when compared with ΔΣM or ΔΔΣM ADCs because the specifications of the main building blocks (metastability, comparator's accuracy and comparison times, noise, etc.) have strong repercussions on ADC performance. However, they present very good energy efficiency, increasing their attractiveness. ΔΣM or ΔΔΣM ADCs are also popular for high-resolution applications. However, the integrator design can, in some cases, be problematic for circuit stability and efficiency.

The magnitude of S_{max} can show different behaviors depending on the architecture; therefore, despite the conclusions depicted in Table 1, this aspect should be taken into account to ensure that it does not represent a strong limitation for the specific IoT application.

Table 1. Comparison of the most-suited ADC architectures for IoT applications.

ADC Architecture	Complexity	Resolution	Energy Efficiency
ΔM	Low	Moderate	Low
SAR-ADC with NS	Moderate	Moderate/high	Very good
First-order $\Delta\Sigma$M	Low	Moderate/high	Good
$\Delta\Delta\Sigma$M	Moderate	Moderate/high	Good
Digital-ΔM with NS	Moderate	Moderate/high	Very good

3. Standard-Cell-Based Active Building Blocks

The implementation of the different architectures presented earlier demands different specific circuits to implement the distinct functional blocks that each topology requires. Typically, integrator synthesis encompasses the design of OTAs, and quantizers involve comparator design. Among others, these are fundamental building blocks of converters.

Over the years, different standard-cell-based circuits, recurring to automated digital design flows and standard cells, have been proposed to implement these well-known analog functions, enabling faster design, and synthesis and layout automation based on standard cells.

Despite the importance of the DAC in all architectures, its design has preferably been passive, facilitating the converter porting between different nodes or technologies. Furthermore, the passive characteristics allow for good energy efficiency, which is extremely relevant for IoT applications. For these reasons, this building block is not described here.

3.1. Dynamic Comparators Using Standard Logic Circuitry

A fully synthesizable dynamic voltage comparator was proposed by Weaver et al. in [19]. As depicted in Figure 7, the circuit relies on a two cross-coupled 3-input digital NAND gates and, when two NANDs are connected, assuming that the common-mode voltage of the input signal is high enough to cut off the input *PMOS* devices, an analog-input comparator is created. When the clock is low, the outputs are reset to the positive supply rail, V_{DD}, and when the clock goes high, the outputs start to discharge through the *NMOS* devices. Since the discharging rate is proportional to the input, once one of the outputs achieves a value below than the threshold voltage, the cross-coupled connection forces the outputs to assume the supply rail values. A static SR latch is also used to hold the output decision and it is buffered by an inverter to reduce the memory effect.

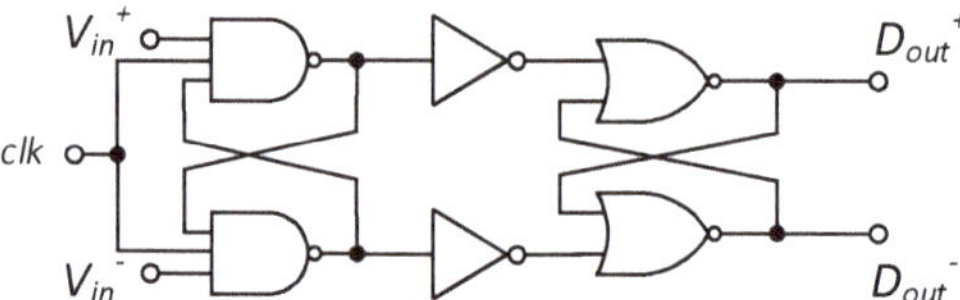

Figure 7. GATE-based comparator proposed by Weaver et al. [19].

In spite of being suitable for an all-digital implementation, this circuit is sensitive to the input common-mode range. Consequently, the usage of this comparator is restricted to stochastic ADCs [20].

Replacing NAND gates with NOR gates, as shown in Figure 8, the comparator only operates correctly if the input common-mode voltage is close to the ground. Thus, merging the 3-input NAND with 3-input NOR solutions, a rail-to-rail dynamic voltage comparator was proposed in [20]. In this case, NAND gates operate correctly for the portion of the common-mode towards V_{DD}, while NOR gates work properly for the portion towards the ground.

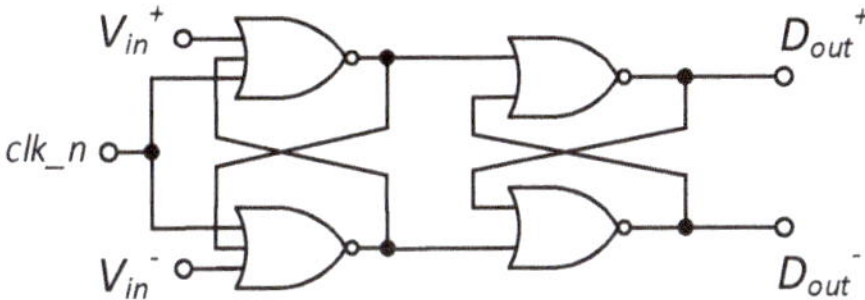

Figure 8. NOR-based comparator that was merged with the NAND-based circuit (shown in Figure 7), producing the proposed rail-to-rail dynamic voltage comparator by Aiello et al. described in [20].

Ojima et al. proposed an NAND-based 4-input clocked comparator to achieve a fully synthesizable SAR ADC [21]. As Figure 9 shows, the four 3-input NAND gates define the preamplifier and the first latch stage (the output of one pair of preamplifiers is fed back to the input of the other pair), while the following 2-input NAND gates form the second latch stage, enhancing the comparator gain and reducing the comparison time. In this scheme, the comparison is carried out on the basis of $(V_{IN}{}^+ + V_{DAC}{}^+)$ and $(V_{IN}{}^- + V_{DAC}{}^-)$.

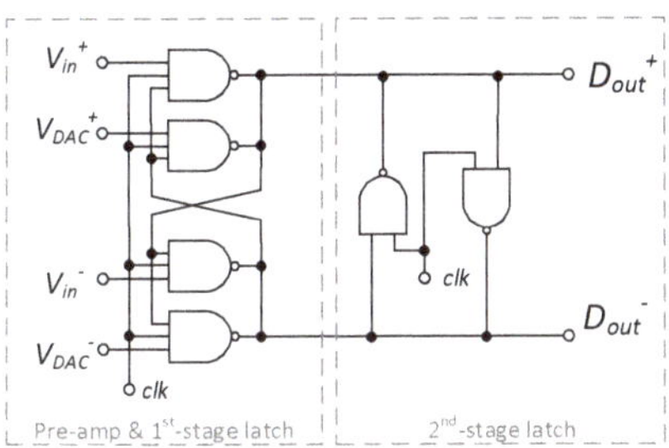

Figure 9. NAND-based four-input clocked comparator proposed by Ojima et al. [21].

In the previous scheme, when the V_{in} was low and the *clk* was disabled, the reset path of the *NMOS* of the preamplifier was cut off. Consequently, a residue voltage remained at the drain node that could be amplified during the next comparison, generating an error output. To resolve that, it was proposed to replace the NAND gates with OR–AND inverter (OAI) cells [6]. Thus, an explicit reset is performed on the drain nodes, eliminating the residue voltages and thereby reducing the probability of a wrong output.

On the basis of the described 4-input solution [6], a 2-input comparator based on the same OAI cells was designed (Figure 10). In addition to the obvious reduction in complexity and power dissipation, because fewer transistors are used, this topology presents satisfactory characteristics (comparison time, noise, and output error probability) for simple ADC topologies such as SAR-ADCs and digital-ΔM, both with NS.

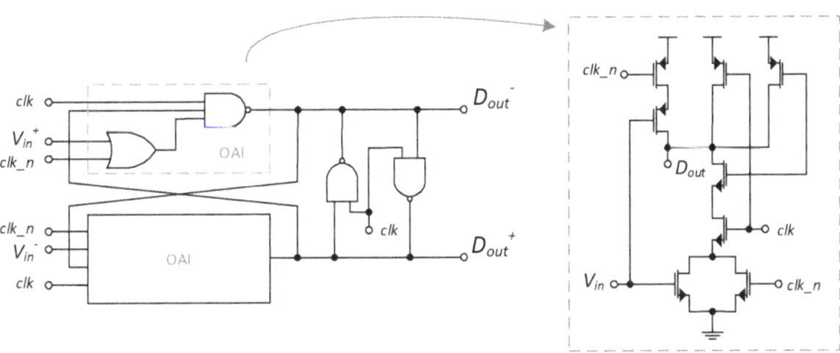

Figure 10. OAI-based comparator with 2 inputs.

Recently, different works have been proposed with the goal of achieving rail-to-rail dynamic voltage comparators with good energy efficiency [22,23].

3.2. Inverter-Based OTA Topologies

Amplifiers are also difficult to design and to port between technologies. Thus, standard-cell-based synthesizable solutions have been drawing attention in recent years. Inverter-based switched-capacitor (SC) circuits are one possibility that has been deeply studied due to the inherent simplicity and capability to operate with low V_{DD}, in contrast with other operational transconductance amplifiers (OTAs).

A simple inverter allows for a push–pull operation, a large output swing (OS), and good energy efficiency. Furthermore, both devices contribute to global transconductance [24]. However, taking into consideration that the inverter does not have an explicit reference virtual ground, different cancellation techniques have been investigated to compensate the offset voltage, V_{off}, and reduce its impact [25].

The technique proposed by Nagaraj et al. in [26] is one of the most used approaches. Besides the offset impact reduction, it requires a lower gain specification, facilitating its design. In this scheme, depicted in Figure 11, while capacitors C_S and C_F perform the integration, the C_{NAG} is used to compensate for the finite gain error and V_{off} [25–27].

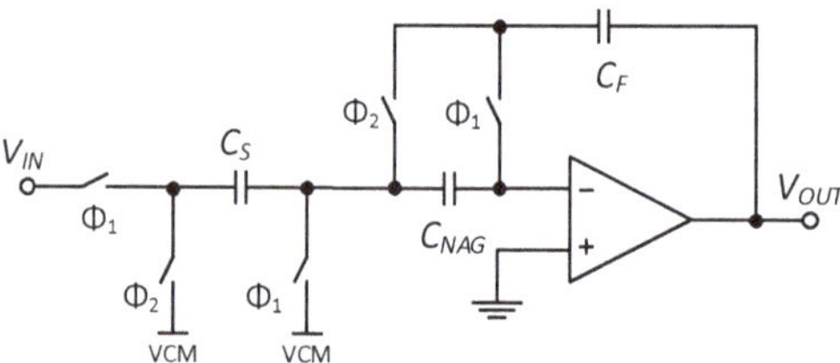

Figure 11. Scheme of the SC integrator proposed by Nagaraj et al. [26].

There are some relevant specifications with which OTAs need to comply in order to render them good candidates for employment in ADCs, namely, the gain and its linearity over V_{out}, since it affects ADC linearity, low complexity, and good energy efficiency. These are fundamental requirements to be observed, especially for IoT and fully synthesizable applications.

Thus, the key performance parameters of three different OTA topologies were evaluated in [28]. The circuits, *OTA 1*, *OTA 2* and *OTA 3*, can be described as follows:

- *OTA 1*: a single pseudodifferential Nagaraj integrator with a fully passive SC common-mode feedback (CMFB) circuit, as illustrated in Figure 12 [24].
- *OTA 2*: a pseudodifferential with a three-stage multipath inverter-based amplifier using a RC network as CMFB, as shown in Figure 13 [29].
- *OTA 3*: a single-path three-stage pseudodifferential Nagaraj integrator using a fully passive SC CMFB, as shown in Figure 14.

As summarized in Table 2, there are significant differences between these three OTA circuits. Since a cascade of inverters was used in *OTA 2* and *OTA 3*, a higher DC gain was achieved. However, this also increased the complexity. Furthermore, the RC network utilized in *OTA 2* as CMFB increased the current consumption. Regarding the linearity over V_{out}, significant differences were also noticed with *OTA 2* and *OTA 3* being the best options when the linearity of the OTA is extremely important in the system, such as in the case of ADCs.

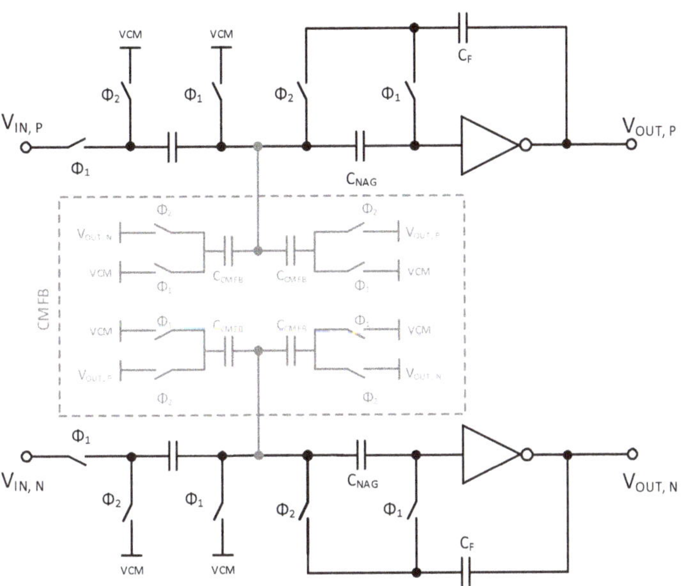

Figure 12. *OTA 1*: a pseudodifferential inverter-based Nagaraj integrator with a fully passive SC CMFB circuit [24].

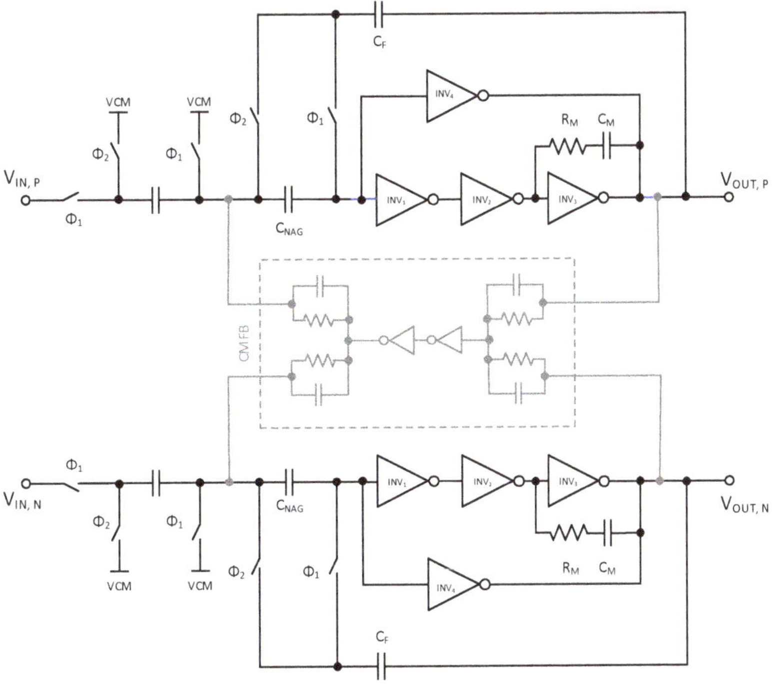

Figure 13. *OTA 2*: a pseudodifferential with a three-stage multipath inverter-based Nagaraj integrator [29].

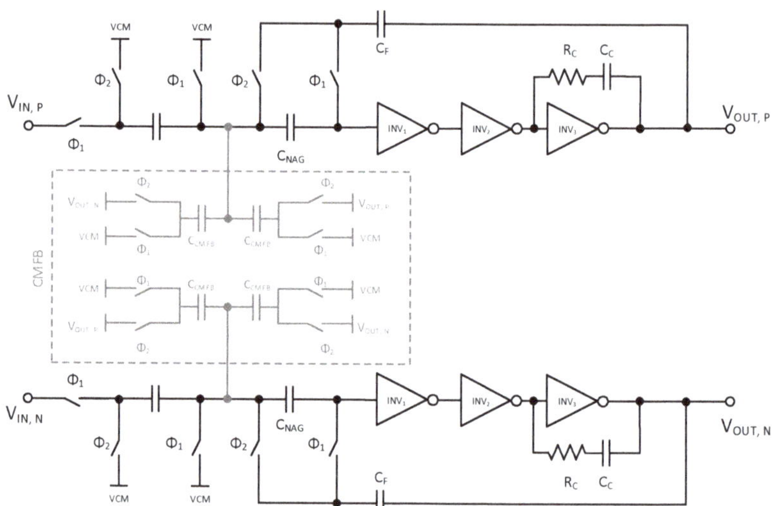

Figure 14. *OTA 3*: a single path three-stage pseudodifferential Nagaraj integrator using a fully passive SC CMFB circuit [28].

Table 2. Dynamic performance for the three described inverter-based OTA topologies.

	OTA 1	*OTA 2*	*OTA 3*
DC gain	Low	High	High
GBW	High	High	Moderate
Linearity [1]	Low	Moderate	Moderate
Current consumption	Low	High	Moderate
Circuit complexity	Two inverters	Ten inverters	Six inverters

[1] Considering 2/3 of the full-scale output.

4. A Standard-Cell-Based Digital–Delta (Δ) Modulator (ΔM) with Noise Shaping (NS)

A complete and differential electrical scheme of the proposed ADC topology, a digital-ΔM employing NS, is depicted in Figure 15, complementing the description in Section 2.5. It comprises a split-capacitor DAC with embedded S/H, a pseudodifferential inverter-based Nagaraj integrator with a fully passive SC CMFB circuit (shown in Figure 12), an OAI-based comparator (whose circuit is shown in Figure 10), an accumulator, and a phase generator. Thus, a fully synthesizable ADC is demonstrated, only recurring to passive and standard-cell-based circuitry.

As described in Figure 16, its operation is based on two different frequencies. The sampling and NS function at F_S, while the delta modulation is performed at a higher frequency.

After the DAC is reset, the sampling of the V_{in} and the noise-shaping voltage, V_{NS}, is simultaneously performed in the most significant bit (MSB) section of the DAC and on the dedicated capacitor, C_{NS}, respectively. After that, delta modulation starts: the comparator makes a decision that is transmitted to the accumulator to reconfigure the DAC for the next comparator decision. This action is performed during M averages, and the accumulator output is lastly ready. Before the new reset of the DAC, residue voltages $V_{RES,P}$ and $V_{RES,N}$ are integrated by the pseudodifferential inverter-based Nagaraj integrator scheme; consequently, $V_{NS,P}$ and $V_{NS,N}$ are updated. Then, a new sampling of the V_{in} and the V_{NS} is performed, and the process continues repeatedly.

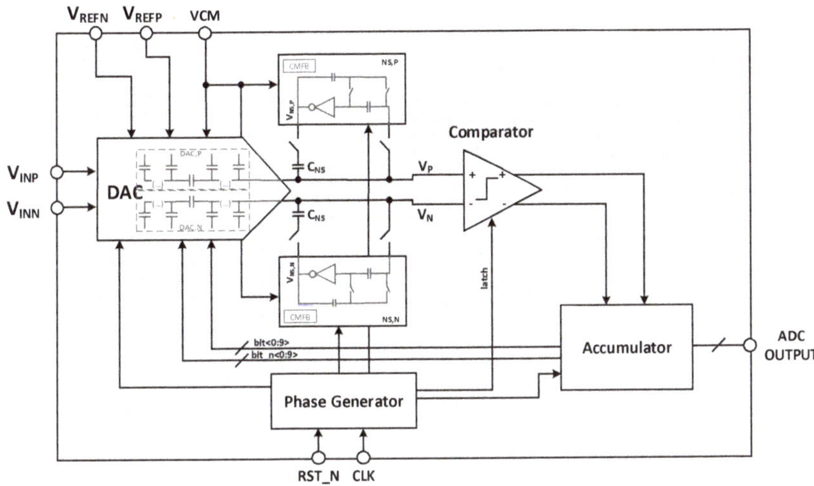

Figure 15. Scheme of the proposed standard-cell-based digital-ΔM with NS employing a split-capacitive DAC, an inverter-based OTA topology, to perform NS and an OAI-based comparator.

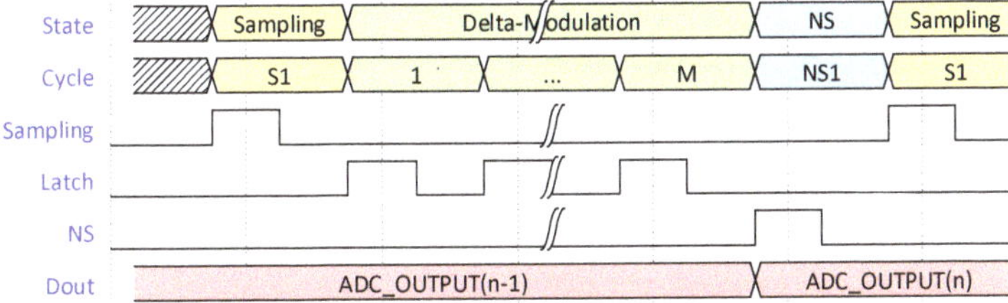

Figure 16. Illustrative timing of the proposed digital-ΔM ADC employing NS.

The simulated output spectrum of the proposed converter, fully implemented in a 28 nm CMOS technology, is shown in Figure 17. With an oversampling ratio (OSR) of 32 and a 10-bit DAC, a peak of 72.5 dB in the signal-to-noise and distortion ratio (SNDR) was achieved for a ≈113 kHz input signal and a 1 MHz BW. Table 3 summarizes the simulation results. The converter dissipated ≈112 μW, which could be translated into a Walden figure of merit, FoM$_{\text{Walden}}$, of 16.2 fJ/conv.-step.

These results are promising, allowing for a fully synthesizable ADC that is capable of achieving both high resolutions and good energy efficiency.

Table 3. Summary of simulated results of the proposed standard-cell-based digital-ΔM employing NS using a 28 nm CMOS technology.

Parameter	Unit	Digital-DM with NS
F_S	MHz	64
BW	MHz	1
OSR		32
SNDR	dB	72.50
ENOB	−bit	11.8
V_{DD}	V	0.9
Power dissipation	μW	112
FoM$_{\text{Walden}}$	fJ/conv.-step	16.2

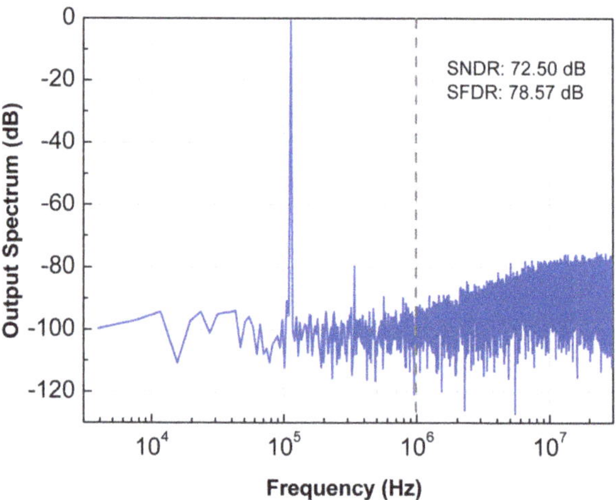

Figure 17. Simulated output spectrum of the schematic of the proposed standard-cell-based digital-ΔM employing NS. This result was achieved using 2^{14} points, $M = 8$, a BW of 1 MHz (OSR of 32), and a Fin of 113 kHz.

5. Conclusions

The most-suited high-resolution ADC topologies for IoT applications were described and compared in terms of complexity, overall performance, and energy efficiency. ΔM was described because it is the basis of some well-known topologies and has inspired others, such as SAR-ADC with NS or $\Delta\Delta\Sigma$M ADC. Taking into consideration the advantages of standard-cell-based synthesizable schemes, some schematics of comparators and amplifiers were reported, and their key performance parameters were compared. An innovative topology, a digital-ΔM ADC employing NS, was detailed employing passive and standard-cell-based circuitry. An SNDR of 72.5 dB was achieved for a 1 MHz BW (OSR of 32) with an estimated FoM$_{\text{Walden}}$ of 16.2 fJ/conv.-step. Thus, a fully synthesizable ADC that is compatible with IoT applications was clearly demonstrated.

Author Contributions: Conceptualization: A.C., V.G.T. and J.G.; methodology: A.C., V.G.T. and J.G.; investigation: A.C., V.G.T. and J.G.; writing—original draft preparation: A.C.; writing—review and editing: all authors; supervision: V.G.T., P.B. and J.G. All authors have read and agreed to the published version of the manuscript.

Funding: This work was funded by National Funds through FCT—Fundação para a Ciência e a Tecnologia, I.P., through a Ph.D. Grant (SFRH/BD/137519/2018) and the following project references: UIDB/50025/2020-2023, SMART-E-PTDC/CTM-PAM/04012/2022, IDS-PAPER-PTDC/CTM-PAM/4241/2020 and PEST (CTS/UNINOVA)-UIDB/00066/2020. This work also received funding from the European Community's H2020 program [Grant Agreement No. 716510 (ERC-2016-StG TREND) and 952169 (SYNERGY, H2020-WIDESPREAD-2020-5, CSA)].

Data Availability Statement: The data presented in this study are contained within the article.

Conflicts of Interest: The authors declare no conflict of interest.

References

1. O'Riordan, N. Industrial IoT. In *Circuits and Systems for the Internet of Things: CAS4IoT*; IEEE CAS, River Publishers: Gistrup, Denmark, 2017.
2. Jiang, D.; Sin, S.W.; Qi, L.; Wang, G.; Martins, R.P. Recent Advances in High-Resolution Hybrid Discrete-Time Noise-Shaping ADCs. *IEEE Open J. Solid-State Circuits Soc.* **2021**, *1*, 129–139. [CrossRef]
3. Fredenburg, J.A.; Flynn, M.P. A 90-MS/s 11-MHz-bandwidth 62-dB SNDR noise-shaping SAR ADC. *IEEE J. Solid-State Circuits* **2012**, *47*, 2898–2904. [CrossRef]
4. Schreier, R.; Temes, G.C. *Understanding Delta-Sigma Data Converters*; IEEE Press: Piscataway, NJ, USA, 2005; Volume 74.

5. Ren, J.; Sarwana, S.; Sahu, A.; Talalaevskii, A.; Inamdar, A. Low-pass delta-delta-sigma ADC. *IEEE Trans. Appl. Supercond.* **2014**, *25*, 1–6. [CrossRef]

6. Xu, Z.; Ojima, N.; Li, S.; Iizuka, T. An all-standard-cell-based synthesizable SAR ADC with nonlinearity-compensated RDAC. *IEEE Trans. Very Large Scale Integr. Syst.* **2021**, *29*, 2153–2162. [CrossRef]

7. Park, J.E.; Hwang, Y.H.; Jeong, D.K. A 0.5-V fully synthesizable SAR ADC for on-chip distributed waveform monitors. *IEEE Access* **2019**, *7*, 63686–63697. [CrossRef]

8. Aiello, O.; Crovetti, P.; Alioto, M. Fully synthesizable low-area analogue-to-digital converters with minimal design effort based on the dyadic digital pulse modulation. *IEEE Access* **2020**, *8*, 70890–70899. [CrossRef]

9. Deloraine, E.M.; Van Mierlo, S.; Derjavitch, B. Methode et Systéme de Transmission par Impulsions. French Patent 932,140, 10 August 1946.

10. Abate, J.E. Linear and Adaptive Delta Modulation. Ph.D. Thesis, Newark College of Enginering, Newark, NJ, USA, 1967.

11. Zrilic, D.G. *Circuits and Systems Based on Delta Modulation: Linear, Nonlinear and Mixed Mode Processing*; Springer Science & Business Media: Heidelberg, Germany; New York, NY, USA, 2005.

12. Carusone, T.; Johns, D.; Martin, K. *Analog Integrated Circuit Design*, 2nd ed.; Wiley: Hoboken, NJ, USA, 2011.

13. Razavi, B. *Principles of Data Conversion System Design*; IEEE Press: New York, NY, USA, 1995.

14. Rabuske, T.; Fernandes, J. *Charge-Sharing SAR ADCs for Low-Voltage Low-Power Applications*; Springer: Berlin/Heidelberg, Germany, 2017.

15. Guo, W.; Sun, N. A 12b-ENOB 61 μW noise-shaping SAR ADC with a passive integrator. In Proceedings of the ESSCIRC Conference 2016: 42nd European Solid-State Circuits Conference, Lausanne, Switzerland, 12–15 September 2016; pp. 405–408.

16. Li, S.; Qiao, B.; Gandara, M.; Sun, N. A 13-ENOB 2nd-order noise-shaping SAR ADC realizing optimized NTF zeros using an error-feedback structure. In Proceedings of the 2018 IEEE International Solid-State Circuits Conference-(ISSCC), San Francisco, CA, USA, 11–15 February 2018; pp. 234–236.

17. Liu, J.; Wang, X.; Gao, Z.; Zhan, M.; Tang, X.; Hsu, C.K.; Sun, N. A 90-dB-SNDR calibration-free fully passive noise-shaping SAR ADC with 4× passive gain and second-order DAC mismatch error shaping. *IEEE J. Solid-State Circuits* **2021**, *56*, 3412–3423. [CrossRef]

18. Correia, A.; Barquinha, P.; Marques, J.; Goes, J. A High-resolution Δ-Modulator ADC with Oversampling and Noise-shaping for IoT. In Proceedings of the 2018 14th Conference on Ph. D. Research in Microelectronics and Electronics (PRIME), Prague, Czech Republic, 2–5 July 2018; pp. 33–36.

19. Weaver, S.; Hershberg, B.; Moon, U.K. Digitally synthesized stochastic flash ADC using only standard digital cells. *IEEE Trans. Circuits Syst. I Regul. Pap.* **2013**, *61*, 84–91. [CrossRef]

20. Aiello, O.; Crovetti, P.; Alioto, M. Fully synthesizable, rail-to-rail dynamic voltage comparator for operation down to 0.3 V. In Proceedings of the 2018 IEEE International Symposium on Circuits and Systems (ISCAS), Florence, Italy, 27–30 May 2018; pp. 1–5.

21. Ojima, N.; Xu, Z.; Iizuka, T. A 0.0053-mm^2 6-bit fully-standard-cell-based synthesizable SAR ADC in 65 nm CMOS. In Proceedings of the 2019 17th IEEE International New Circuits and Systems Conference (NEWCAS), Munich, Germany, 23–26 June 2019; pp. 1–4.

22. Aiello, O.; Crovetti, P.; Toledo, P.; Alioto, M. Rail-to-rail dynamic voltage comparator scalable down to pW-range power and 0.15-V supply. *IEEE Trans. Circuits Syst. II Express Briefs* **2021**, *68*, 2675–2679. [CrossRef]

23. Li, X.; Zhou, T.; Ji, Y.; Li, Y. A 0.35 V-to-1.0 V synthesizable rail-to-rail dynamic voltage comparator based OAI&AOI logic. *Analog. Integr. Circuits Signal Process.* **2020**, *104*, 351–357.

24. Christen, T. A 15-bit 140-μW Scalable-Bandwidth Inverter-Based ΔΣ Modulator for a MEMS Microphone With Digital Output. *IEEE J. Solid-State Circuits* **2013**, *48*, 1605–1614. [CrossRef]

25. Ki, W.H.; Temes, G.C. Offset-compensated switched-capacitor integrators. In Proceedings of the IEEE International Symposium on Circuits and Systems, New Orleans, LA, USA, 1–3 May 1990; pp. 2829–2832.

26. Nagaraj, K.; Vlach, J.; Viswanathan, T.; Singhal, K. Switched-capacitor integrator with reduced sensitivity to amplifier gain. *Electron. Lett.* **1986**, *21*, 1103–1105. [CrossRef]

27. Chae, Y.; Han, G. Low voltage, low power, inverter-based switched-capacitor delta-sigma modulator. *IEEE J. Solid-State Circuits* **2009**, *44*, 458–472. [CrossRef]

28. Correia, A.; Tavares, V.G.; Barquinha, P.; Goes, J. Trade-offs and Limitations in Energy-Efficient Inverter-based CMOS Amplifiers. In Proceedings of the XXXVI Conference on Design of Circuits and Integrated Systems (DCIS), Vila do Conde, Portugal, 24–26 November 2021.

29. Neofytou, M.; Zhou, M.; Bolatkale, M.; Liu, Q.; Zhang, C.; Radulov, G.; Baltus, P.; Breems, L. A 1.9 mW 250 MHz Bandwidth Continuous-Time ΣΔ Modulator for Ultra-Wideband Applications. In Proceedings of the 2018 IEEE International Symposium on Circuits and Systems (ISCAS), Florence, Italy, 27–30 May 2018; pp. 1–5.

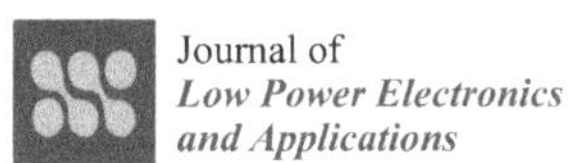

Journal of
*Low Power Electronics
and Applications*

A Fully-Differential CMOS Instrumentation Amplifier for Bioimpedance-Based IoT Medical Devices

Israel Corbacho, Juan M. Carrillo, José L. Ausín, Miguel Á. Domínguez, Raquel Pérez-Aloe and J. Francisco Duque-Carrillo *

Departamento de Ingeniería Eléctrica, Electrónica y Automática, Universidad de Extremadura, Avenida de Elvas s/n, 06006 Badajoz, Spain
* Correspondence: duque@unex.es

Abstract: The implementation of a fully-differential (FD) instrumentation amplifier (IA), based on indirect current feedback (ICF) and aimed to electrical impedance measurements in an Internet of Things (IoT) biomedical scenario, is presented. The IA consists of two FD transconductors, to process the input signal and feed back the output signal, a summing stage, used to add both contributions and generate the correcting current feedback signal, and a common-mode feedback network, which controls the DC level at the output nodes of the circuit. The transconductors are formed by a voltage-to-current conversion resistor and two voltage buffers, which are based on a super source follower cell in order to improve the overall response of the circuit. As a result, a compact single-stage structure, suitable for achieving a high bandwidth and a low power consumption, is obtained. The FD ICF IA has been designed and fabricated in 180 nm CMOS technology to operate with a 1.8-V supply and provide a nominal gain of 4 V/V. Experimental results show a voltage gain of 3.78 ± 0.06 V/V, a BW of 5.83 MHz, a CMRR at DC around 70 dB, a DC current consumption of 266.4 µA and a silicon area occupation of 0.0304 mm^2.

Keywords: CMOS; fully-differential; indirect current feedback; instrumentation amplifier; low-voltage; wide bandwidth

Citation: Corbacho, I.; Carrillo, J.M.; Ausín, J.L.; Domínguez, M.Á.; Pérez-Aloe, R.; Duque-Carrillo, J.F. A Fully-Differential CMOS Instrumentation Amplifier for Bioimpedance-Based IoT Medical Devices. *J. Low Power Electron. Appl.* **2023**, *13*, 3. https://doi.org/10.3390/jlpea13010003

Academic Editors: Orazio Aiello and Andrea Acquaviva

Received: 28 October 2022
Revised: 7 December 2022
Accepted: 25 December 2022
Published: 30 December 2022

1. Introduction

Currently, an increase of the life expectancy in the population of developed countries is taking place. Therefore, new habits for healthy lifestyles are being adopted, many of them trying to implement preventive health programs and early detection of diseases, as the most effective way to improve the effectiveness of treatments and therapies and ensure, as far as possible, a high quality of life and a healthy aging. The Internet of Things (IoT) that allows data to be collected and analysed at any time and from anywhere, is called to play a fundamental role to offer a solving strategy in healthcare [1]. In IoT-based healthcare, sensors and devices are developed for a variety of objectives, such as monitoring the medical conditions of people, assisting in the treatment of diseases, and providing access to patient information. In this context, wearable devices are seamlessly connected to improve information delivery and the care-giving process in healthcare services [2]. Given the large-scale challenges caused by chronic diseases, very low cost and effective wearable devices for telemedicine have become of higher importance.

Electrical bioimpedance (EBI), or simply bioimpedance, joins the attributes to become a promising sensor technology in the IoT environment. EBI is a well-established physical concept in which an object's impedance to an applied alternating current over increasing frequencies can be measured, to assess tissue composition [3]. In addition to being economic, lightweight, easy-to-use, and noninvasive, bioimpedance can be used for a wide range of clinical applications, ranging from examine body composition in healthy people to monitoring various types of diseases such as diabetes, hypertension, and others. Therefore, in recent years, a pronounced trend towards the integration of EBI in wearable systems has been observed.

In practice, for detecting some transient physiological events, bioimpedance spectroscopy (BIS) is used. As with any spectroscopy technique, BIS implies the measurement of the bioimpedance spectrum in a determined frequency range, for which a sequential sweep of analysis varying the frequency is carried out. Typical frequencies in BIS are in the range from several hundreds of Hz to 1 MHz, also known as the β-dispersion range. Therefore, the use of such a broad signal spectrum puts several challenges for the full integration of wearable bioimpedance-based devices into the clinical health care system. In particular, a CMOS integrated BIS system in the IoT horizon requires a great circuit optimization not only in size but also in energy consumption.

The block diagram of a bioimpedance-based IoT system for medical applications is illustrated in Figure 1. The source of power, which can be a battery or an energy harvester, is controlled by a power management unit (PMU), which optimizes and regulates the signals used to supply the rest of the blocks. The bioimpedance under test, Z_{BIO}, is excited by an AC signal, usually a current in order to avoid any damage on the biological sample, and the resulting voltage is acquired and conditioned by the analog front-end (AFE). Then, signals are efficiently processed in the digital domain, by a digital signal processor (DSP), and can be locally stored or transmitted by means a wireless protocol. The user interface allows control of the operation of the overall system.

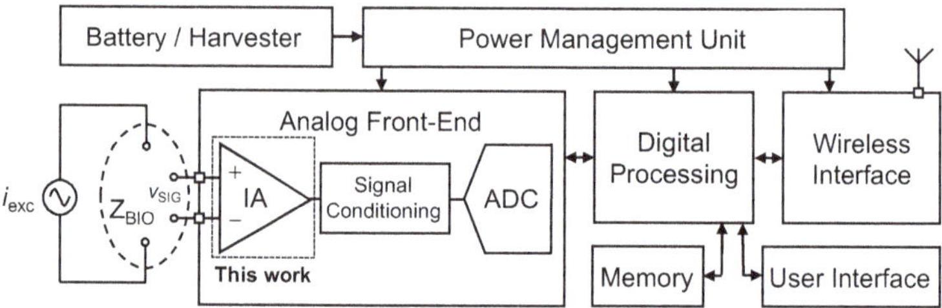

Figure 1. Conceptual block diagram of a bioimpedance-based IoT system for biomedical applications.

The IA is a critical constituent block of the system previously described [4–29]. Indeed, an appropriate signal acquisition is required, which includes a demanding performance in terms of differential-mode (DM) signals amplification, common-mode (CM) signals rejection, and noise, among others, whereas the overall power consumption has to be kept to a minimum extent, which is particularly a challenge in applications that require the processing of signals contained in a wide frequency range and with a relatively large amplitude. The indirect current feedback (ICF) technique results suitable to design a monolithic IA with low-voltage capability [5,22,29]. In addition, a single-stage ICF IA provides compactness and the possibility of achieving operation over a broad frequency range [11,12,22,26,29].

The overall performance of an analog system in general, and of an IA in particular, can be enhanced by adopting a fully differential (FD) implementation [23,25,30]. There are well-known advantages associated to this solution, such as the extension of the signal range, due to the availability of two output terminals, the increase of the linearity, thanks to the ideal cancellation of even-order harmonics, and the decrease of the effects of undesired noises coming from the supply, which can be considered as CM signals. There are also disadvantages related to the use of a FD circuit, such as the increase of the circuitry to obtain a fully symmetrical structure, with the consequent increase in area and power consumption, or the need of a CM feedback (CMFB) network, to control the CM component of the output signal. Therefore, all the pros and cons must be considered and a design tradeoff has to be established.

A FD IA, relying on the ICF technique and suitable for bioimpedance analysis in an IoT biomedical application, is presented in this contribution. An analysis of the main characteristics of the proposed circuit is provided, which is confirmed by means of simulated and experimental results. In addition, the solution is compared in terms of circuit structure to other differential IA previously reported [29], whereas a performance comparison with similar solutions in the literature is also carried out. The circuit has been designed and fabricated in 180 nm CMOS technology to operate with a single-supply voltage of 1.8 V. The experimental characterization illustrates the robustness of the proposed solution. The rest of the manuscript has been organized as follows. Section 2 deals with the block diagram and the transistor level implementation of the IA, whereas different design considerations are discussed in Section 3. Measurement results are reported in Section 4 and conclusions are drawn in Section 5.

2. Principle of Operation

The block diagram of the proposed FD IA is described in order to clearly understand the role of each constitutive section. In addition, the transistor level implementation of both the core of the IA and the CMFB network are also detailed.

2.1. Block Diagram

Different approaches to implement a differential IA have been previously reported [9,17–20,23–25,29]. Among them, there are solutions based on the ICF technique, as the pseudo-differential (PD) IA proposed in [29], the block diagram of which is illustrated in Figure 2a. The sections G_{mI} and G_{mO} are an input and an output (or feedback) transconductor, used, respectively, to process the input signal and establish the current feedback. When the input signal, $v_{I,DM}$, is applied to the transconductor G_{mI}, a current i_I is generated. Similarly, an output current i_O is produced when the voltage $v_{SENSE} - V_{REF}$ is applied to the input terminals of the voltage-to-current (V-to-I) converter G_{mO}. The voltage v_{SENSE} is used as feedback signal and V_{REF} is a reference voltage used to set the DC component of v_O to the intended level. In the particular case of Figure 2a, a single-stage structure is represented, in which an unitary feedback loop is established. Indeed, the output voltages, v_O^+ and v_O^-, are shorted to the feedback terminals, v_{SENSE}^+ and v_{SENSE}^-, whereas two copies of the block G_{mO} are required to stablish the differential feedback loop. The feedback action around each output transconductor controls individually the DC level at the two output terminals and; hence, no CMFB is needed.

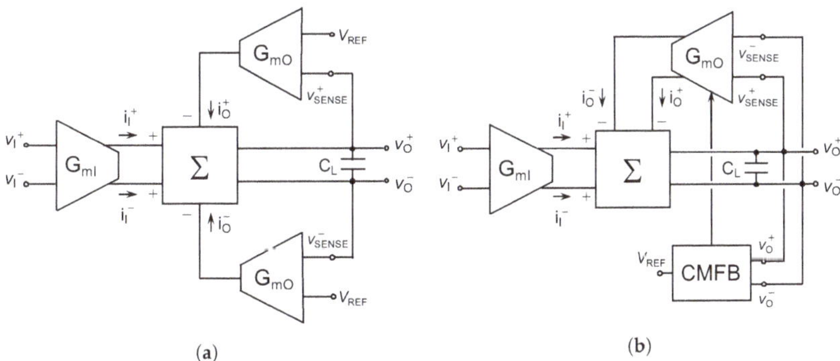

(a) (b)

Figure 2. Block diagram of (**a**) a pseudo-differential and (**b**) a fully-differential ICF IA.

The block diagram of the proposed FD IA is depicted in Figure 2b. As observed, the feedback network is implemented differentially and, hence, only one output transconductor is required. Nevertheless, it is well known that the establishment of a differential feedback loop relies on the assistance of a CM control network, in order to dynamically set the CM component of the output voltage to the intended level. With this purpose, the CMFB section illustrated in Figure 2b has been included. As observed, the DC component of the output voltage is induced to be equal to V_{REF} by the CMFB circuit, rather than being applied to the output transconductor, as it is done in the PD structure in Figure 2a. The existence of well-differenced signal paths for the DM and CM components in the FD approach allows the individual optimization of their response, which is not possible in the PD solution, where the control of the output CM voltage is embedded in the implementation of the output section of the circuit.

A hand analysis of the block diagram in Figure 2b led to the following transfer function for the system:

$$H(s) \equiv \frac{v_o(s)}{v_i(s)} = \frac{G_{mI} \cdot \left(R_{out} \parallel \frac{1}{sC_L} \right)}{1 + G_{mO} \left(R_{out} \parallel \frac{1}{sC_L} \right)} \tag{1}$$

where R_{out} and C_L are the output resistance and the load capacitance, respectively, of the summing stage. Assuming a high gain for the loop around the transconductor G_{mO}, the voltage gain, A_v, and the BW of the IA are inferred from (1) and can be expressed as:

$$A_v \equiv \frac{v_o}{v_{i,dm}} = \frac{G_{mI}}{G_{mO}} \tag{2}$$

$$BW = \frac{G_{mO}}{C_L} \tag{3}$$

The voltage gain of the IA is adjusted by means of the ratio of G_{mI} and G_{mO}. In addition, a proper value of C_L has to be selected in order to ensure an optimal phase margin and, hence, appropriate frequency and time responses.

2.2. Transistor Level Implementation

The transistor level implementation of the proposed FD IA is illustrated in Figure 3, where the different circuit sections are labelled at the bottom. The *V*-to-*I* conversion at the input (output) transconductor is carried out by a resistor and two voltage followers. The input (output) voltage is applied to resistor R_I (R_O) through two super-source-follower (SSF) sections, which act as voltage buffers. The SSF block incorporates an implicit feedback loop, implemented by transistors MDI and MFI (MDO and MFO), that reduces the effective output resistance of the block and makes its voltage gain very close to unity, regardless of the value of the linearization resistor. As a result, the value of R_I (R_O) can be greatly reduced without hardly affecting the operation of the SSF sections, which allows a reduction in the noise contribution of the resistor to be made, as well as the silicon area occupied by this passive component. The SSF structures are biased by means of devices MSUI and MSDI (MSUO and MSDO), which are single-transistor current sources providing tail currents $2I_B$ and I_B, respectively. The gate terminals of these transistors are connected to the corresponding bias voltage, V_{BN} or V_{BP}, in the biasing network represented at the left of Figure 3. Capacitors C_{C1} to C_{C4} are used to optimize the phase margin of the feedback loop inherent in each SSF cell. The effective transconductance of the input and output *V*-to-*I* cells is equal to:

$$G_{m,eff} \equiv \frac{i}{v_{DM}} = \frac{2}{R} \frac{1}{\left[1 + \left(1 + \frac{2}{R} \frac{1}{g_{m,MD}} \right) \left(\frac{g_{o,MD} + g_{o,MSD}}{g_{mF}} \right) \right]} \approx \frac{2}{R} \tag{4}$$

where $g_{m,Mi}$ and $g_{o,Mi}$ are the transconductance and output conductance, respectively, of transistor *Mi*, at the input and the output transconductor, R is the linearization, or source degeneration, resistor (R_I or R_O), and $g_m \gg g_o$ has been assumed. The factor of 2 in (4) indicates that the current signal generated in the input and the output transconductor, i_I and i_O, respectively, is conveyed to the output terminals of the IA by the two branches of the circuit section. In addition, the second term in (4), multiplying the main contribution $2/R$, represents the load regulation effect of resistor R on the voltage buffers. In first order of approximation, the effective transconductance of each *V*-to-*I* converter is approximately equal to two times the inverse of the linearization resistor.

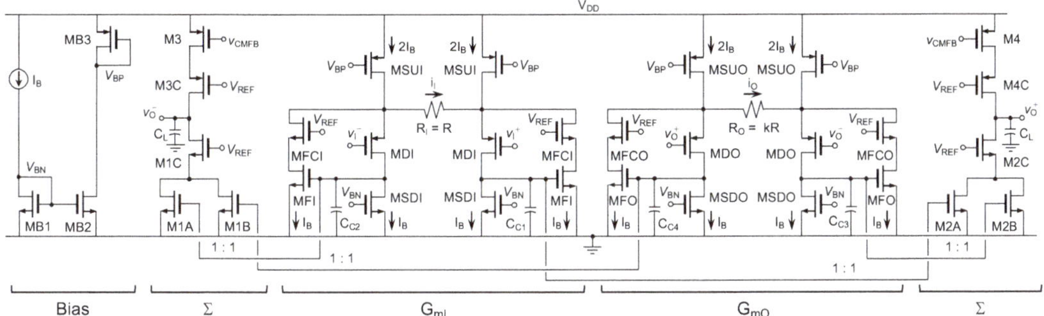

Figure 3. Transistor level implementation of the proposed fully-differential IA.

The current signals generated at G_{mI} and G_{mO} are mirrored to the output nodes of the IA by using current mirrors with gains $1:1$. Cascode transistors are used in the output branches in order to increase the output resistance and, hence, the open-loop voltage gain. In addition, cascode devices MFCI and MFCO are used in G_{mI} and G_{mO}, respectively, to ideally cancel out the systematic offset in the current reflections. Additional design flexibility to adjust the voltage gain and bandwidth of the IA to the intended values can be obtained by sizing the current mirrors with a gain different from unity [29]. The voltage gain and the BW of the proposed IA can be specified by considering the general expressions (2) and (3), along with the equation of the effective transconductance in (4), and can be rewritten as:

$$A_v \approx \frac{R_O}{R_I} \tag{5}$$

$$BW \approx \frac{2}{C_L R_O} \tag{6}$$

The input CM voltage of the FD IA in Figure 3 can be adjusted over a reasonably wide range. Indeed, the operation for input signals around the midsupply is ensured by adequately setting the aspect ratio of the input devices, so that the upper current source transistors, MSUI, can operate in saturation. Thus, the maximum level of the input CM voltage that can be achieved close to V_{DD} is constrained by the operation in saturation of transistors MSUI. Furthermore, the operation for $v_{I,CM}$ around ground can be easily achieved by proper sizing of transistors MFI. Indeed, the voltage at the drain of transistors MDI, which could force their operation in the triode region, can be reduced to an appropriate level by increasing the aspect ratio of transistors MFI, thus ensuring the operation of the input drivers in saturation.

The structure of the CMFB network used to control the DC level of the output voltage is depicted in Figure 4. A current-mode approach, based on generating a CM current signal that is a function of the output CM voltage, has been followed. The output voltages of the FD IA are used as input signals in the CMFB section and are applied to the inputs of two cross-coupled differential pairs. The other two input terminals of the CMFB are connected to the reference voltage V_{REF}. Assuming the voltage difference $v_O^+ - v_O^-$ small and, hence, the differential pairs operating within their linear region, a current signal i_{cm}, proportional to the output CM voltage, is generated. This current, superimposed to a DC level nominally equal to $2I_B$, is mirrored by a NMOS and a PMOS current mirror and injected into the FD IA through the terminal v_{CMFB}. The CM loop is closed through the output branches of the IA, which are connected to the input of the CMFB network. The action of the feedback loop forces the CM component of the output voltage to be equal to V_{REF}, setting the DC level of the output voltage to this value. The dominant pole of the feedback loop established for the CM signal is the same to that of the DM loop and is determined by the load capacitor. The secondary poles in both cases, DM and CM signals, are associated to low impedance nodes, that is, the corresponding time constants are the product of a low resistance, in general the inverse of a transconductance, and a parasitic capacitance. As a consequence, the frequency compensation of the DM and CM loops in the FD IA can be easily achieved by properly setting the value of the load capacitor. Indeed, the value of C_L must be adjusted to have a phase margin higher than 60° in both the CM and the DM feedback loop.

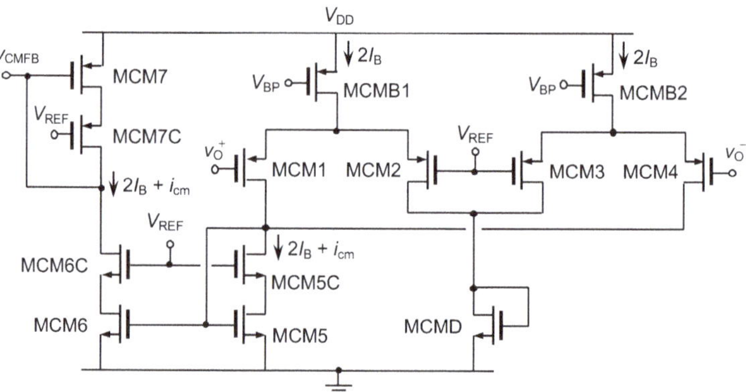

Figure 4. Transistor level implementation of the CMFB network.

3. Design Considerations

The main features of the FD IA proposed are analysed and discussed in view of the fundamental performance, in order to facilitate the design procedure. In a FD implementation, the CM signal must be processed at least with the same accuracy and speed as the DM signal. Therefore, the CMFB network, in particular, and the CM feedback loop, in general, must be designed so that the open-loop gain (LG) and gain-bandwidth product (LGBW) of both components are similar [30]. This requirement can be analytically expressed as

$$LG_{CM} \simeq LG_{DM} \tag{7a}$$

$$LGBW_{CM} \simeq LGBW_{DM} \tag{7b}$$

Therefore, it is recommendable to provide similar paths to the DM and the CM signal in order to accomplish these requisites. In the case of the IA represented in Figures 3 and 4, the output branch of the core circuit (Figure 3) is common to both the DM and the CM section. Nevertheless, the differential input stage of each loop, and hence the corresponding effective transconductance, is different in every case. Indeed, for the DM signal, the input transconductance is given by (4), whereas for the CM component the transconductance is equal to the individual transconductances of transistors MCM1 to MCM4 in Figure 4. As the linearization carried out in G_{mI} implies a reduction of the transconductance value, it is expected that effective input transconductance of the DM loop is lower as compared to the CM loop. This fact ensures that an appropriate treatment of the DM signal will result in an adequate processing of the CM signal.

Regarding the signal processing of the FD IA, only the DM component gives rise to an output current in the input and output V-to-I converters, being the CM signal rejected by the differential structure of these stages. However, a CM signal can also produce an output current, given that the presence of mismatches is unavoidable in a real implementation. In order to evaluate the impact of the join action of a CM signal and the mismatches on the output current produced, the residual transconductance of the input and the output transconductor in the IA, defined as $\Delta G_m \equiv \frac{i}{v_{CM}}$, has been analytically calculated. With this purpose, each small signal parameter g_i has been assumed to have values equal to $g_i + \Delta g_i/2$ and $g_i - \Delta g_i/2$ for a given pair of ideally matched transistors. In addition, the contributions to the residual transconductance, due to considering mismatches in every pair of transistors, have been evaluated individually. The corresponding expressions were obtained by means of a hand analysis and the main terms were determined by simulations, resulting to be dominant the responses associated to mismatches in the transconductance (Δg_m) and output conductance (Δg_o) of the input driver transistors, MDI and MDO. The corresponding expressions are:

$$\Delta G_m|_{\Delta g_{m,MD}} \approx \frac{2}{R} \cdot \frac{\Delta g_{m,MD} g_{o,MD}}{g_{m,MD}^2} \cdot \frac{1}{\left[1 + \frac{2}{R}\frac{1}{g_{m,MD}}\left(\frac{g_{o,MD}+g_{o,MSD}}{g_{m,MF}}\right)\right]} \tag{8a}$$

$$\Delta G_m|_{\Delta g_{o,MD}} \approx \frac{2}{R} \cdot \frac{\Delta g_{o,MD}}{g_{m,MF}} \cdot \frac{1}{\left[1 + \frac{2}{R}\frac{1}{g_{m,MD}}\left(\frac{g_{o,MD}+g_{o,MSD}}{g_{m,MF}}\right)\right]} \tag{8b}$$

where MD represents the driver transistors in G_{mI} and G_{mO}. The impact of the transconductance and output conductance mismatches of other transistors on ΔG_m is negligible and, hence, is not reported here for the sake of conciseness.

The use of the CM rejection ratio (CMRR) is a very widespread habit in order to compare the magnitude of the CM gain with respect to the DM gain. As the proposed IA has a single-stage structure, the voltage gain for DM and CM signals will be given by the product of the input transconductance and the output impedance. Assuming the same output impedance for both signal components, the CMRR of the IA can be expressed in terms of the ratio of the effective and the residual transconductance, given respectively by (4) and (8b), as:

$$CMRR \equiv \frac{G_{mI}}{\Delta G_{mI}} = \frac{1}{\left(\frac{\Delta g_{m,MDI} g_{o,MDI}}{g_{m,MDI}^2} + \frac{\Delta g_{o,MDI}}{g_{m,MFI}}\right)} \cdot \frac{\left[1 + \frac{2}{R_I}\frac{1}{g_{m,MDI}}\left(\frac{g_{o,MDI}+g_{o,MSDI}}{g_{m,MFI}}\right)\right]}{\left[1 + \left(1 + \frac{2}{R_I}\frac{1}{g_{m,MDI}}\right)\left(\frac{g_{o,MDI}+g_{o,MSDI}}{g_{m,MFI}}\right)\right]} \tag{9}$$

The most-right term in (9) represents the ratio of the load regulation effects of resistor R for the CM and the DM signals, respectively. Thanks to the improved response of the SSF cell, the value of these terms is very close to unity, which allows the expression of the CMRR as a function of the different mismatches in the actual implementation of the circuit. At this point it is worth to mention that, as observed in Figure 2, the structure of the input section of both the PD IA and the FD IA is the same and, hence, both structures present a similar rejection to CM signals form the architecture point of view.

Another key parameter for an IA is the noise, as it indicates the minimum signal level that can be processed. In the case of an IA for bioimpedance spectroscopy, the signal bandwidth required is usually wide and, hence, thermal noise is dominant. The spectral density of the input referred thermal noise has been analytically determined, assuming that the main contributions are due to the input *V*-to-*I* converter, and can be expressed as:

$$\frac{v_{iN,th}^2}{\Delta f} = \left[1 + \left(1 + \frac{2}{R_I}\frac{1}{g_{m,MDI}}\right)\left(\frac{g_{o,MDI} + g_{o,MSDI}}{g_{m,MFI}}\right)\right]^2 \cdot 4kTR_I \cdot \left[1 + \frac{4}{3}(g_{m,MDI} + g_{m,MSUI})R_I\right] \tag{10}$$

where k and T are Boltzmann's constant and the absolute temperature, respectively. The first term in (10) represents the conversion factor for referring the noise from the resistor to the input of the circuit, and is the inverse of the load regulation effect of resistor R_I on the SSF cell (see Equation (4)), the second factor is the thermal noise of resistor R_I, and the last term includes the main thermal noise contributions of the devices involved in the circuit implementation of the input *V*-to-*I* converter, G_{mI}. It can be inferred from (10) that the noise of the IA can be decreased by reducing the value of the source degeneration resistor R_I, which is possible until a certain limit thanks to the use of SSF sections.

The fact of linearizing the *V*-to-*I* converters in the IA by means of a resistor, requires a given level of biasing current to achieve a given input DM voltage range with a determined linearity. In each SSF cell in the input and output transconductors, the bias current $2I_B$ is split into two branches corresponding to the input and feedback transistors. As the tail current of the driver devices is fixed to I_B by the lower current sources, a current equal to I_B is steered towards the feedback transistors. Consequently, the maximum input DM signal that can be processed by each V-to-I converter is that leading to a current equal to zero through one of the feedback transistors. This condition can be expressed for G_{mI} as

$$v_{I,DM_{max}} = \pm R_I \cdot I_B \tag{11}$$

where the voltage gain of the SSF cells has been assumed to be equal to unity. Nevertheless, this is an extreme situation that leads to switching off one of the branches of the input transconductor. Instead, a specific criterion, such as considering a given total harmonic distortion (THD) level, is assumed in a practical case to determine the value of $v_{I,DM_{max}}$ in an objective way.

4. Experimental Results

The fully-differential IA illustrated in Figure 3, along with the CMFB section in Figure 4, has been designed and fabricated in 180 nm CMOS technology to operate with a single-supply voltage of 1.8 V. The microphotograph of the chip, including details on the layout, is depicted in Figure 5a, and the aspect ratios of the main transistors in the circuit are reported in Table 1. The measurements have been carried out over 10 samples of the silicon prototype. The testbench implemented for the experimental characterization is represented in Figure 5b, where the on-chip and the PCB levels have been highlighted. An on-chip differential voltage buffer, referred to as $\times 1$, has been included for test purposes in order to isolate the output terminals of the FD IA from heavy loads. The buffer consists of two PMOS source followers including low-V_{th} transistors, so that operation with the general 1.8-V supply is possible. Auxiliary circuits AD8475 and AD8429 in Figure 5b are used to carry out, respectively, a single-to-differential signal conversion at the input of the IA and a differential-to-single signal conversion at the output in order to facilitate measurements. Even though these commercial components have been selected with a bandwidth higher that the circuit under test, their influence on the measurement procedure is unavoidable. The value of the reference voltage V_{REF} used to set the DC level of the output voltage was set to 0.9 V. In addition, this voltage is also used to bias the gate terminal of the cascode transistors. The biasing current of each V-to-I converter, i.e., G_{mI} and G_{mO}, was adjusted as $I_B = 10\ \mu A$. The source degeneration resistors R_I and R_O were implemented with non-salicided high-resistance polysilicon having values equal to $R_I = 5$ kΩ and $R_O = 20$ kΩ, thus leading to a nominal voltage gain of 4 V/V (12.04 dB).

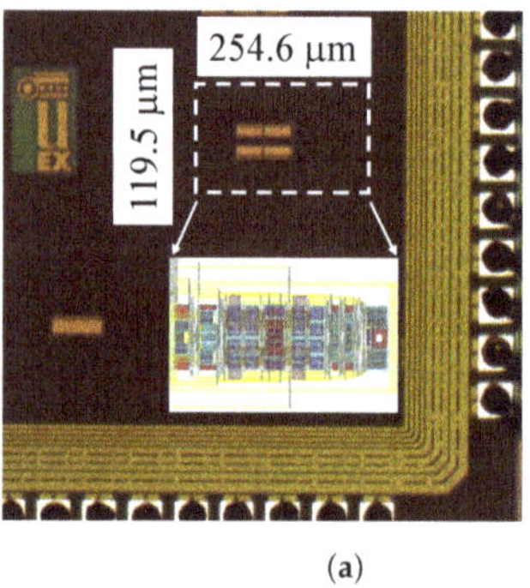

(a)

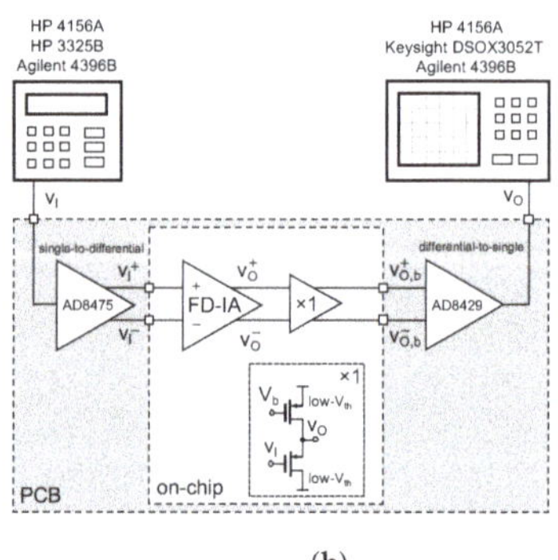

(b)

Figure 5. (a) Chip microphotograph and (b) measurement setup.

Table 1. Transistor aspect ratios (μm/μm) for the FD IA (Figure 3).

Device	W/L (μm/μm)	Device	W/L (μm/μm)
MDI	200/1	MDO	200/1
MFI	80/0.5	MFO	80/0.5
MFCI	20/0.5	MFCO	20/0.5
MSDI	16/1	MSDO	16/1
MSUI	48/1	MSUO	48/1
M1A, M2A	80/0.5	M1B, M2B	80/0.5
M1C	20/0.5	M2C	20/0.5
M3, M4	30/0.5	M3C, M4C	60/0.5

The load capacitors, C_L, were built on-chip as metal-insulator-metal devices to make stable the feedback loop established around the transconductor G_{mO}. The design criterion selected was to ensure a phase margin of 60° considering the nominal value of the load capacitors, equal to 1.33 pF each, and the parasitic capacitance also connected to the output terminals due to the test buffer. In addition, it is worth to point out that the effective value of the parasitic capacitance introduced by the test buffer slightly relies on the value of the total external capacitance, associated to the PCB and the test probe used for measurements. This external capacitance has been estimated to be around 30 pF in most of the test configurations followed. Under these conditions, the open-loop frequency response of the DM and CM signal paths has been simulated and is represented in Figure 6. For the DM signal LG_{DM} = 58.0 dB and $LGBW_{DM}$ = 5.9 MHz with a phase margin of 52.8° and a gain margin of 17.6 dB, whereas the CM signal response provides LG_{CM} = 64.2 dB and $LGBW_{DM}$ = 18.1 MHz with a phase margin of 75.5° and a gain margin of 14.3 dB. These results show the stability of both the DM and the CM feedback loop and confirm the requirements imposed in (7a) and (7b) to the CM signal path. The bandwidth of the CM signal is noticeably higher than that of the DM component. This is due to the fact that the linearization carried out in the input differential structure of the IA leads to a lower effective transconductance as compared to the CMFB section, which results in a narrower frequency range.

The DC measurements on the 10 available samples allowed to obtain an average DC supply current for the IA equal to 266.4 μA, with a standard deviation of 2.6 μA. The DC voltage level shift introduced by the on-chip buffer did not allow characterizing the actual output voltage of the IA, expected to be very close to V_{REF}. Hence, only the standard deviation of the buffered output voltage, equal to 3.63 mV, is reported in order to determine the variability of the output voltage among the different samples. The experimental $v_I - v_O$ DC transfer characteristic of the IA is represented in Figure 7. The CM level of the output voltage, defined as $(v_O^+ + v_O^-)/2$, has been used to shift all plots from their original DC level down to zero, so that results can be more easily interpreted. A linear voltage range at DC larger than ± 50 mV can be inferred for the differential output response. As observed in Figure 7, the non-linearity appreciable in v_O^+ and v_O^- is cancelled out when the overall output signal is obtained as the difference of the individual responses, i.e., $v_O = v_O^+ - v_O^-$.

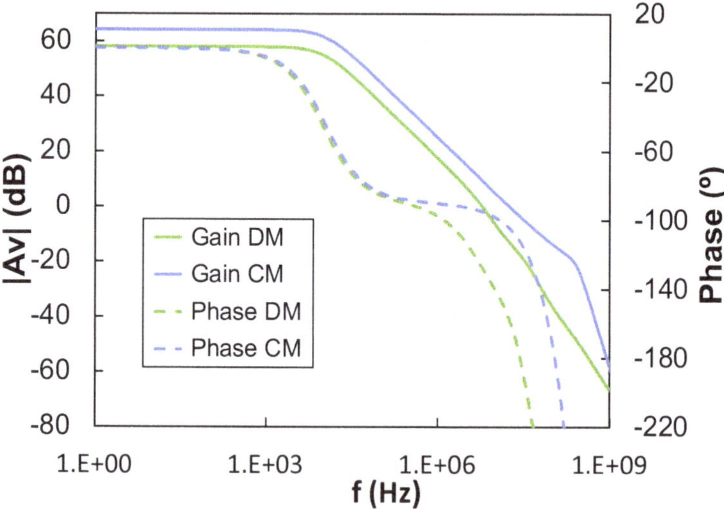

Figure 6. Simulated open-loop frequency response of the DM and CM signal paths in the proposed IA.

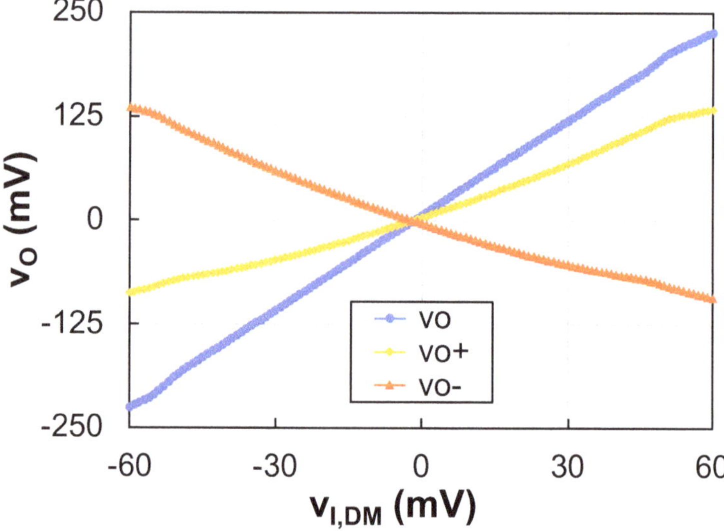

Figure 7. Input–output DC transfer characteristic.

The simulated and experimental frequency response of the IA is illustrated in Figure 8, where the magnitude of the DM voltage gain is depicted. From the experimental response the voltage gain in the passband, A_v, and the BW of the IA can be extracted, obtaining values equal to 3.78 V/V (11.4 dB) and 5.83 MHz, respectively. The gain value is in close agreement with the design value of 4 V/V or 12.4 dB (relative error of 5.0%) and with the simulated value of 3.69 V/V or 11.34 dB (relative error of 2.4%), whereas the measured BW deviates from the corresponding simulated value, equal to 7.76 MHz (relative error of 24.8%). The difference between the simulated and the experimental responses in Figure 8 has two possible reasons. On the one hand, it has been found that the on-chip voltage buffer is more sensible to external load capacitors than expected from simulations. On the other hand, the BW of the IA is determined by the on-chip load capacitors illustrated in Figure 3, the value of which can suffer important absolute variations during the fabrication process. The nominal simulated value of the BW has been complemented with the result extracted from a 1000-run Montecarlo analysis, considering mismatch and process variations, which has been found to be equal to 10.27 ± 4.70 MHz. Considering the standard deviation as a suitable error margin, the lower bound of the statistically simulated BW encloses the values of both the nominally simulated and the measured BW. The time response of the proposed IA, depicted in Figure 9, has been used to confirm its stability. In particular, a 100-mV$_{pp}$ input signal (yellow plot) is applied and an appropriate establishment of the output voltage (green plot) can be observed.

The response to CM signals has also been obtained. The CMRR has been simulated and measured as a function of the frequency of the input signal and is shown in Figure 10. In the simulated plot (in green color), the average value, extracted from a 1000-run Montecarlo analysis including mismatch and process variations, is represented, whereas the error bars indicate the standard deviation, σ. As observed, the experimental CMRR lays below the error margin when the standard deviation is considered, but it has been proved that is enclosed by a 3-σ error region. The measured CMRR at low frequencies and at the frequency of the BW is equal to 73.3 dB and 42.0 dB, respectively. Furthermore, the impact of process, voltage, and temperature (PVT) variations on the CMRR at DC has been determined by nesting a 100-run Montecarlo analysis and a corner analysis. In particular, typical-typical (*tt*), slow-slow (*ss*), fast-fast (*ff*), fast-n-slow-p (*fs*), and slow-n-fast-p (*sf*) corners were considered for the active devices, whereas the temperature was set to values (0,27,80) °C and the supply voltage was adjusted to (1.62,1.8,1.98) V, i.e., a variation equal to $\pm 10\%$ was assumed. The corresponding results are summarized in Figure 11, where in the axis corresponding to the temperature the considered range has been replicated for each corner of the active devices. As observed, the CMRR varies between 74.8 dB and 90.5 dB.

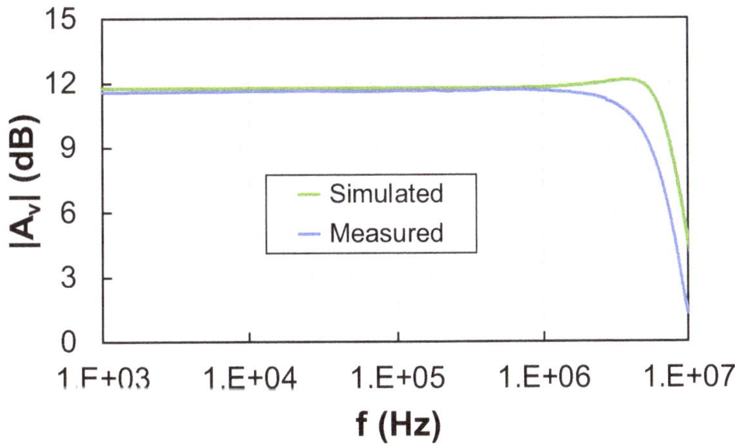

Figure 8. Simulated and measured frequency response of the proposed IA.

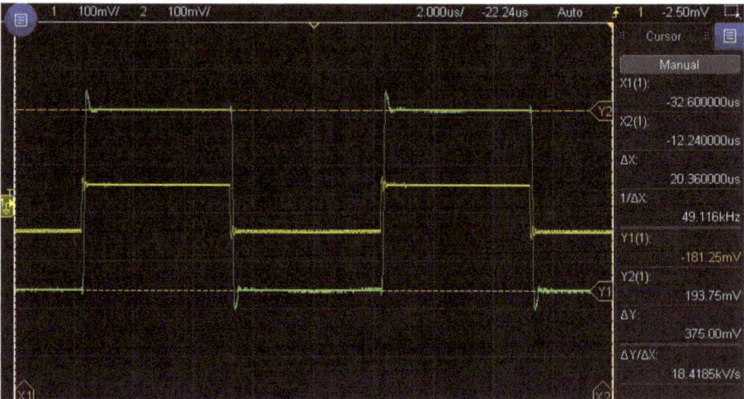

Figure 9. Transient response of the IA output voltage (green) to a 100-mV$_{pp}$ input square wave (yellow).

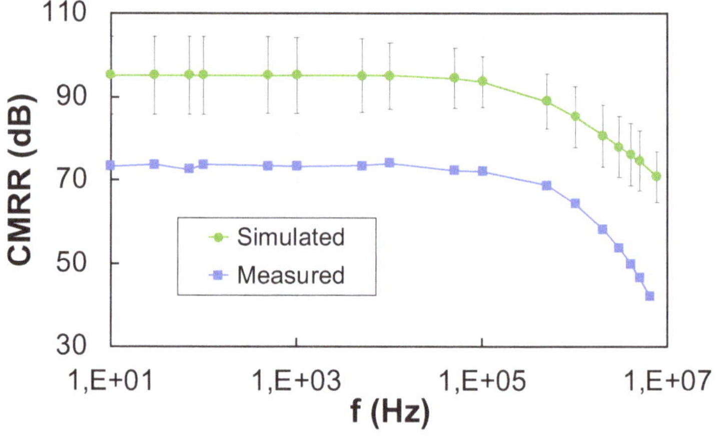

Figure 10. Simulated and measured CMRR vs. frequency.

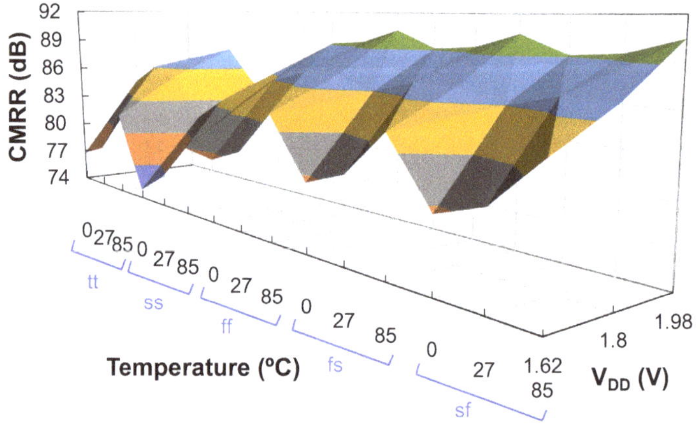

Figure 11. CMRR extracted from a Montecarlo analysis for the different corners.

The noise response of the FD IA has also been characterized. In particular, the spectral density of noise has been simulated and measured and is depicted in Figure 12. In addition, the noise has been integrated over a frequency band between 100 Hz and the frequency of the BW, obtaining a value equal to 86.4 μV_{rms}. The calculated experimental noise is slightly higher than the actual value, due to the finite approximation followed to integrate the noise. In any case, the simulated noise, equal to 74.7 μV_{rms} (relative error of 15.7%), is much lower. The reason of the noise increase in measurements is ascribed to the experimental setup and to the contributions of the different auxiliary circuits used for the test, as illustrated in Figure 5b and already indicated at the beginning of this section. The THD has been used to asses the linearity of the dynamic response of the FD IA. In Figure 13 the simulated and experimental THD of the output voltage is represented as a function of the input DM signal amplitude for frequencies of 1 kHz and 10 kHz. The simulated THD is reduced as compared to the experimental response for small values of the input signal due to the lower noise floor level in simulations. Nevertheless, for high input signals the measured response results even more linear. Using the widespread criterion of considering the 1%-THD as a limit to determine the maximum input signal that can be processed with reasonable linearity, experimental values of 59.6 mV and 57.6 mV were obtained for input frequencies of 1 kHz and 10 kHz, respectively.

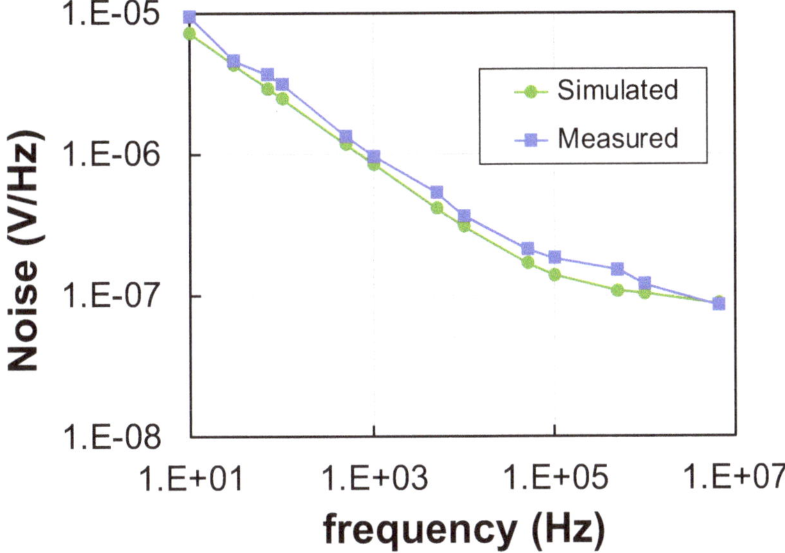

Figure 12. Spectral density of noise vs. frequency: simulated (green) and measured (blue) responses.

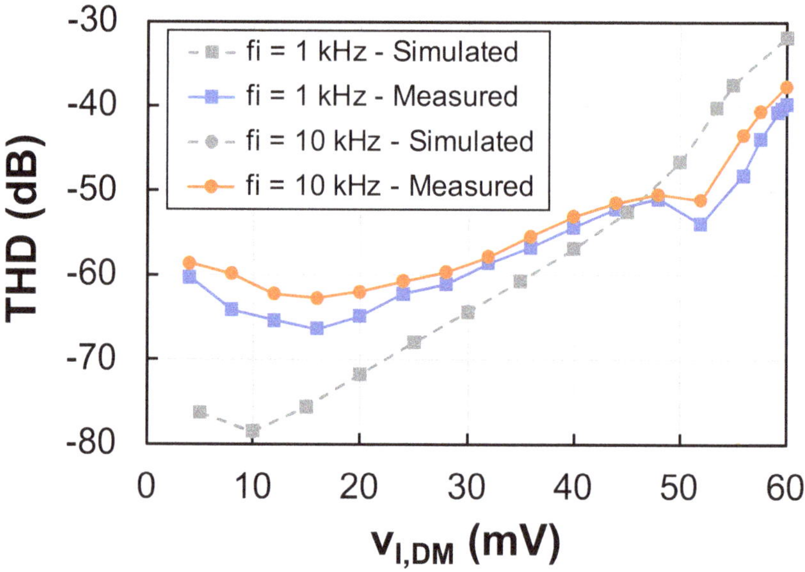

Figure 13. Simulated and experimental THD vs. $v_{I,DM}$ for f_I equal to 1 kHz and 10 kHz.

The performance of the designed and fabricated FD IA is summarized in Table 2, where simulated and measured results are reported. The data expressed as the mean value plus/minus the standard deviation were obtained from a 1000-run Montecarlo analysis with mismatch and process variations in the case of simulations and from the measurements on 10 samples in the case of experimental results. In general, there is a good agreement between the simulated and the measured metrics, being the corresponding differences due to the variations of the process parameters during fabrication. One exception is the case of the noise, which, as discussed previously, greatly increases in measurements with respect to simulations.

The comparison of the previous metrics for different IAs is done usually in terms of a widespread figure-of-merit (FoM) known as noise efficiency factor (NEF) [4]. This parameter indicates how large is the noise of a system as compared to the white noise of a single MOS transistor with the same drain current and bandwidth, and is defined as:

$$NEF = V_{iN,rms} \sqrt{\frac{2I_{DD}}{\pi V_T 4kTBW}} \tag{12}$$

where I_{DD} and V_T are the supply current of the IA and the thermal voltage, respectively. Nevertheless, this parameter does not take into account the amplitude of the signals to be processed. Indeed, when large input signals must be handled, a high biasing current is required, thus resulting in a penalty in therms of NEF. In this case, the dynamic range (DR), defined as

$$DR = 20 \cdot \log \left(\frac{v_{I,DM_{max}}}{V_{iN,rms}} \right) \tag{13}$$

can be used as a complementary FoM for performance comparison.

Table 2. Simulated vs. experimental performance of the FD IA (Technology: 180 nm CMOS, V_{DD} = 1.8 V, $A_{v,nom}$ = 4 V/V).

Parameter	Simulated	Measured
Voltage gain (V/V)	3.69 ± 0.07	3.78 ± 0.06
Voltage gain error (%)	-7.7	-5.5
BW (MHz)	10.27 ± 4.70	5.83
$\sigma(v_O)$ (mV)	5.14	3.63
$v_I\|_{THD=-40\,dB}$ @ 1 kHz (mV)	53.5	59.6
$v_I\|_{THD=-40\,dB}$ @ 10 kHz (mV)	53.5	57.6
$v_I\|_{THD=-40\,dB}$ @ 100 kHz (mV)	53.2	59.0
$v_I\|_{THD=-40\,dB}$ @ 1 MHz (mV)	44.8	38.0
SR^+/SR^- (V/μs)	10.4/10.4	8.3/8.3
CMRR @ DC (dB)	95.1 ± 9.2	73.3
CMRR @ BW (dB)	70.8 ± 6.2	42.0
$V_{iN,rms}$ [100 Hz-BW] (μV_{rms})	74.7	86.4
I_{DD} (μA)	199.1	266.4

The FD IA presented is compared in Table 3 to other works previously reported and with similar characteristics, i.e., based on current feedback and presenting a wide bandwidth. The work by Worapishet et al. [11] presents very good values of NEF and DR, especially considering that measured results are given, but the BW is more limited than in the other solutions. The IAs in [12,22] have a good response in general, even tough they are solutions supported by simulated results. In [26] a very high bandwidth is achieved but no data regarding the size of the processed signals and the noise are reported. The IA proposed in [29] has also a differential structure and achieves a higher BW than the IA proposed here, but the signal processed are smaller and the noise is higher, thus resulting in a higher NEF and a lower DR. The proposed IA has a BW suitable for electrical bioimpedance analysis and is able to process the largest input differential signals for similar supply currents. In addition, it is a compact solution in terms of silicon area as compared to most of the other solutions, especially considering that it has a FD structure. Finally, it is worth to point out that the increase of the experimental noise, previously indicated, leads to a noticeable reduction of the measured DR and to an increase of the experimental value of the NEF. Indeed, the simulated characterization of the IA reported values for the NEF and the DR equal to 14.6 and 57.1, respectively.

Table 3. Performance comparison of the proposed IA with other works previously reported.

Parameter	[11] TCAS-I'11	[12] IMCSSD'12	[22] IJEC'20	[26] TCAS-II'21	[29] Electronics'22	This Work
Technology	0.35-µm CMOS	0.35-µm CMOS	0.35-µm CMOS	0.18-µm CMOS	0.18-µm CMOS	0.18 µm CMOS
Technique [*]	LCF	LCF	ICF	G_m-TI	ICF	ICF
Results	Meas.	Sim.	Sim.	Sim.	Meas.	Meas.
V_{DD} (V)	3	2	3	1.8	1.8	1.8
I_{DD} (μA)	285	240	250.6	162	219.3	266.4
Gain (dB)	34	8	34	0/40	11.4	11.4
BW (MHz)	2.0	4.0	7.6	$6.7 \times 10^{-6}/87.0$	8.0	5.83
CMRR (dB)	>90 @ DC	80 @ 1 MHz	99.5 @ DC	164.4 @ 100 kHz	80.6 @ DC	73.3 @ DC
THD (dB) @ v_I (mV$_{pp}$)	−56.2 @ 10	N.A.	−57.4 @ 10	N.A.	−61.6 @ 20	−64.9 @ 20
$v_{I,max}$ (mV)	30	N.A.	8	N.A.	53	59.6
$V_{iN,rms}$ (μV_{rms})	16	36	32.4	N.A.	92.0	86.4
Area (mm^2)	0.068	0.037	—	0.0569	0.0291	0.0304
NEF	5.9	10.8	7.2	N.A.	26.3	21.3
DR	65.5	N.A.	47.9	N.A.	52.2	56.8

[*] LCF: local current feedback; ICF: indirect current feedback; G_m-TI: transconductance-transimpedance.

5. Conclusions

An IA suitable for bioimpedance-based IoT applications and based on the ICF technique has been presented. The SSF structure has been incorporated in the design of the IA in order to reduce input referred noise and silicon area. In addition, a FD implementation has been selected to enhance the overall performance of the circuit. The proposed ICF FD IA been designed and fabricated in 180 nm CMOS technology to operate with a supply voltage of 1.8 V and provide a voltage gain of 4 V/V. Measurements on 10 different samples of the silicon prototype showed wide bandwidth, high CMRR and linear signal processing, thus confirming the suitability of the proposed solution for the intended application.

Author Contributions: Conceptualization, I.C., J.M.C. and J.L.A.; methodology, I.C., J.M.C. and J.L.A.; software, I.C. and J.M.C.; formal analysis, J.M.C. and R.P.-A.; investigation, I.C., J.M.C. and J.L.A.; resources, I.C., J.M.C. and M.Á.D.; data curation, I.C., J.M.C., M.Á.D. and R.P.A.; writing—original draft preparation, I.C., J.M.C. and J.L.A.; writing—review and editing, I.C., J.M.C., J.L.A., M.Á.D., R.P.-A. and J.F.D.-C.; visualization, I.C. and J.M.C.; supervision, J.M.C. and J.L.A.; project administration, J.F.D.-C.; funding acquisition, J.L.A. and J.F.D.-C. All authors have read and agreed to the published version of the manuscript.

Funding: Work funded by projects RTI2018-095994-B-I00, from MCIN/AEI/10.13039/501100011033, and IB18079, from *Junta de Extremadura* R&D Plan, and by Fondo Europeo de Desarrollo Regional (FEDER) Una manera de hacer Europa. Silicon samples granted by EUROPRACTICE MPW and design tool support.

Conflicts of Interest: The authors declare no conflict of interests.

References

1.   Minhee, K.; Eunkyoung, P.; Hwan, C.B.; Kyu-Sung, L. Recent Patient Health Monitoring Platforms Incorporating Internet of Things-Enabled Smart Devices. *Int. Neurourol. J.* **2018**, *22*, S76–S82.
2.   Gope, P.; Gheraibia, Y.; Kabir, S.; Sikdar, B. A Secure IoT-Based Modern Healthcare System With Fault-Tolerant Decision Making Process. *IEEE J. Biomed. Health Inf.* **2021**, *25*, 862–873. [CrossRef] [PubMed]
3.   Grimnes, S.; Martinsen, V.G. *Bioimpedance and Bioelectricity Basics*, 3rd ed.; Academic Press: Cambridge, MA, USA, 2015.
4.   Steyaert, M.S.J.; Sansen, W.M.C. A micropower low-noise monolithic instrumentation amplifier for medical purposes. *IEEE J. Solid-State Circuits* **1987**, *22*, 1163–1168. [CrossRef]
5.   van den Dool, B.J.; Huijsing, J.K. Indirect current feedback instrumentation amplifier with a common-mode input range that includes the negative rail. *IEEE J. Solid-State Circuits* **1993**, *28*, 743–749. [CrossRef]
6.   Martins, R.; Selberherr, S.; Vaz, F.A. A CMOS IC for portable EEG acquisition systems. *IEEE Trans. Instrum. Meas.* **1998**, *47*, 1191–1196. [CrossRef]
7.   Harrison, R.R.; Charles, C. A low-power low-noise CMOS amplifier for neural recording applications. *IEEE J. Solid-State Circuits* **2003**, *38*, 958–965. [CrossRef]
8.   Zhao, Y.Q.; Demosthenous, A.; Bayford, R.H. A CMOS instrumentation amplifier for wideband bioimpedance spectroscopy systems. In Proceedings of the 2006 IEEE International Symposium on Circuits and Systems, Kos, Greece, 21–24 May 2006; pp. 5079–5082.
9.   Yazicioglu, R.F.; Merken, P.; Puers, R.; Van Hoof, C. A 60 µW 60 nV/$\sqrt{\text{Hz}}$ readout front-end for portable biopotential acquisition systems. *IEEE J. Solid-State Circuits* **2007**, *42*, 1100–1110. [CrossRef]
10. Denison, T.; Consoer, K.; Santa, W.; Avestruz, A.; Cooley, J.; Kelly, A. A 2 µW 100 nV/$\sqrt{\text{Hz}}$ chopper-stabilized instrumentation amplifier for chronic measurement of neural field potentials. *IEEE J. Solid-State Circuits* **2007**, *42*, 2934–2945. [CrossRef]
11. Worapishet, A.; Demosthenous, A.; Liu, X. A CMOS instrumentation amplifier with 90-dB CMRR at 2-MHz using capacitive neutralization: Analysis, design considerations, and implementation. *IEEE Trans. Circuits Syst. I Regul. Pap.* **2011**, *58*, 699–710. [CrossRef]
12. Ramos, J.; Ausín, J.L.; Duque-Carrillo, J.F.; Torelli, G. Wideband low-power current-feedback instrumentation amplifiers for bioelectrical signals. In Proceedings of the International Multi-Conference on Systems, Signals and Devices, Chemnitz, Germany, 20–23 March 2012; pp. 1–5.
13. Abdelhalim, K.; Jafari, H.M.; Kokarovtseva, L.; Velazquez, J.L.P.; Genov, R. 64-channel UWB wireless neural vector analyzer SOC with a closed-loop phase synchrony-triggered neurostimulator. *IEEE J. Solid-State Circuits* **2013**, *48*, 2494–2510. [CrossRef]
14. Ong, G.T.; Chan, P.K. A power-aware chopper-stabilized instrumentation amplifier for resistive Wheatstone bridge sensors. *IEEE Trans. Instrum. Meas.* **2014**, *63*, 2253–2263. [CrossRef]
15. Van Helleputte, N.; Konijnenburg, M.; Pettine, J.; Jee, D.; Kim, H.; Morgado, A.; Van Wegberg, R.; Torfs, T.; Mohan, R.; Breeschoten, A.; et al. A 345 µW multi-sensor biomedical SoC with bio-impedance, 3-channel ECG, motion artifact reduction, and integrated DSP. *IEEE J. Solid-State Circuits* **2015**, *50*, 230–244. [CrossRef]
16. Worapishet, A.; Demosthenous, A. Generalized analysis of random common-mode rejection performance of CMOS current feedback instrumentation amplifiers. *IEEE Trans. Circuits Syst. I Regul. Pap.* **2015**, *62*, 2137–2146. [CrossRef]
17. Chang, C.; Zahrai, S.A.; Wang, K.; Xu, L.; Farah, I.; Onabajo, M. An analog front-end chip with self-calibrated input impedance for monitoring of biosignals via dry electrode-skin interfaces. *IEEE Trans. Circuits Syst. I Regul. Pap.* **2017**, *64*, 2666–2678. [CrossRef]
18. Rezaeiyan, Y.; Zamani, M.; Shoaei, O.; Serdijn, W.A. A 0.5 µA/channel front-end for implantable and external ambulatory ECG recorders. *Microelectron. J.* **2018**, *74*, 79–85. [CrossRef]
19. Nasserian, M.; Peiravi, A.; Moradi, F. A fully-integrated 16-channel EEG readout front-end for neural recording applications. *AEU–Int. J. Electron. Commun.* **2018**, *94*, 109–121. [CrossRef]
20. Lee, C.; Song, J. A chopper stabilized current-feedback instrumentation amplifier for EEG acquisition applications. *IEEE Access* **2019**, *7*, 11565–11569. [CrossRef]
21. Psychalinos, C.; Minaei, S.; Safari, L. Ultra low-power electronically tunable current-mode instrumentation amplifier for biomedical applications. *AEU–Int. J. Electron. Commun.* **2020**, *117*, 153120. [CrossRef]
22. Carrillo, J.M.; Domínguez, M.A.; Pérez-Aloe, R.; de la Cruz Blas, C.A.; Duque-Carrillo, J.F. Low-power wide-bandwidth CMOS indirect current feedback instrumentation amplifier. *AEÜ–Int. J. Electron. Commun.* **2020**, *123*, 153299. [CrossRef]
23. Kwon, Y.; Kim, H.; Kim, J.; Han, K.; You, D.; Heo, H.; Cho, D.i.; Ko, H. Fully differential chopper-stabilized multipath current-feedback instrumentation amplifier with R-2R DAC offset adjustment for resistive bridge sensors. *Appl. Sci.* **2020**, *10*, 63. [CrossRef]
24. Han, K.; Kim, H.; Kim, J.; You, D.; Heo, H.; Kwon, Y.; Lee, J.; Ko, H. A 24.88 nV/$\sqrt{\text{Hz}}$ Wheatstone bridge readout integrated circuit with chopper-stabilized multipath operational amplifier. *Appl. Sci.* **2020**, *10*, 399. [CrossRef]
25. Matthus, C.D.; Buhr, S.; Kreißig, M.; Ellinger, F. High gain and high bandwidth fully differential difference amplifier as current sense amplifier. *IEEE Trans. Instrum. Meas.* **2021**, *70*, 1–11. [CrossRef]
26. Pérez-Bailón, J.; Sanz-Pascual, M.T.; Calvo, B.; Medrano, N. Wide-band compact 1.8 V-0.18 µm CMOS analog front-end for impedance spectroscopy. *IEEE Trans. Circuits Syst. II Express Briefs* **2022**, *69*, 764–768. [CrossRef]
27. Pérez-Bailón, J.; Calvo, B.; Medrano, N. 1.0 V-0.18 µm CMOS tunable low pass filters with 73 dB DR for on-chip sensing acquisition systems. *Electronics* **2021**, *10*, 563. [CrossRef]

28. Ashayeri, M.; Yavari, M. A front-end amplifier with tunable bandwidth and high value pseudo resistor for neural recording implants. *Microelectron. J.* **2022**, *119*, 105333. [CrossRef]
29. Corbacho, I.; Carrillo, J.M.; Ausín, J.L.; Domínguez, M.A.; Pérez-Aloe, R.; Duque-Carrillo, J.F. Compact CMOS wideband instrumentation amplifiers for multi-frequency bioimpedance measurement: A design procedure. *Electronics* **2022**, *11*, 1668. [CrossRef]
30. Banu, M.; Khoury, J.; Tsividis, Y. Fully differential operational amplifiers with accurate output balancing. *IEEE J. Solid-State Circuits* **1988**, *23*, 1410–1414. [CrossRef]

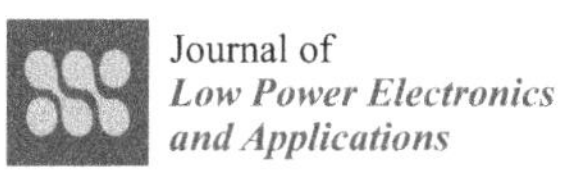

Journal of
Low Power Electronics and Applications

Article

Wideband Cascaded and Stacked Receiver Front-Ends Employing an Improved Clock-Strategy Technique

Arash Abbasi *,† and Frederic Nabki †

École de Technologie Supérieure (ÉTS), Montreal, QC H3C 1K3, Canada
* Correspondence: arash.abbasi.1@ens.etsmtl.ca
† Current address: Department of Electrical Engineering, École de Technologie Supérieure ÉTS, Montreal, QC H3C 1K3, Canada.

Abstract: A wideband cascaded receiver and a stacked receiver using an improved clock strategy are proposed to support the software-defined radio (SDR). The improved clock strategy reduces the number of mixer switches and the number of LO clock paths required to drive the mixer switches. This reduces the dynamic power consumption. The cascaded receiver includes an inverter-based low-noise transconductance amplifier (LNTA) using a feed-forward technique to enhance the noise performance; a passive mixer; and an inverter-based transimpedance amplifier (TIA). The stacked receiver architecture is used to reduce the power consumption by sharing the current between the LNTA and the TIA from a single supply. It utilizes a wideband LNTA with a capacitor cross-coupled (CCC) common-gate (CG) topology, a passive mixer to convert the RF current to an IF current, an active inductor (AI) and a $1/f$ noise-cancellation (NC) technique to improve the noise performance, and a TIA to convert the IF current to an IF voltage at the output. Both cascaded and stacked receivers are simulated in 22 nm CMOS technology. The cascaded receiver achieves a conversion-gain from 26 dB to 36 dB, a double-sideband noise-figure (NFDSB) from 1.4 dB to 3.9 dB, $S_{11} < -10$ dB and an IIP3 from -7.5 dBm to -10.5 dBm, over the RF operating band from 0.4 GHz to 12 GHz. The stacked receiver achieves a conversion-gain from 34.5 dB to 36 dB, a NFDSB from 4.6 dB to 6.2 dB, $S_{11} < -10$ dB, and an IIP3 from -21 dBm to -17.5 dBm, over the RF operating band from 2.2 GHz to 3.2 GHz. The cascaded receiver consumes 11 mA from a 1 V supply voltage, while the stacked receiver consumes 2.4 mA from a 1.2 V supply voltage.

Keywords: wireless receiver; wideband; cascaded; stacked; harmonic recombination; N-path receiver; LNTA

Citation: Abbasi, A.; Nabki, F. Wideband Cascaded and Stacked Receiver Front-Ends Employing an Improved Clock-Strategy Technique. *J. Low Power Electron. Appl.* **2023**, *13*, 14. https://doi.org/10.3390/jlpea13010014

Academic Editor: Orazio Aiello

Received: 24 November 2022
Revised: 27 January 2023
Accepted: 30 January 2023
Published: 2 February 2023

1. Introduction

Wireless standards operate over a wide frequency spectrum spanning tens of GHz and employ various modulation schemes. Wireless applications have led to the rapid growth of wireless devices in all sectors of the internet of things (IoT), such as health monitoring, agriculture, and smart cities. A wideband system such as the software defied radio (SDR) is a well-suited architecture to address several wireless standards in a single receiver module. Conventional SDRs required a high specification analog-to-digital converters (ADCs), which increased the power consumption and complexity of the design [1]. In [2], down conversion is proposed to reduce the power consumption.

In wideband operation, the wanted signal at the local oscillator (LO) frequency down-converts to the baseband, along with other components of the LO harmonics. This degrades the error vector magnitude (EVM) performance. Harmonic recombination using the N-path receiver architecture is employed to overcome this problem [3–6]. One of the drawbacks of N-path receivers is that they require a high-frequency driving clock (e.g., 8× the LO frequency) to generate the clock phases that are needed to down-convert the signal at the desired LO frequency, and they reject the harmonics of the LO signal . Similarly, the

N-path passive mixer-first topologies that offer input matching without using external components and high-quality filtering can be used [7–9]. However, they consume high power to achieve a low noise figure (NF). In addition, mixer-first topologies are not suitable for wideband applications. An N-path ultra-low-power mixer-first receiver is presented in [10]. Although very low power consumption is achieved, it requires an off-chip inductor and achieves a low modulation bandwidth of 3.5 MHz. A feed-forward technique with tuned LO phase was employed in [11] to reject the LO harmonics. However, the phase-correction circuit increases the complexity of the design and reduces accuracy. A harmonic recombination technique that down-converts the signal at $3\times$ the LO frequency is used in [12]. This technique removes all of the other harmonics at the LO frequencies such as the 1st, 2nd, 4th, and 5th harmonics. However, it consumes a significant amount of power in the baseband harmonic recombination circuitry. In addition, it uses a low noise transconductance amplifier (LNTA) with two inductors that occupy a relatively large area. Our earlier work [13] overcame the problems mentioned above. It employed a clock strategy technique to down-convert the signal at $4\times$ the LO frequency while removing all of the other harmonics at the LO frequencies (i.e., 1st , 2nd, 3rd, 5th, 6th, and 7th harmonics). To reduce the power consumption, a current-reuse receiver topology was used. It employed the common-gate LNTA topology with a single to differential balun. In addition, an active inductor was used to improve the receiver sensitivity at higher RF. However, the clock strategy technique proposed in that prior work can be improved to reduce the mixer design complexity and the number of switches. In addition, the technique suffers from low-frequency noise due to the direct coupling of the LNTA noise to the output through the active inductor (AI).

To overcome the limitations mentioned above, this work proposes an improved clock strategy technique that reduces the number of mixer switches and the number of LO clock paths required to drive the mixer switches. This reduces the dynamic power consumption. The clock strategy technique down-converts an RF signal at $4\times$ the LO frequency. The proposed clock strategy is verified through simulations in both cascaded and stacked receiver front-ends. In the cascade receiver front-end, very high RF bandwidth, low noise figure, and good linearity are achieved compared to the stacked receiver front-end at the cost of higher power consumption. The $1/f$ noise problem of [13] is resolved in this work by using a $1/f$ noise-cancellation (NC) technique. Current mode harmonic recombination is used to reduce the power consumption by avoiding the use of additional harmonic recombination circuitry.

The paper presents the clock strategy in Section 2, the cascaded receiver front-end design and its simulation results in Section 3, the stacked receiver front-end design and its simulation results in Section 4, and the comparison and discussion in Section 5. This is followed by a conclusion.

2. Clock Strategy Technique

The harmonic recombination technique has been used for wideband receiver front-ends to suppress the harmonics of the LO frequency that they down-convert to the baseband along with unwanted signals and noise. A harmonic rejection mixer using a parallel mixing path with a gain ratio of $1 : \sqrt{2} : 1$ is proposed in [14] to reject the third and fifth harmonics at the cost of using two frequency generator circuits that consume area and power. Another approach [15] achieved higher harmonic rejection using a digital adaptive-interference-cancelling (AIC) technique to enhance harmonic rejection. However, it requires high power. In addition, it requires $4\times$ the LO frequency to generate 8-phase clocks to down-convert the signal at the LO. In [16], a 32-phase non-overlapping LO clock is used to achieve very good harmonic rejection (HR) after LO clock-phase calibration. However, it consumes 30 mW, and it requires harmonic selective TIAs, which increases the area and power consumption. Figure 1a shows the conventional harmonic recombination technique that down-converts a wanted signal at f_{LO} and rejects all of the LO harmonics (i.e., $2\times f_{LO}$, $3\times f_{LO}$, ..., where $f_{LO} = CLKIN/4$). This requires a CLKIN signal that is equal to $4\times$

of the LO frequency, increasing the complexity and power draw of the clock generation circuit, due to the high clock frequencies required. To overcome these issues, this work proposes a harmonic recombination technique, shown in Figure 1b, that employs a clock strategy to down-convert the wanted signal at $4 \times f_{LO}$ and reject signals at f_{LO}, $2 \times f_{LO}$, $3 \times f_{LO}$, $5 \times f_{LO}$ etc. For instance, using a CLKIN at 10 GHz, an RF input at 10 GHz is down-converted to baseband. This relaxes the requirements to design the LO clock generation circuits. Figure 1c shows the circuit diagram of the clock divider to generate 8-phase clocks (PH0, PH45, ..., PH316) for conventional harmonic recombination. OR-gates are used to combine the mentioned clocks to generate LO_1 and LO_2. The proposed clock strategy technique reduces the number of switches in the mixer to two in comparison to eight in the conventional harmonic recombination technique. This reduces the dynamic power consumption in the LO clock paths. In addition, the simplified LO routing on the chip reduces clock signal leakage to the substrate and improves signal integrity.

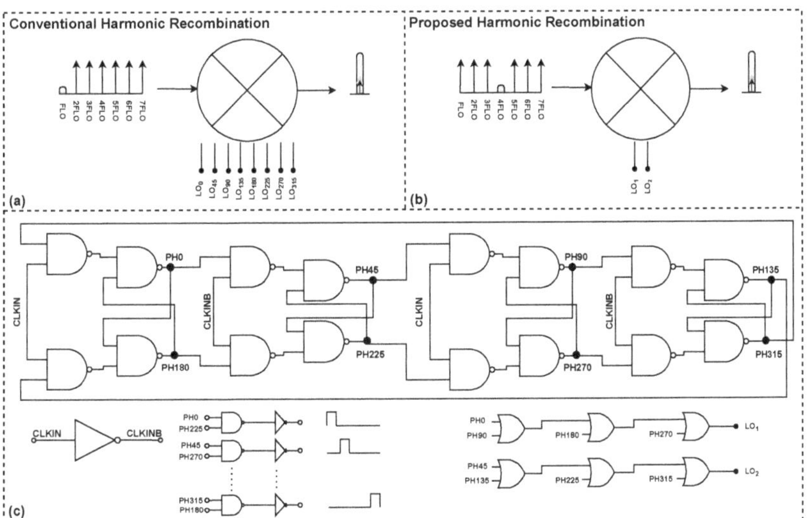

Figure 1. Harmonic recombination techniques: (**a**) cascaded approach, (**b**) stacked approach, and (**c**) clock generation circuitry.

To evaluate the proposed recombination strategy, the Fourier series coefficients are calculated using

$$
\begin{cases}
a_0 = \frac{2}{N} \sum_{k=1}^{N} s[k] \\
a_n = \frac{2}{N} \sum_{k=1}^{N} s[k].cos[\frac{2\pi}{N}nk] \\
b_n = \frac{2}{N} \sum_{k=1}^{N} s[k].sin[\frac{2\pi}{N}nk]
\end{cases}
\tag{1}
$$

where N is the pulse period, k is the sample number, n is the harmonic number, and $s[k]$ is the signal given by

$$
s[k] = \sum_{m=1}^{M} (-1)^m P_m[k] = \sum_{m=1}^{M} (-1)^m (u[k - \frac{(m-1)\pi}{4}] - u[k - \frac{m\pi}{4}])
\tag{2}
$$

where $u[k]$ is the step function and M is the number of each shifted single pulses. The coefficients are calculated based on (1), where a_0 and a_n are zero for all harmonics, and b_n for harmonics $n = 4, 12, ..., 4(2i-1)$ is calculated by $\frac{16}{n\pi}$. Table 1 shows the calculated Fourier series coefficients of the proposed harmonic recombination technique for seven harmonics. It also presents the example scenario that a signal at 4 GHz is down-converted while the other LO harmonics are rejected.

Table 1. Fourier series coefficients of the proposed method.

b_n	b_1	b_2	b_3	b_4	b_5	b_6	b_7
Value (dB)	$-$Inf	$-$Inf	$-$Inf	2.2	$-$Inf	$-$Inf	$-$Inf
Freq. (GHz)	1	2	3	4	5	6	7

3. Cascaded Receiver Front-End Using the Proposed Clock Strategy

The functionality of the proposed harmonic recombination technique is verified using a cascaded receiver front-end architecture where the LNTA, passive mixer, and TIA are cascaded. Despite previous N-path receiver architectures that use complex and power-consuming circuits to combine the signals at the mixer output [4,17], the proposed receiver performs harmonic recombination in current-mode at the mixer output followed by a single TIA shown in Figure 2.

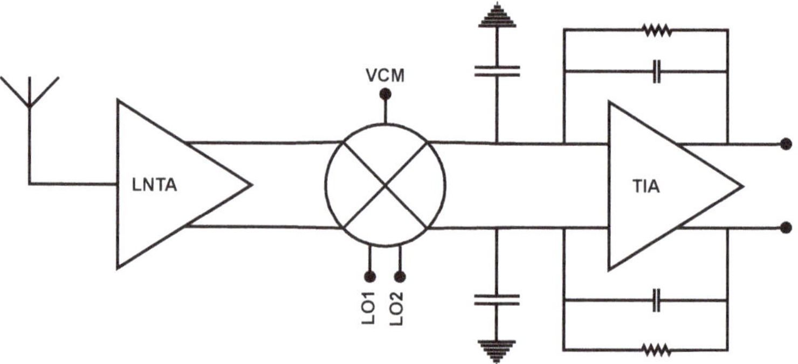

Figure 2. Cascaded receiver diagram.

3.1. LNTA Design

Conventionally, a wideband low noise amplifier (LNA) is used followed by a g_m-stage, down-conversion mixer and transimpedance amplifier (TIA). This helps reduce the NF but also amplifies blockers, which can saturate the following stages of the receiver[18]. The LNTA is an alternative that can be used to convert the RF voltage to an RF current that is then down-converted to the IF or BB current through a passive mixer. In this fashion, the receiver is not compressed by blockers due to the inherent low-voltage gain [15]. This work employs a low noise and wideband LNTA proposed in [19], shown in Figure 3.

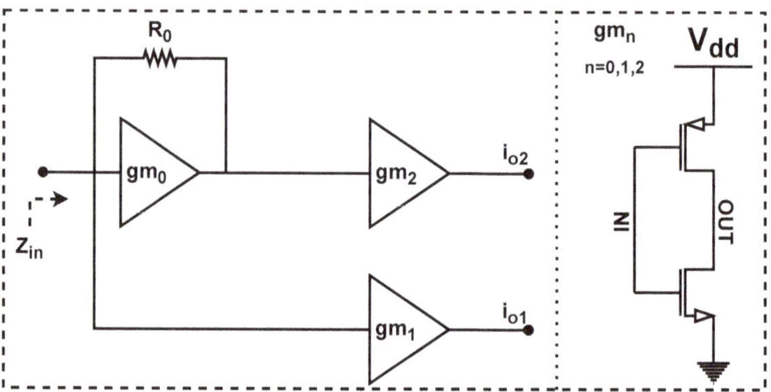

Figure 3. LNTA circuit diagram.

The first transconductance stage (gm_0) provides the input impedance matching, which is given by

$$Z_{in} \cong \frac{1}{gm_0}(1 + R_0/r_0)$$ (3)

where R_0 is the feedback resistor and r_0 is the output resistance of the gm_0 circuit.

Thanks to the feed-forward technique that provides noise-cancellation, the LNTA can achieve a low NF given by

$$NF \cong 1 + \frac{gm_2}{gm_1} + \frac{\lambda}{gm_1 R_s}$$ (4)

where λ is the short-channel effect coefficient, which can be reduced by increasing the transistor lengths. It can be seen the NF is independent of gm_0 and can be reduced by increasing the gm_1 value.

The total transconductance gain is approximated and given by

$$G_m \cong \frac{1}{2}\left[\left(\frac{gm_0 R_0 - 1}{1 + R_0/r0}\right)gm_2 + gm_1\right]$$ (5)

Transconductance gm_0 provides gate biasing for the gm_1 and gm_2 inverters along with bulk biasing of the PMOS and NMOS transistors since flip-well devices are used in a fully depleted silicon on insulator (FDSOI) CMOS technology. This reduces the area and parasitic capacitance of the AC-coupling capacitors at the input of gm_1 and gm_2.

3.2. Mixer Design

The mixer is responsible for down-converting the RF signal to the IF signal using the LO signal. There are two well-known mixer architectures: the passive and the active mixer. The passive mixer is preferred over the active mixer due to its high linearity performance yielded by the current-mode operation. In the passive mixer, switches are biased in a linear region. The gate of the switches are biased to make sure that the LO signal is able to turn on the mixer switches when it toggles between 0 V and the supply voltage. The circuit diagram of the mixer is shown in Figure 4. The mixer input is ac-coupled using a capacitor to separate the LNTA biasing and block low-frequency noise. The drain and source of the mixer switches are biased by the TIA common-mode voltage.

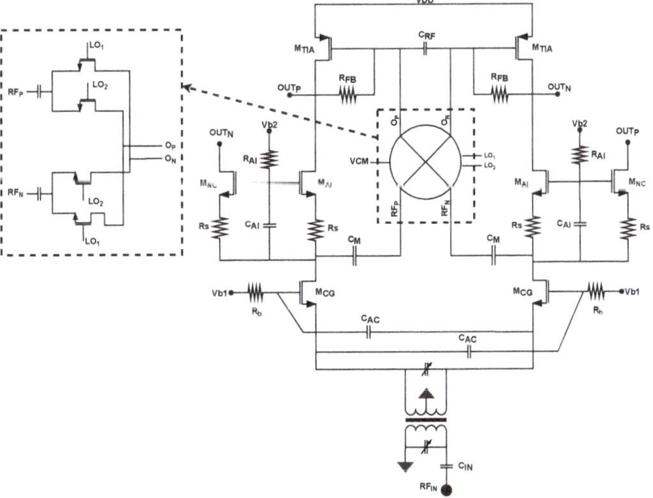

Figure 4. Stacked receiver circuit diagram.

3.3. TIA Design

The transimpedance amplifier (TIA) is used after the passive mixer to convert the IF current to an IF voltage at the output. In addition, it provides low-input impedance that improves the linearity. This work employs an inverter-based TIA using a feedback resistor to control the gain and a capacitor bank to define the IF bandwidth, shown in Figure 5. The current reuse inverter using both PMOS and NMOS enhances the overall transconductance without consuming extra power.

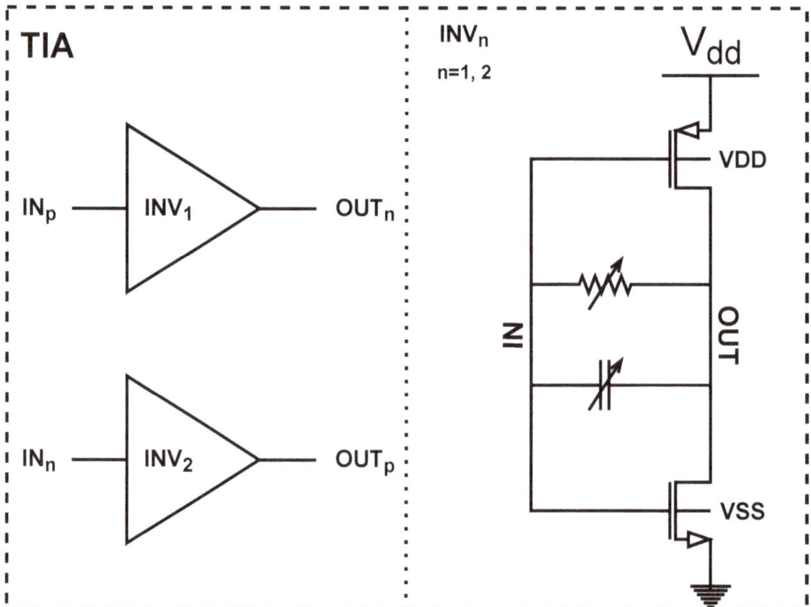

Figure 5. TIA circuit diagram.

The input impedance looking into the INV_n input is given by

$$Z_{in} = \frac{R}{g_m R_{out}} + \frac{1}{g_m}$$ (6)

where R is the feedback resistor and g_m is given by $g_{mp} + g_{mn}$.

The conversion-gain of the proposed receiver can be calculated as

$$CG \cong \frac{2}{\pi} \frac{Sin(\pi d)}{2d} g_m R_{FB}$$ (7)

where d is the clock duty-cycle that is 12.5% in this work.

3.4. Simulation Results of the Cascaded Receiver Front-End

The wideband receiver front-end using the clock strategy was designed and simulated in a 22 nm CMOS technology. The receiver consumes 11 mA from a 1 V supply voltage. The LNTA and TIA consume 4 mA and 7 mA, respectively.

The wideband input matching (S_{11}) of the LNTA is shown in Figure 6, showing an S_{11} of less than −10 dB at up to 13 GHz, which is suitable for ultra wideband applications.

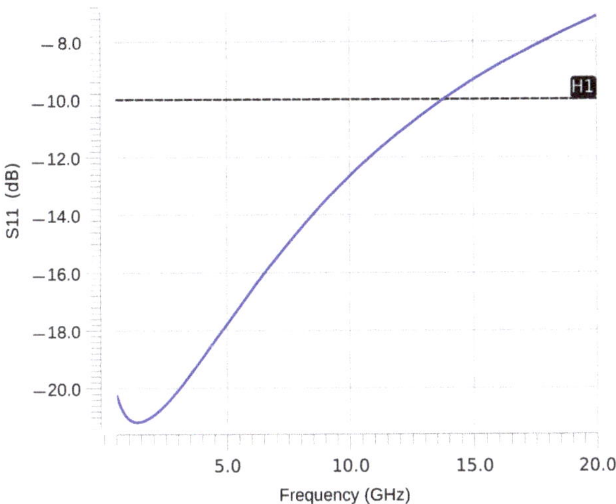

Figure 6. The cascaded received input matching (S11) performance versus RF.

The receiver-performance, integrated double-sideband noise-figure (NFDSB) from 10 kHz to 100 MHz, conversion gain and input-referred third-order intercept point (IIP3) versus f_{LO}, which is equivalent to an input RF signal having a frequency of $4 \times f_{LO}$, from 400 MHz to 12 GHz, is shown in Figure 7 . It shows the NFDSB is increasing in frequency from almost 1.4 dB to 3.9 dB. On the other hand, the conversion gain reduces from almost 36 dB to 26 dB as f_{LO} is increased. A constant value of feedback resistor is used in the TIA. To perform the IIP3 simulation, a two-tone signal at $4 \times f_{LO} + 10$ MHz and $4 \times f_{LO} + 11$ MHz is applied at the input of the LNTA. This generates two third-order intermodulation products at 9 MHz and 12 MHz along with two fundamental products at 10 MHz and 11 MHz. The IIP3 performance varies over f_{LO} from -10.5 dBm to -7.5 dBm.

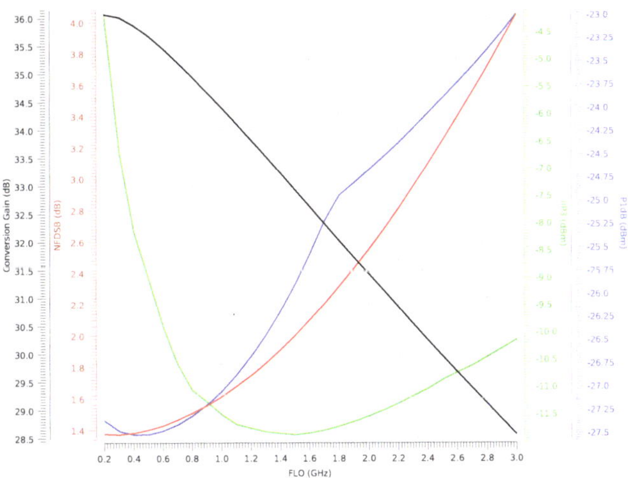

Figure 7. The cascaded receiver NFDSB, conversion gain, and IIP3 performances versus f_{LO}.

The TIA bandwidth can be configured with four settings using two capacitors. Figure 8 shows the receiver bandwidth can be configured from 250 MHz to almost 1 GHz. This can be changed using different values of feedback capacitors and also the shunt capacitors after the mixer, which are 2 pF in this work. It shows that the receiver is suitable for very wideband baseband modulation.

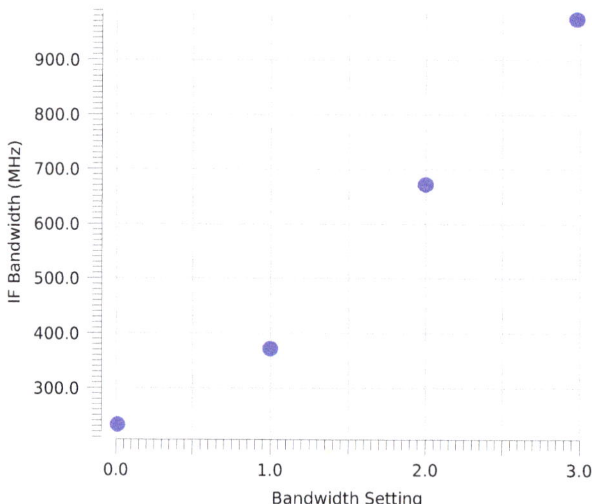

Figure 8. The cascaded receiver bandwidth versus the bandwidth settings.

The harmonic rejection can be affected by the transistor process and mismatch variations. The effect of the transistor process and mismatch variation is verified using Monte-Carlo simulation over 100 runs, and the results are shown in Figure 9. The HR1, HR2, ... HRn (n = 7) are the 1st, 2nd, ... nth harmonics rejected relative to the 4th harmonic, which is the wanted signal in this work. This shows that very good harmonic rejection is achieved at all harmonics with a minimum rejection of 134 dB in HR7 using a 3× sigma calculation.

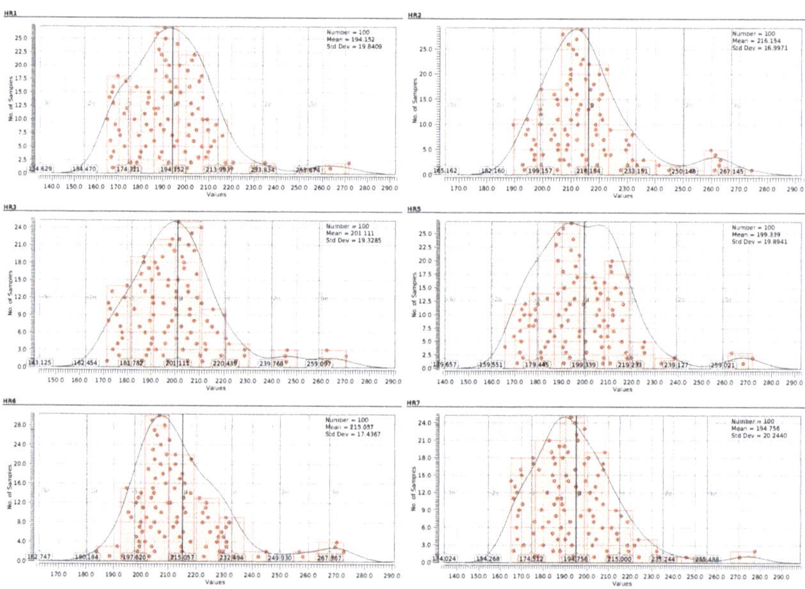

Figure 9. HR_n performance of the cascaded receiver over 100 runs.

4. Stacked Receiver Front-End Using the Proposed Clock Strategy

By scaling down the CMOS technology node , the threshold voltage (V_{TH}) is lowered, enhancing the frequency of operation and enabling new low-power design techniques that have emerged. One effective low-power design technique is the current-reuse or stacked technique by means of stacking different circuits such as LNTA, mixer, and TIA to share the biasing current from a single supply. Thus, in the stacked receiver front-end, the LNTA, mixer, and TIA are stacked.

Although this reduces power consumption, it still has drawbacks. The LNA, mixer, and voltage controlled oscillator (VCO) are staked in [20] to improve the power efficiency, but the circuit may suffer from the injection locking of the VCO due to the large blockers. Moreover, it has high NF. An unbalanced single to differential LNA, active mixer, and baseband circuitry are cascoded to reduce power consumption in [21]. However, the active mixer degrades the linearity performance and increases the voltage headroom requirements. A simultaneous input matching and $1/f$ NC technique is employed in [22] that results in a very low NF of less than 2 dB. However, the use of the common-source (CS) LNTA topology reduces the RF bandwidth. In addition, due to the receiver topology that connects the mixer input to the output node, the receiver is not able to operate at high frequency. Our earlier works [13,23,24], overcome the problems mentioned above. However, the mixer circuit in [13] needs to be improved to reduce the number of mixer switches. In addition, it suffers from $1/f$ noise that does not allow the receiver to operate at low frequency with good NF performance.

To overcome the limitations mentioned above, this work proposes a stacked receiver front-end, shown in Figure 4. It includes an on-chip balun to convert the single-ended antenna to a differential signal at the input of the LNTA, a capacitive cross-coupled common-gate (CG) LNTA topology to convert the RF voltage to an RF current, an active inductor (AI) and a $1/f$ noise-cancellation (NC) technique to isolate the mixer input and enhance low-frequency noise performance, a passive mixer to down-convert the RF current to an IF current, and a TIA to convert the IF current to an IF voltage at the output. The TIA and LNTA share the current using a single supply, reducing the power consumption.

4.1. LNTA Design

Two well-known LNTA topologies, common-gate (CG) and common-source (CS), can be used. The CS LNTA is suitable for very-low-noise applications at the cost of a narrow RF bandwidth. It is also very susceptible to non-idealities related to fabrication and packaging. On the other hand, the CG topology is used for wideband matching, and it is more reliable. The input impedance looking into the CG LNTA is 1/gm. Very high current and large device sizes are required to achieve the 1/gm = 50 Ω. To overcome this, a capacitor cross-coupled (CCC) technique is used to boost the effective g_m by two times without consuming extra power. It also improves the NF. The LNTA circuit is formed by M_{CG}, R_b, and C_{AC}. The noise factor of the LNTA is given by

$$F = 1 + \frac{\gamma}{R_S \times g_m}, \tag{8}$$

where R_S is the source impedance and γ is the short-channel effect coefficient, and can be reduced by increasing the transistor lengths.

4.2. Active-Inductor and $1/f$ Noise-Cancellation Design

In [22], the mixer input is connected to the output node. This increases the RF signal loss, and it does not allow the circuit to maintain its performance at high RF. To overcome this, the proposed AI circuit isolates the mixer input from the output. The impedance looking into the AI circuit is low at DC, while it increases at RF. In this case, the signal loss is then limited to parasitic capacitors. M_{AI}, C_{AI}, and R_{AI} form the AI circuit. The small R_s is used to boost the impedance at high frequencies with minimal impact at low frequencies. The impedance looking into the AI circuit is approximately given by

$$Z_{AI}(s) \cong \frac{g_{m,AI}R_S(R_{AI}C_{AI}s + 1) + R_{AI}C_{AI}s}{g_{m,AI}R_S C_{AI}s + g_{m,AI} + C_{AI}s} \| \frac{1}{sC_{par}}, \tag{9}$$

where C_{par} is the parasitic capacitance at the mixer input.

The stacked receiver front-end suffers from high low-frequency noise due to the direct coupling of the noise through the AI circuit. To overcome this issue, a $1/f$ NC circuit is formed by the M_{NC} transistors. It provides the signal path to the output with the opposite polarity to cancel the low-frequency noise and push the $1/f$ corner to a lower frequency. The functionality of the $1/f$ NC circuit is being verified in Figure 10. It shows the $1/f$ noise corner is pushed to a very low IF when the $1/f$ NC circuit is enabled, while the thermal noise remains almost constant.

4.3. TIA Design

The transimpedance amplifier (TIA) is used to convert the IF current to an IF voltage at the output. The TIA circuit is formed by transistors M_{TIA} and feedback resistors R_{FB}, where an additional harmonic recombination circuit is not required. The length of M_{TIA} should be large enough to enhance the output impedance. The conversion gain of the proposed stacked receiver can be calculated using Equation (7).

4.4. Simulation Results of the Stacked Receiver Front-End

The wideband stacked receiver front-end using a clock strategy was designed and simulated in a 22 nm CMOS technology. The receiver consumes 2.4 mA from a 1.2 V supply voltage.

The wideband input matching (S_{11}) of the LNTA is shown in Figure 11, showing an S_{11} of less than -10 dB over a wide frequency range by switching the capacitor bank at the input balun. The capacitor bank uses 16 binary weighted codes. The rest of the simulations in this work are verified using code 8.

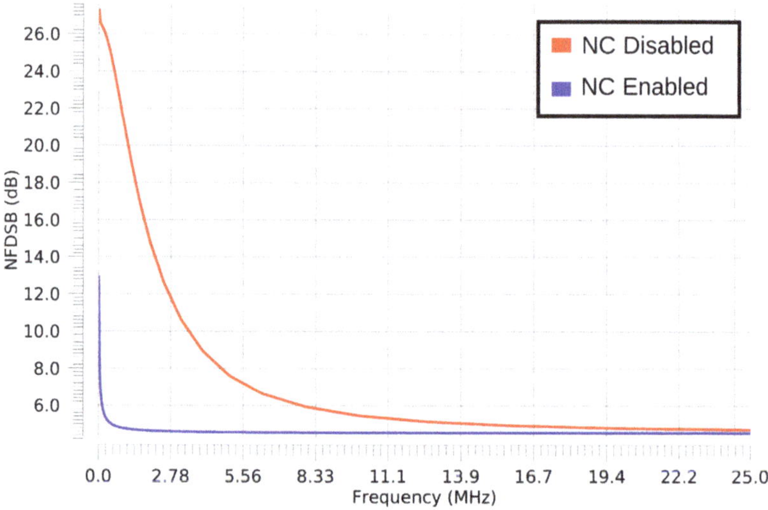

Figure 10. NFDSB of the stacked receiver with the $1/f$ NC circuit enabled and disabled.

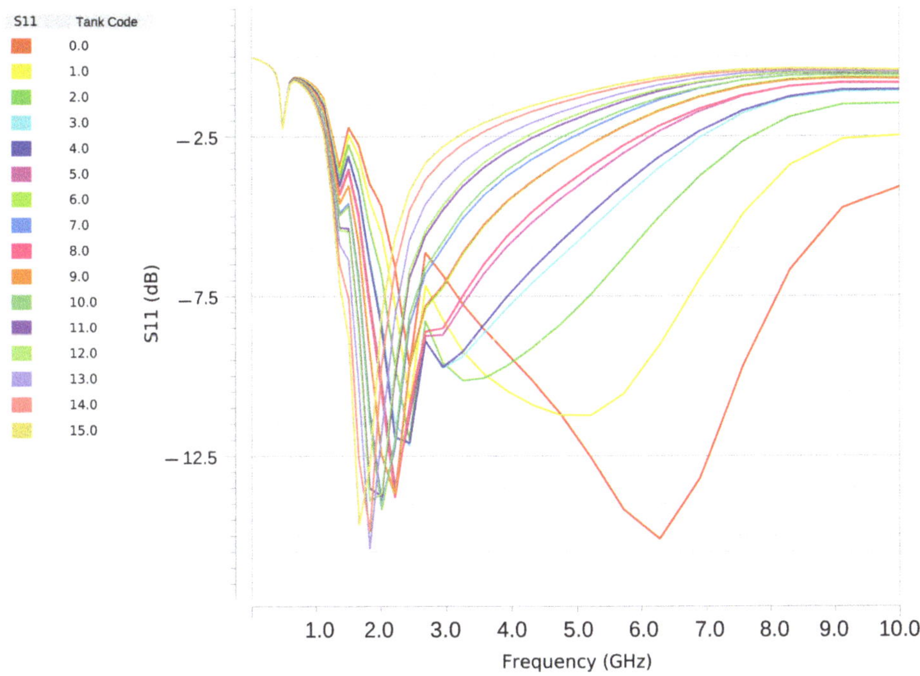

Figure 11. The stacked receiver input matching (S11) performance versus RF.

The receiver-performance, integrated NFDSB from 100 kHz to 50 MHz, conversion gain and input-referred third-order intercept point (IIP3) versus f_{LO}, which is equivalent to an RF input signal of $4 \times f_{LO}$, from 2.2 GHz to 3.2 GHz, is shown in Figure 12 . It shows that the NFDSB varies in terms of frequency from almost 4.5 dB to 6.3 dB. On the other hand, the conversion gain varies from almost 34.5 dB to 36 dB as f_{LO} is increased. A constant feedback resistor is used in the TIA. To perform IIP3 simulation, a two-tone signal at $4 \times f_{LO} + 10$ MHz and $4 \times f_{LO} + 11$ MHz is applied at the input of the LNTA. This generates two third-order intermodulation products at 9 MHz and 12 MHz, along with two fundamental products at 10 MHz and 11 MHz. The IIP3 performance varies over f_{LO} from −21 dBm to −17.5 dBm.

The harmonic rejection can be affected by transistor and layout mismatch. The effect of the transistor process and mismatch variation is verified using Monte-Carlo simulation over 100 runs, and the results are shown in Figure 13. The HR1, HR2, ... HRn (n = 7) are the 1st, 2nd, ... nth harmonics rejected relative to the 4th harmonic, which is the wanted signal in this work. The harmonic rejection in the stacked receiver front-end architecture is much less than the cascaded receiver architecture. The minimum rejection is achieved in HR5 with a 61 dB rejection using a $3\times$ sigma calculation.

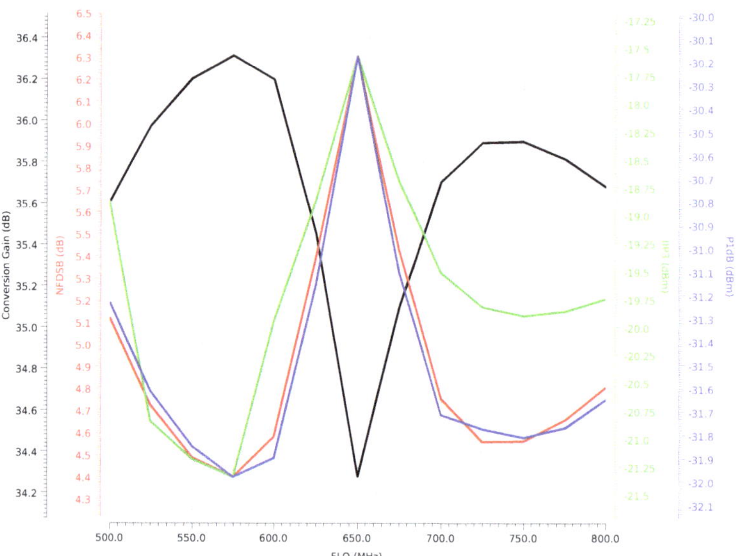

Figure 12. The stacked receiver NFDSB, conversion gain, and IIP3 performances versus f_{LO}.

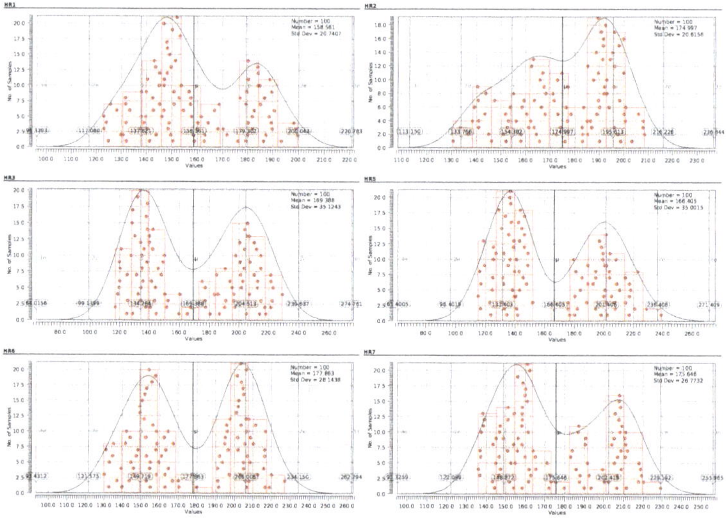

Figure 13. HR_n performance of the stacked receiver over 100 runs.

5. Discussion

Table 2 provides a performance summary and comparison of the cascaded and stacked receiver front-end using the clock strategy proposed in this work and compare them to the state-of-the-art. The cascaded receiver front-end with the clock strategy achieves a higher RF bandwidth, IIP3, and lower NF compared to the staked receiver front-end proposed in this work, while consuming almost four times current but operating at a slightly lower supply voltage of 1 V. The cascaded receiver architecture also achieved a higher RF bandwidth than work in [3,9,16] while consuming much lower power. Overall, both circuits presented in this work are suitable for wide modulation bandwidth application. The NF, IIP3, and bandwidth performance compare well with the state-of-the-art considering the power consumption of the presented receivers. The harmonic recombination technique verified by both receiver architectures validates the viability of the technique for different receiver architectures. The minimum harmonic rejection ratio of the cascaded, stacked receiver, refs [3,4,12,16] are approximately 134, 61, 35, 51, 80, and 52 dB, respectively. The stacked receiver architecture is better suited to very low power wireless applications with relaxed performance requirements such as Bluetooth Low Energy, while the cascaded receiver architecture can be used for a wide range of higher performance applications.

Table 2. Performance summary and comparison.

Parameters	This Work Cascaded [⊖]	This Work Stacked [⊖]	[12] [⊖]	[4] [⊕]	[16] [⊕]	[10] [⊕]	[9] [⊕]	[3] [⊕]
Process node	22 nm CMOS	22 nm CMOS	65 nm CMOS	65 nm CMOS	28 nm CMOS	28 nm CMOS	65 nm CMOS	28 nm CMOS
Freq. (GHz)	0.4–13	2–6	5.7–7.2	0.15–0.85	0.5–3	1–2	0.5–2	0.1–3.3
S11 (dB)	<-10	<-10	<-10	<-10	<-10	<-10	<-10	N/A
Gain (dB)	26–36	34.5–36	36.4	51	42	29.4	36	N/A
NF (dB)	1.4–3.9	4.6–6.2	4.4	5.4	2.4–5	5.7	2.2–4.2	1.7
IIP3 (dBm)	-10.5–7.5	-21–17.5 *	-18.9 *	-12 *	4 [+]	-10 *	-11 *	11.5 [+]
P_{DC}	11	2.9	13	7.5	21	0.141	41–65	36.8–62.4

* In-band IIP3 [+] out-of-band IIP3; [⊕] measurement results; and [⊖] simulation results.

6. Conclusions

Wideband cascaded and stacked receiver front-ends employing a clock strategy to down-convert an RF signal at $4\times$ the f_{LO} frequency were designed in a 22-nm CMOS process for SDR applications. The simulation results are presented showing the benefits of both architectures. The cascaded receiver front-end achieved higher bandwidth, lower NF, and better linearity performance than the stacked receiver front-end at the cost of higher power consumption. In the cascaded receiver front-end, low NF was achieved thanks to the feed-forward noise cancelling technique of the LNTA. The LNTA used by the cascaded receiver front-end operates over a frequency range from 0.4 GHz to 13 GHz. In the stacked receiver front-end, the low power consumption was achieved by sharing the current between the TIA and the LNTA using a single supply. The noise performance was also improved by using an AI and $1/f$ noise-cancellation technique.

Thanks to the current mode harmonic recombination, both receivers do not require additional circuits for harmonic recombination, reducing the power consumption. Dynamic power consumption is ultimately reduced thanks to the clock strategy technique that down-converts an RF signal at $4 \times f_{LO}$, reducing the clock frequency requirements. In the stacked receiver architecture, the CCC technique boosts g_m by two times without consuming additional power. The LNTA and balun can be tuned over an input frequency range from 2 GHz to 6 GHz.

The wideband operation and performance metrics of the proposed front-ends make them very suitable for SDR receivers that require a wideband frequency response and good harmonic rejection performance.

Author Contributions: Conceptualization, A.A. and F.N.; methodology, A.A.; software, A.A.; validation, A.A.; formal analysis, A.A.; investigation, A.A.; resources, A.A.; data curation, A.A.; writing—original draft preparation, A.A.; writing—review and editing, A.A. and F.N.; visualization, A.A.; supervision, A.A. and F.N.; project administration, A.A.; and funding acquisition, F.N. All authors have read and agreed to the published version of the manuscript.

Funding: This research received funding from the Natural Sciences and Engineering Research Council of Canada.

Institutional Review Board Statement: Not applicable

Informed Consent Statement: Not applicable

Data Availability Statement: Not applicable

Acknowledgments: The author would like to thank CMC Microsystems for providing access to the EDA tools.

Conflicts of Interest: The authors declare no conflict of interest.

Abbreviations

The following abbreviations are used in this manuscript:

ADC	Analog-to-digital converter
AI	Active inductor
AIC	Adaptive interference cancelling
CG	Common-gate
CCC	Capacitor cross-coupled
CMOS	Complementary metal–oxide semiconductor
CS	Common-source
ÉTS	École de technologie supérieure
FDSOI	Fully depleted silicon on insulator
HR	Harmonic rejection
IF	Intermediate frequency
IIP3	Third-order intercept point
IoT	Internet of things
LNA	Low-noise amplifier
LNTA	Low-noise transconductance amplifier
LO	Local-oscillator
NF	Noise figure
NC	Noise cancellation
NFDSB	Double side-band noise figure
RF	Radio frequency
SDR	Software-defined radio
TIA	Transimpedance amplifier
VCO	Voltage-controlled oscillator
V_{TH}	Voltage-threshold
g_m	Tranconductance

References

1. Mitola, J. The software radio architecture. *IEEE Commun. Mag.* **1995**, *33*, 26–38. [CrossRef]
2. Ke, Y.; Gao, P.; Craninckx, J.; Van der Plas, G.; Gielen, G. A 2.8-to-8.5 mW GSM/bluetooth/UMTS/DVB-H/WLAN fully reconfigurable CT $\Delta\Sigma$ with 200kHz to 20MHz BW for 4G radios in 90nm digital CMOS. In Proceedings of the 2010 Symposium on VLSI Circuits, Honolulu, HI, USA, 16–18 June 2020; IEEE: Piscataway Township, NJ, USA, 2010; pp. 153–154.
3. Murphy, D.; Darabi, H.; Xu, H. A noise-cancelling receiver resilient to large harmonic blockers. *IEEE J. -Solid-State Circuits* **2015**, *50*, 1336–1350. [CrossRef]
4. Lin, F.; Mak, P.I.; Martins, R.P. An RF-to-BB-current-reuse wideband receiver with parallel N-path active/passive mixers and a single-MOS pole-zero LPF. *IEEE J. -Solid-State Circuits* **2014**, *49*, 2547–2559. [CrossRef]
5. Elmi, M.; Tavassoli, M.; Jalali, A. A wideband receiver front-end using 1st and 3rd harmonics of the N-path filter response. *Analog. Integr. Circuits Signal Process.* **2018**, *94*, 451–467. [CrossRef]
6. Zinjanab, A.P.; Jalali, A.; Farshi, H.T. A standard and harmonic blocker tolerant receiver front-end using a harmonic rejection differential N-path notch filter and blocks withstand to possible variations. *AEU-Int. J. Electron. Commun.* **2020**, *125*, 153356. [CrossRef]
7. Andrews, C.; Molnar, A.C. A passive mixer-first receiver with digitally controlled and widely tunable RF interface. *IEEE J. -Solid-State Circuits* **2010**, *45*, 2696–2708. [CrossRef]
8. Lien, Y.C.; Klumperink, E.A.; Tenbroek, B.; Strange, J.; Nauta, B. Enhanced-selectivity high-linearity low-noise mixer-first receiver with complex pole pair due to capacitive positive feedback. *IEEE J. -Solid-State Circuits* **2018**, *53*, 1348–1360. [CrossRef]

9. Zolkov, E.; Cohen, E. A Mixer-First Receiver With Enhanced Matching Bandwidth by Using Baseband Reactance-Canceling LNA. *IEEE Solid-State Circuits Lett.* **2021**, *4*, 109–112. [CrossRef]

10. Mohammadpour, A.; Manstretta, D.; Castello, R. A 140-μW Front-End With 5.7-dB NF and+ 10-dBm OOB-IIP3 Using Voltage-Mode Boosting Mixer. *IEEE Microw. Wirel. Components Lett.* **2021**, *31*, 729–732. [CrossRef]

11. Bazrafshan, A.; Taherzadeh-Sani, M.; Nabki, F. A 0.8–4-GHz software-defined radio receiver with improved harmonic rejection through non-overlapped clocking. *IEEE Trans. Circuits Syst. Regul. Pap.* **2018**, *65*, 3186–3195. [CrossRef]

12. Shams, N.; Nabki, F. Analysis and Comparison of Low-Power 6-GHz N-Path-Filter-Based Harmonic Selection RF Receiver Front-End Architectures. *IEEE Trans. Very Large Scale Integr. (VLSI) Syst.* **2022**, *30*, 253–266. [CrossRef]

13. Abbasi, A.; Moshrefi, A.H.; Nabki, F. A Wideband Low-Power Current-Reuse RF-to-BB Receiver Using a Clock Strategy Technique. In Proceedings of the 2022 20th IEEE Interregional NEWCAS Conference (NEWCAS), Quebec, QC, Canada, 19–22 June 2022; IEEE: Piscataway Township, NJ, USA, 2022; pp. 275–279.

14. Weldon, J.A.; Narayanaswami, R.S.; Rudell, J.C.; Lin, L.; Otsuka, M.; Dedieu, S.; Tee, L.; Tsai, K.C.; Lee, C.W.; Gray, P.R. A 1.75-GHz highly integrated narrow-band CMOS transmitter with harmonic-rejection mixers. *IEEE J. -Solid-State Circuits* **2001**, *36*, 2003–2015. [CrossRef]

15. Ru, Z.; Moseley, N.A.; Klumperink, E.A.; Nauta, B. Digitally enhanced software-defined radio receiver robust to out-of-band interference. *IEEE J. -Solid-State Circuits* **2009**, *44*, 3359–3375. [CrossRef]

16. Wu, H.; Murphy, D.; Darabi, H. A harmonic-selective multi-band wireless receiver with digital harmonic rejection calibration. *IEEE J. -Solid-State Circuits* **2019**, *54*, 796–807. [CrossRef]

17. Xu, Y.; Zhu, J.; Kinget, P.R. A Blocker-Tolerant RF Front End With Harmonic-Rejecting N-Path Filter. *IEEE J. -Solid-State Circuits* **2017**, *53*, 327–339. [CrossRef]

18. Bagheri, R.; Mirzaei, A.; Chehrazi, S.; Heidari, M.E.; Lee, M.; Mikhemar, M.; Tang, W.; Abidi, A.A. An 800-MHz–6-GHz software-defined wireless receiver in 90-nm CMOS. *IEEE J. -Solid-State Circuits* **2006**, *41*, 2860–2876. [CrossRef]

19. Manetakis, K.; McKay, T.G. Wideband Low Noise Amplifier Having DC Loops with Back Gate Biased Transistors. US Patent 10,700,653. 30 June 2020.

20. Tedeschi, M.; Liscidini, A.; Castello, R. Low-power quadrature receivers for ZigBee (IEEE 802.15. 4) applications. *IEEE J. -Solid-State Circuits* **2010**, *45*, 1710–1719. [CrossRef]

21. Lin, Z.; Mak, P.I.; Martins, R.P. A 2.4 GHz ZigBee Receiver Exploiting an RF-to-BB-Current-Reuse Blixer+ Hybrid Filter Topology in 65 nm CMOS. *IEEE J. -Solid-State Circuits* **2014**, *49*, 1333–1344. [CrossRef]

22. Kim, S.; Kwon, K. A Low-Power RF-to-BB-Current-Reuse Receiver Employing Simultaneous Noise and Input Matching and $1/f$ Noise Reduction for IoT Applications. *IEEE Microw. Wirel. Components Lett.* **2019**, *29*, 614–616. [CrossRef]

23. Abbasi, A.; Nabki, F. A Design Methodology for Wideband Current-Reuse Receiver Front-Ends Aimed at Low-Power Applications. *Electronics* **2022**, *11*, 1493. [CrossRef]

24. Abbasi, A.; Moshrefi, A.H.; Nabki, F. A Wideband Low-Power RF-to-BB Current-Reuse Receiver Using an Active Inductor and 1 /f Noise-Cancellation for L-Band Applications. *IEEE Access* **2022**, *10*, 95839–95848. [CrossRef]

Journal of
*Low Power Electronics
and Applications*

Article

A 1.1 V 25 ppm/°C Relaxation Oscillator with 0.045%/V Line Sensitivity for Low Power Applications

Yizhuo Liao and Pak Kwong Chan *

School of EEE, Nanyang Technological University, Singapore 639798, Singapore
* Correspondence: epkchan@ntu.edu.sg; Tel.: +65-67904513

Abstract: A fully-integrated CMOS relaxation oscillator, realized in 40 nm CMOS technology, is presented. The oscillator includes a stable two-transistor based voltage reference without an operational amplifier, a simple current reference employing the temperature-compensated composite resistor, and the approximated complementary to absolute temperature (CTAT) delay-based comparators compensate for the approximated proportional to absolute temperature (PTAT) delay arising from the leakage currents in the switches. This relaxation oscillator is designed to output a square wave with a frequency of 64 kHz in a duty cycle of 50% at a 1.1 V supply. The simulation results demonstrated that the circuit can generate a square wave, with stable frequency, against temperature and supply variation, while exhibiting low current consumption. For the temperature range from −20 °C to 80 °C at a 1.1 V supply, the oscillator' output frequency achieved a temperature coefficient (T.C.) of 12.4 ppm/°C in a typical corner in one sample simulation. For a 200-sample Monte Carlo simulation, the obtained T.C. is 25 ppm/°C. Under typical corners and room temperatures, the simulated line sensitivity is 0.045%/V with the supply from 1.1 V to 1.6 V, and the dynamic current consumption is 552 nA. A better figure-of-merit (FoM), which equals 0.129%, is displayed when compared to the representative prior-art works.

Keywords: relaxation oscillator; voltage reference; composite resistor; current reference; temperature compensation; cross-coupled pair; delay drift compensation

Citation: Liao, Y.; Chan, P.K. A 1.1 V 25 ppm/°C Relaxation Oscillator with 0.045%/V Line Sensitivity for Low Power Applications. *J. Low Power Electron. Appl.* **2023**, *13*, 15. https://doi.org/10.3390/jlpea13010015

Academic Editor: Orazio Aiello

Received: 11 November 2022
Revised: 16 January 2023
Accepted: 19 January 2023
Published: 7 February 2023

1. Introduction

With the rapid development of wearable electronics and IoT (Internet of things), the demand of an on-chip and low-power oscillator has received much attention in the research. Low power consumption is especially important for IoT devices because of the real-time clock which has to stay awake all the time, even when other circuits are in sleep mode [1]. Although the crystal oscillator can provide an accurate signal with high stability, it is relatively expensive and occupies a large area with high current consumption [2]. For small size, the on-chip oscillators, such as ring oscillators and relaxation oscillators (ROSC), are widely used. Regarding to the ring oscillator, despite its simple architecture and low-power consumption under low oscillation frequency, the circuit is sensitive to process, supply, and temperature (PVT) variation [3]. This leads to a significant variation in the output frequency. Although other ring oscillators [4–6] can achieve relatively low sensitivity for output frequency, the power consumption is large. Hence, it may not be suitable for providing a stable clock using the standalone ring oscillator topology. Several reported works [7–11] have shown that the relaxation oscillator can provide a good tradeoff between frequency stability, temperature variation, and supply variation while occupying at reasonably small area. Thus, the relaxation oscillator is preferred as on-chip oscillator for those applications that require good stability with low cost and moderate precision. For example, the switched-capacitor based sensor interfaces [12,13] usually employ a low-frequency clock to control the sampling and charge transfer action in the circuits. Moreover, for the design of an instrumentation amplifier, ROSC can be applied for chopping amplifiers [14,15] to provide the chopping signal for modulating the signal to high frequency for amplification and translating it back to baseband for analog signal processing.

Figure 1 depicts the plot of the T.C. and line sensitivity of representative relaxation oscillators against the power consumption. As can be seen, it shows a tradeoff relationship among the performance parameters in context of the frequency stability and the power consumption [16]. The same goes for line sensitivity. In order to realize a lower T.C., one previous work [17] adopted the error feedback to achieve temperature-dependent delay cancellation, while another one [18] utilized the second-order compensation with a charge pump and filter. However, these complex designs led to avoidable high power consumption. Although other designs [16,19] reduced the power to a relatively small value, their thermal stability is slightly weakened due to the timing error and the temperature-sensitive current reference, respectively. Regarding the line sensitivity, there are works [8,10] utilizing high-gain operational amplifiers to minimize the effect from the supply variation, but at the cost of higher current consumption. Although other works [9,20] employed lowered the supply voltage as well as the bias current to reduce power, the circuit topologies were subject to higher supply sensitivity. Therefore, it is challenging to achieve a good stability of output frequency with relatively low power consumption in the ROSC design.

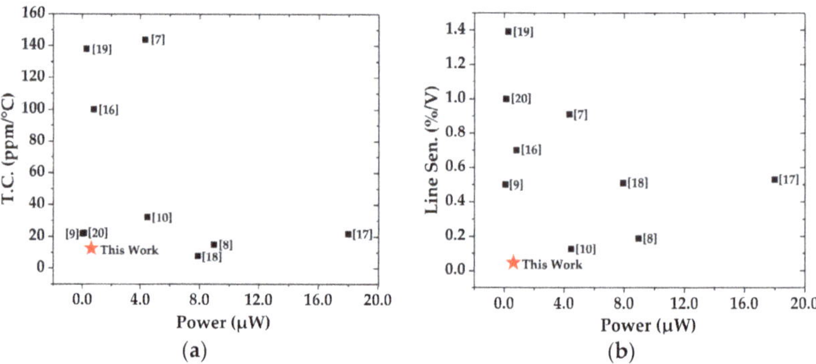

Figure 1. Tradeoff performance of reported relaxation oscillators: (**a**) T.C. versus power; (**b**) line sensitivity versus power.

In this paper, an improved relaxation oscillator with simple circuit topology is presented. As illustrated in Figure 1, the proposed ROSC features excellent stability against temperature and supply variation while achieving relatively low power consumption. This is achieved by using a simple delay drift compensation technique to enhance the thermal stability, in conjunction with the design of a simple V-I converter, which is based on two-transistor-type circuit topology to provide good immunity against the fluctuation of supply voltage change. Section 1 provides the introduction. Several representative prior-art relaxation oscillator designs are described in Section 2. Section 3 describes the design and implementation of the proposed relaxation oscillator. Section 4 presents the results and discussions. This is followed by the conclusion in Section 5.

2. Review of Reported Relaxation Oscillator

A low temperature coefficient relaxation oscillator [8] with a merged window comparator is shown in Figure 2. There are two different reference voltages (V_{REF_H} and V_{REF_L}) in the reported work. V_{REF_H} is connected to the non-inverting input of the comparator, and the voltage (V_{osc}) across the capacitor is initially zero. The capacitor (C_{osc}) is first charged by I_{REF} until V_{osc} reaches V_{REF_H}. Then C_{osc} starts to discharge with constant current I_{REF} while V_{REF_L} is connected to the comparator. When V_{osc} is lower than V_{REF_L}, C_{osc} is charged, and V_{osc} is again compared with V_{REF_H}. The delay generation units prevent the oscillator from entering metastability.

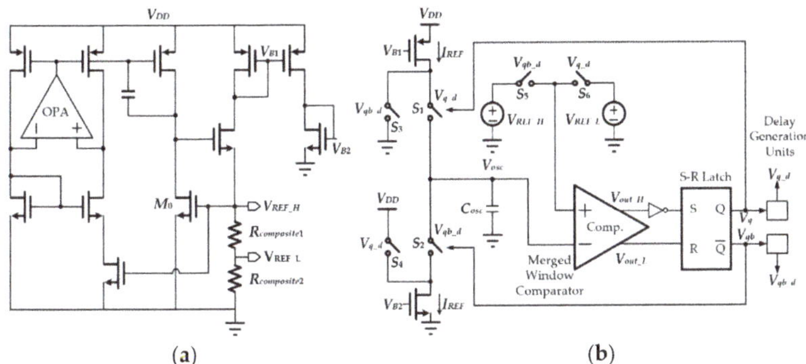

Figure 2. A CMOS relaxation oscillator with a merged window comparator: (**a**) reference generator; (**b**) block diagram.

The voltage reference is based on the architecture improved from the threshold monitoring circuit [21] that can effectively compensate for the temperature effect. Moreover, the reference current is also derived from the same reference voltage, in association with the optimized series/parallel composite resistor. As a result, this leads to the output frequency with low T.C. The merged window comparator is able to cancel out the offset of the comparator arising from the component mismatch effect. As such, this allows the T.C. of output frequency to maintain a good value, even if there is a 10 mV offset in the analog-based comparator, for example. However, the transistor M_0 in the reference generator needs to work in a saturation region, thus, for reliable temperature compensation, the current flowing through it is not allowed to be reduced to a small value. At this juncture, an operational amplifier is also utilized to provide a high loop gain to minimize the circuit sensitivity with respect to the supply variation. This suggests an additional current consumption source. Therefore, high current consumption becomes the main limitation of this circuit technique.

Another relaxation oscillator [9] that provides good a T.C. of output frequency while maintaining low power consumption, is depicted in Figure 3. This oscillator starts when ϕ is logic low. At the beginning, current I_2 flows though resistor R to generate the reference voltage at the non-inverting input of the comparator, and capacitor C_1 is charged by constant current I_1. After the voltage across C_1 becomes bigger than the reference voltage, ϕ transits to logic high; thus, the capacitor C_2 begins to be charged, and the reference voltage is connected to the inverting input of the comparator until the charging operation for C_2 is completed to make ϕ logic low again.

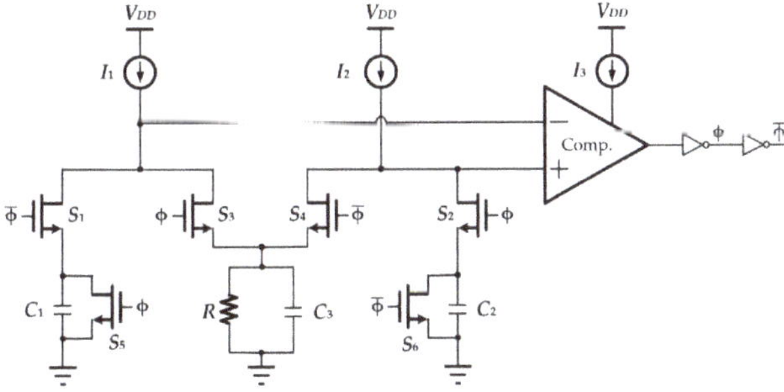

Figure 3. A Relaxation Oscillator with Ultra-Low Power Consumption.

In this work, the currents I_1 and I_2 can be reused at different phases, and only one comparator [22] is needed. All the currents, including the bias current for the comparator, can be achieved from one current source, hence the current consumption is minimized. Because the reference voltage is connected to different inputs of the comparator at different phases, the offset of the comparator increases the period at one phase, while decreasing the period at another phase. Thus, the offset can be cancelled out as long as the two capacitors are identical and the two charging currents are assumed the same. However, because of the restricted drain-to-source voltage headroom for each transistor working in a low supply environment, the transistor is subject to more stressing, resulting in not having good matching characteristics. On top of that, the mismatch effect between I_1 and I_2 is unavoidable. The current mismatch leads to the residual offset of the comparator that cannot be cancelled out completely. Ultimately, an error in the reference voltage exists between the different phases.

Another relaxation oscillator [10], with a self-chopped technique to achieve good stability against temperature and supply variations, is depicted in Figure 4. A current-mode comparator [23] is used to compare the voltage (V_r) across the composite resistors and the voltage (V_c) across the capacitor. Initially, V_c, V_{cmp}, and V_{rst} are logic low, and the capacitor is charged by constant current I_r until V_c becomes larger than V_r to change V_{cmp} from low to high, which causes V_{rst} to transition to high to discharge the capacitor. After the discharge action has been completed, V_{cmp} and V_{rst} become low to allow the current I_r to charge the capacitor again.

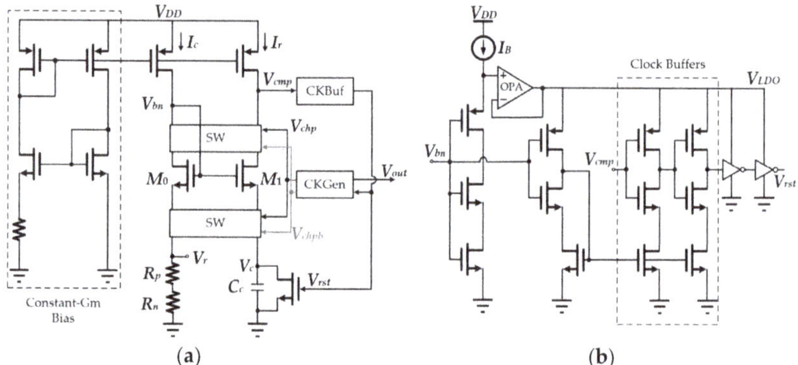

Figure 4. A self-chopped relaxation oscillator with adaptive supply generation: (**a**) block diagram; (**b**) clock buffer.

In this design, the ratio between I_c and I_r is independent of temperature and supply variations, and the temperature effect on the metal-oxide-metal (MoM) capacitor is negligible. In addition, the offset voltage in transistors M_0 and M_1 can be cancelled out by flipping M_0 and M_1 at every half cycle. Therefore, a good T.C. and good line sensitivity of the output frequency can be obtained due to good thermal stability through the use of a stable composite resistor and capacitor, in conjunction with offset cancellation using the chopping technique. However, it may be difficult to reduce the current consumption due to the operational amplifier which exhibits good transient response for powering the fast-switching clock buffers and the need for a complicated replica-biasing circuit. Hence, this design also suffers from the problem of relatively high supply current consumption.

3. Proposed Relaxation Oscillator

3.1. Topology of the Relaxation Oscillator

The relaxation oscillator, which makes use of the reported topology [24], is depicted in Figure 5. The major difference is that of the design and implementation of the current reference I_{REF}, while the lower supply current consumption is further addressed in the proposed work. The oscillator comprises a reference generator, two comparators (Comp. 1 and Comp. 2), an S-R latch, four switches, and two capacitors (C_1 and C_2). The reference generator generates a stable reference voltage (V_{REF}) and a reference current (I_{REF}), charging two capacitors with a constant current.

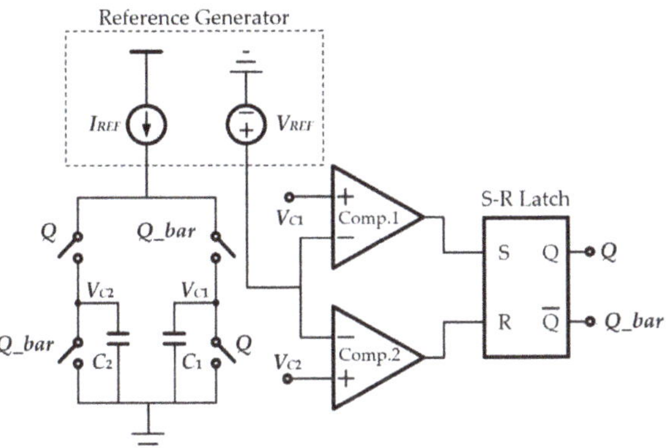

Figure 5. Block diagram of relaxation oscillator.

Note that the relaxation oscillator circuit starts when Q is logic low, and the voltages across two capacitors are initially zero. At the beginning, the outputs of two comparators are low, and C_1 is charged by I_{REF}. After the voltage (V_{C1}) across C_1 reaches V_{REF}, the output of Comp.1 transits to high, which allows Q to become high and Q_bar to become low. Then C_1 starts to discharge, while C_2 is charged by I_{REF}. The inputs of the SR latch are both low when I_{REF} is charging C_2, so that the outputs of the SR latch remain constant until the voltage (V_{C2}) rises to V_{REF}. At this juncture, Q transits to low, while Q_bar transits to high, allowing C_1 to be charged again and C_2 to discharge. The output Q of the SR latch will generate a rail-to-rail square wave with the desired frequency. As capacitors are charged by a constant current, the duration of the high level and low level of the square wave are dependent on the charging time of each capacitor; hence, the 50% duty cycle can be achieved by using two identical capacitors. The output frequency can be expressed as

$$f_{ROSC} = \frac{I_{REF}}{2CV_{REF}} \tag{1}$$

where f_{ROSC} is the output frequency, and C is the capacitance of C_1 and C_2. From (1), the accuracy of the output frequency is dependent on the accuracy of V_{REF} and I_{REF}. From the design considerations, the reference generator and the comparator are the key components in circuit design. Two nonideal effects exist in the relaxation oscillator circuit. They are the delay and offset of the comparator. Regarding the comparator's delay, it is not critical because of the low frequency specification and the moderate precision requirement in its application. With a careful designing of the comparator, the temperature-dependent delay of the comparator can be minimized to cause less impact to the circuit, and ultimately, on the output frequency. Pertaining to the comparator's offset, it is also addressed in the design phase with an appropriate choice of critical device sizes so that the offset effect to the circuit is acceptable, without significantly jeopardizing the oscillator's performance.

3.2. Reference Generator

The reference generator, which provides both the reference voltage (V_{REF}) and the reference current (I_{REF}), is depicted in Figure 6. The reference voltage generator is based on the two-transistor topology [25] and the cascode current mirror. This is achieved by employing voltage-to-current and current-to-voltage conversions to produce V_{REF}. This is then followed by another voltage-to-current converter with a composite resistor [26] and V_{REF} to generate the reference current I_{REF}.

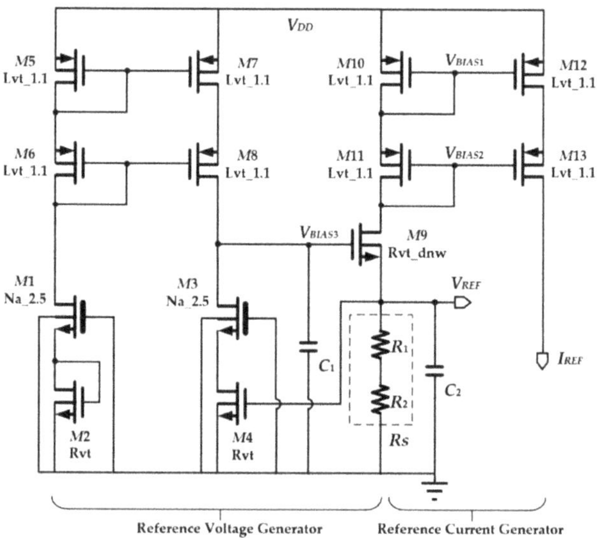

Figure 6. Proposed reference voltage and reference current generator.

Regarding the reference voltage generator, M_1, M_2, M_3, and M_4 work in the weak inversion region, where M_1 and M_3 are identically designed native transistors with a negative threshold voltage, whereas M_2 and M_4 are identical standard transistors. It is given that, for a sub-threshold biased MOSFET, its drain current is

$$I_{sub} = \mu C_{OX}(\eta - 1)V_T{}^2 \frac{W}{L} \exp\left(\frac{V_{GS} - V_{TH}}{\eta V_T}\right)\left[1 - \exp\left(\frac{-V_{DS}}{V_T}\right)\right] \tag{2}$$

where μ is the carrier mobility, C_{ox} is the gate-oxide capacitance, η is the subthreshold slope factor, V_T is the thermal voltage, W is transistor's channel width, L is transistor's channel length, V_{GS} is the gate-to-source voltage, V_{TH} is the threshold voltage, and V_{DS} is the drain-to-source voltage. When V_{DS} is larger than 100 mV (4 V_T), the effect of V_{DS} on I_{sub} is negligible; hence, the current I_{sub} can be approximated as

$$I_{sub} \approx \mu C_{OX}(\eta - 1)V_T{}^2 \frac{W}{L} \exp\left(\frac{V_{GS} - V_{TH}}{\eta V_T}\right) \tag{3}$$

Since the currents in M_1 and M_2 are the same, we can obtain

$$\begin{aligned}
I_{sub} &= \mu_1 C_{OX1}(\eta_1 - 1)V_T{}^2 \frac{W_1}{L_1} \exp\left(\frac{V_{GS_1} - V_{TH1}}{\eta_1 V_T}\right) \\
&= \mu_2 C_{OX2}(\eta_2 - 1)V_T{}^2 \frac{W_2}{L_2} \exp\left(\frac{V_{GS_2} - V_{TH2}}{\eta_2 V_T}\right)
\end{aligned} \tag{4}$$

The current of M_1 and M_2 is the same as that of M_3 and M_4 due to the identical current copying action in the cascode current mirror M_5–M_8. Hence, the V_{GS} of M_2 is identical to that of M_4, which is the V_{REF}. When the gates of M_1 and M_3 are connected to a ground, it suggests that the V_{GS} of M_1 and M_3 are identical negative reference voltages. Thus, V_{REF} can be obtained as

$$V_{REF} = -V_{GS1} = V_{GS2} = \frac{\eta_1 \eta_2}{\eta_1 + \eta_2}(V_{TH2} - V_{TH1}) + \frac{\eta_1 \eta_2}{\eta_1 + \eta_2}V_T \ln\left(\frac{\mu_1 C_{OX1}W_1 L_2}{\mu_2 C_{OX2}W_2 L_1}\right) \tag{5}$$

In this design, the first-order temperature effect on V_{TH} is given as [7]

$$V_{TH} = V_{TH0} - \kappa T \tag{6}$$

where V_{TH0} is the threshold voltage at room temperature (300 K), and κ is the temperature coefficient of the threshold voltage. Therefore, the T.C. of V_{REF} is as follows:

$$TC_{V_{REF}} = \frac{1}{V_{REF}} \frac{\partial V_{REF}}{\partial T} = \frac{(\kappa_1 - \kappa_2) + \frac{k}{q} \ln\left(\frac{\mu_1 C_{OX1} W_1 L_2}{\mu_2 C_{OX2} W_2 L_1}\right)}{(V_{TH20} - V_{TH10}) + (\kappa_1 - \kappa_2)T + \frac{k}{q}T \ln\left(\frac{\mu_1 C_{OX1} W_1 L_2}{\mu_2 C_{OX2} W_2 L_1}\right)} \tag{7}$$

where k is the Boltzman constant, and q is the electronic charge. In (7), the temperature effect on μ is ignored. By selecting appropriate aspect ratios of M_1 and M_2, while M_3 and M_4 remain the same size as M_1 and M_2, respectively, the temperature compensation can be achieved to permit V_{REF} in the first-order temperature compensation. Finally, it yields

$$V_{REF} = \frac{\eta_1 \eta_2}{\eta_1 + \eta_2}(V_{TH20} - V_{TH10}) \tag{8}$$

In addition, the V_{REF} has a good power supply rejection (PSR) at low frequency. Since the effect of ΔV_{DD} on the flowing currents, M_2 and M_4 are negligible, as long as their V_{DS} values are larger than 100 mV, while the transistors have a long channel length to reduce the drain-induced barrier lowering (DIBL) effect on $V_{TH1} \sim V_{TH4}$. Besides, the negative feedback formed by M_3, M_4, and M_9 can further stabilize V_{REF}.

For the reference current, it is produced by V_{REF} driving a temperature-compensated composite resistor (R_s). Of particular note, R_s comprises the series connection of an n-poly resistor (R_n) and a p-poly resistor (R_p), where R_n is PTAT and R_p is CTAT. The T.C. of R_p, R_n, and R_s are given as follows:

$$TC_{Rp} = \frac{1}{R_p} \frac{\partial R_p}{\partial T} \tag{9}$$

$$TC_{Rn} = \frac{1}{R_p} \frac{\partial R_n}{\partial T} \tag{10}$$

$$TC_{Rs} = \frac{1}{R_p + R_n} \frac{\partial (R_p + R_n)}{\partial T} \tag{11}$$

Substituting (9) and (10) into (11), TC_{Rs} can be rewritten as

$$TC_{Rs} = \frac{R_p}{R_p + R_n} TC_{Rp} + \frac{R_n}{R_p + R_n} TC_{Rn} \tag{12}$$

where TC_{Rp} is negative and TC_{Rn} is positive. Thus, TC_{Rs} can be made zero when choosing R_p/R_n equal to $|TC_n/TC_p|$. This indicates that R_s can be independent of the first-order temperature effect. Therefore, the temperature-insensitive reference current (I_{REF}) can be obtained with the temperature-insensitive voltage and the composite resistor.

$$I_{REF} = \frac{V_{REF}}{R_s} = \frac{\eta_1 \eta_2 (V_{TH20} - V_{TH10})}{(\eta_1 + \eta_2)(R_p + R_n)} \tag{13}$$

Moreover, since R_s is independent of V_{DD}, V_{REF} is insensitive to the change in V_{DD}. As a result, I_{REF} is also insensitive to the supply variations.

As seen in Figure 6, the capacitor C_1 is used as a frequency compensation for the negative feedback loop which is formed by M_3, M_4, and M_9. In addition, the capacitor C_2 is used to stabilize V_{REF} when the switches in Figure 4 are turned on and off. This is because the voltage change will be coupled to the gate of M_9 by the parasitic capacitors.

The current mirror pairs M_5–M_8 and M_{10}–M_{13} have a long channel length to reduce the current mismatch.

The 1.1 V supply voltage of this reference generator can ensure that all transistors still work in the proper region when there is a 10% supply voltage drop, but if the supply continuously decreases below 1 V, there will not be adequate V_{DS} headroom for the current mirror pair in Figure 6 at the SS corner under a low temperature, due to the increase in V_{TH}.

Finally, the size of each component pertaining to Figure 5 in the reference generator is listed in Table 1.

Table 1. Size of components in the reference generation.

Component	Size	Component	Size
$M_{1,3}$	10/15 (μm/μm)	$M_{10,12}$	28/12 (μm/μm)
M_2	2.04/1 (μm/μm)	$M_{11,13}$	7/2 (μm/μm)
M_4	2.05/1 (μm/μm)	R_1	433.3 kΩ
$M_{5,7}$	5/4 (μm/μm)	R_2	1.366 MΩ
$M_{6,8}$	6/2 (μm/μm)	C_1	1.5 pF
M_9	1.5/1 (μm/μm)	C_2	4 pF

3.3. Comparator

The comparator in Figure 7 shows an OTA topology using dual cross-coupled load pairs and a cascode arrangement to boost the overall gain. The front differential stage, which makes use of the cross-coupled load pairs, M_3–M_6 and M_{13}–M_{16}, is used to produce gain enhancement as well as to reduce the delay in the comparator, the outputs of which are followed by the current mirror high-gain stage consisting of M_7–M_{12} and M_{15}–M_{20}. Finally, the CMOS inverter, formed by M_{23} and M_{24}, aims to sharpen the output square waveform and provide the driving capability of the comparator.

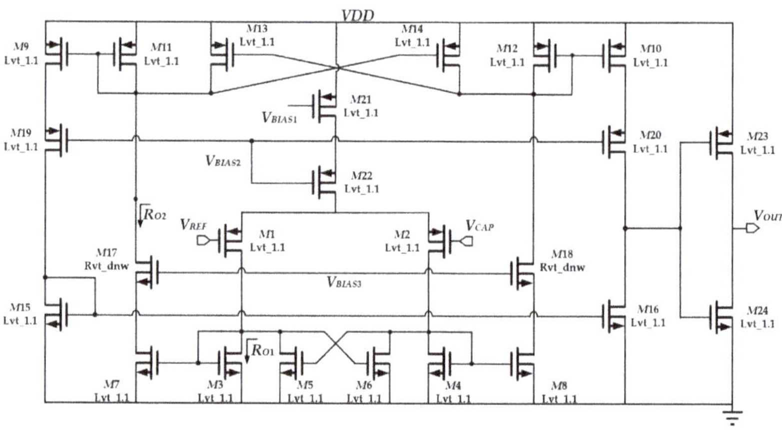

Figure 7. Proposed comparator.

Considering the cross-coupled pairs M_3–M_6, the aspect ratio of M_3 is larger than that of M_5. In small-signal analysis, the output impedance is obtained as

$$R_{O1} = \frac{r_{O3} + r_{O5}}{(g_{m3} - g_{m5})(r_{O3} + r_{O5}) + 1} \approx \frac{1}{g_{m3} - g_{m5}} \tag{14}$$

where g_{mi} is the respective transistor's transconductance and r_{Oi} is the respective transistor's output resistance, with i = 3 or 5. From (14), R_{O1} is increased from $1/g_{m3}$ to $1/(g_{m3}-g_{m5})$ to increase the voltage gain because of the positive feedback allowing M_5 to behave as a negative resistance. When there is a voltage change on the drain of M_3, the positive feedback introduced by M_5 can accelerate this change, causing faster output response to reduce the delay in the comparator.

The four cascode transistors M_{17}–M_{20} are used to reduce the effect of supply variation ΔV_{DD} on the delay in the comparator. For M_7 and M_{17} in small-signal analysis, the change in V_{DS7} caused by ΔV_{DD} can be approximated as [27]

$$\Delta V_{DS7} \approx \frac{\Delta V_{DD}}{g_{m17} r_{O17}} \tag{15}$$

As interpreted from (15), it indicates that the change in V_{DS} on M_7–M_{10} can be ignored when V_{DD} varies. This means the current change in each branch caused by channel length modulation and DIBL can be minimized with cascode transistors. Moreover, the bias current for the comparator is directly copied from I_{REF}, with the cascode current mirror in different ratios. This avoids the need for an extra bias branch, which would cause an increase in supply current consumption.

In fact, when the constant bias current is applied to the comparator, the delay in the comparator is reduced with an increasing temperature. From (3) and (6), the V_{GS} of the MOS transistor working under weak inversion is expressed as

$$V_{GS} = \eta V_T \ln \left[\frac{I_{sub}}{\mu C_{OX}(\eta - 1)V_T^2 \frac{W}{L}} \right] + V_{TH0} - \kappa T \tag{16}$$

It can be observed that the V_{GS} exhibits CTAT behavior. M_3 is in the diode connection, meaning that V_{DS3} is equal to V_{GS3}. When the temperature increases, V_{DS3} (or V_{GS3}) in the diode-connected topology is significantly reduced with respect to V_{DS7}. Thus, the mismatch between the drain-to-source voltage of the transistors can lead to a rising current to allow for the delay in the comparator to decrease from 285.3 ns to 278.8 ns as the temperature increases, as shown in Figure 8. This feature is particularly useful for the delay compensation arising from the observation of the increase in delay through the leakage current of the switches, as depicted in Figure 4. As a result, the thermal stability of the oscillator circuit is enhanced. This will be further discussed in the next subsection. Of particularly note, the leakage current in the advanced technology node can be a serious issue.

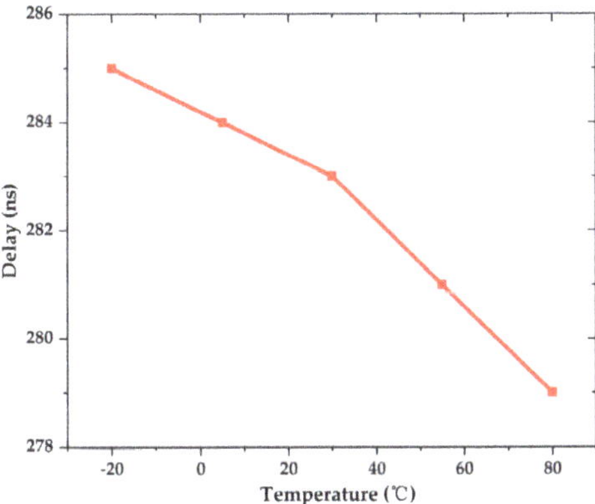

Figure 8. Temperature characteristics of delay in comparators.

Since offset is critical in the comparator design, the Monte Carlo simulation, with 400 samples for the offset evaluation, is shown in Figure 9. This result indicates that the mean offset of the comparator is 0.37 mV, and its standard derivation is 5.63 mV. As observed, the offset is minimized by sizing the input transistor pair and the cross-coupled pairs with a long channel length ($L > 4L_{min}$). Although the parasitic capacitors in the large-size transistors will enlarge the response time, the delay, which is around 0.28 μs, including the hysteresis, contributes 3.6% of the oscillation period. Therefore, it is considered acceptable, with a low output frequency and a moderate precision requirement. Based on the result, the comparator offset cancellation scheme is not implemented in this work. The sizes of each component pertaining to the comparator in Figure 6 are given in Table 2.

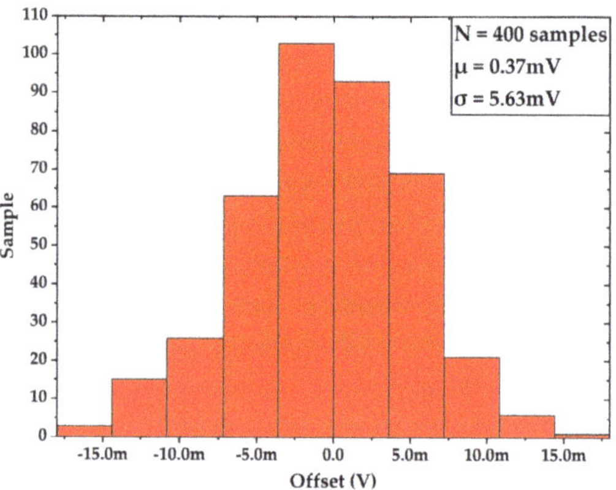

Figure 9. Monte Carlo simulation of the comparator's offset.

Table 2. Size of the components in the comparator.

Component	Size	Component	Size
$M_{1,2}$	1/0.6 (μm/μm)	$M_{15,16}$	0.5/0.6 (μm/μm)
$M_{3,4}$	1/1.2 (μm/μm)	$M_{17,18}$	0.3/0.3 (μm/μm)
$M_{5,6}$	0.8/1.2 (μm/μm)	$M_{19,20}$	0.3/0.3 (μm/μm)
$M_{7,8}$	0.5/0.6 (μm/μm)	M_{21}	20/12 (μm/μm)
$M_{9,10}$	0.5/0.6 (μm/μm)	M_{22}	5/2 (μm/μm)
$M_{11,12}$	1/1.2 (μm/μm)	M_{23}	0.36/0.12 (μm/μm)
$M_{13,14}$	1.2/1.2 (μm/μm)	M_{24}	0.12/0.12 (μm/μm)

3.4. Leakage Current in Switches and Delay Compensation

The four switches controlling the charging and discharging actions, as depicted in Figure 5, are arranged in the inverted style, as shown in Figure 10.

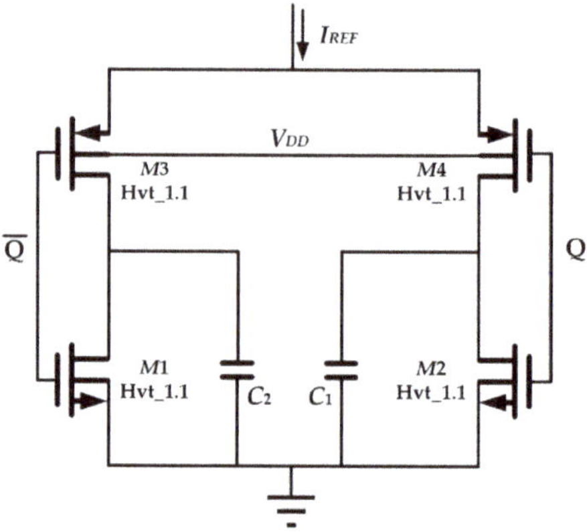

Figure 10. Four transistor switches for charging and discharging the matched capacitor pair.

Since I_{REF} is quite small, the leakage currents in the advanced technology node, flowing through the switch transistors, can be significant when charging C_1 and C_2. With the increase in temperature, the delay caused by the transistor switches is increased from 1.9 ns to 9.8 ns, as shown in Figure 11, and this effect is particularly pronounced. Therefore, this will cause the reduction in the output frequency. Thus, according to the reverse short-channel effect, high threshold-voltage transistors with the smallest channel length can be used as switches. This is also in conjunction with introducing the body effect in pmos to maximize the threshold voltage. Finally, the leakage current effect can be reduced by introducing the CTAT delay, as discussed in Section 3.3, such that the output frequency can be kept constant. The size of each transistor and capacitor shown in Figure 10 are given in Table 3.

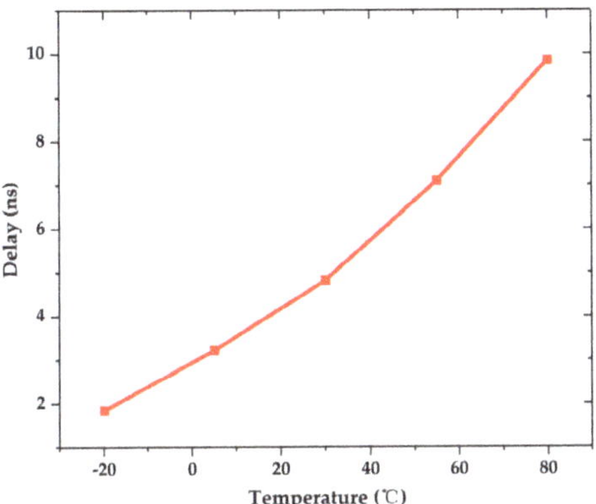

Figure 11. Temperature characteristics of the delay in transistor switches.

Table 3. Size of components in Figure 10.

Component	Size	Component	Size
$M_{1,2}$	0.12/0.04 (μm/μm)	$C_{1,2}$	1.49 pF
$M_{3,4}$	0.36/0.04 (μm/μm)		

4. Results and Discussions

The proposed ROSC, with leakage current compensation, is simulated using TSMC-40 nm CMOS process technology. The output frequency is 64.59 kHz at V_{DD} = 1.1 V under room temperature, and the transient simulation result of the output signal is depicted in Figure 12. All analog-biased transistors in the proposed ROSC work in the subthreshold region, the bias current of the comparators can be made small for low frequency design, and the current derives from the dedicated I_{REF} instead of from the addition of an extra current source. This permits the current consumption of 552 nA at room temperature in a typical corner. However, there is always a performance tradeoff between I_{REF} and low current consumption.

Figures 13 and 14 illustrate the respective simulation results of the output frequency against the temperature variation from 20 °C to 80 °C at different supply voltages and process corners. The T.C. of the output frequency of the proposed ROSC is obtained as 12.4 ppm/°C, 13.3 ppm/°C, and 21.8 ppm/°C at the TT corner, SS corner, and FF corner, respectively, at a 1.1 V supply. Regarding the 1 V supply, the obtained T.C. is 14.3 ppm/°C, 26.7 ppm/°C, and 22.2 ppm/°C, respectively. Of particular note, the T.C. is observed with some degradation at the 1 V supply with respect to that of the 1.1 V supply at the SS corner. This is mainly because the transistors in the reference generator are stressed under limited V_{DS}. Therefore, the proposed ROSC can still work properly when the supply is slightly lower than 1.1 V. Considering the operation margin, 1.1 V is regarded as the minimum supply voltage for the oscillator. Regarding the low T.C. values achieved by the ROSC with respect to those of prior-art works, the T.C. improvement is attributed to the compensation for the delay drift resulting from temperature, as seen in Figures 8 and 11. By calculation, without the delay compensation, this T.C. will increase to about 19.1 ppm/°C.

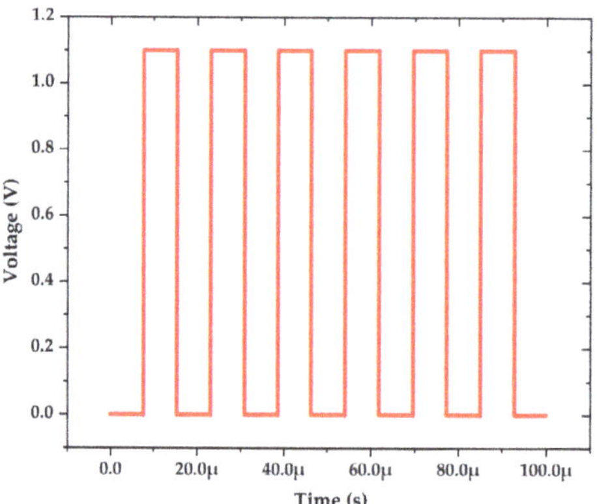

Figure 12. Output signal in time domain.

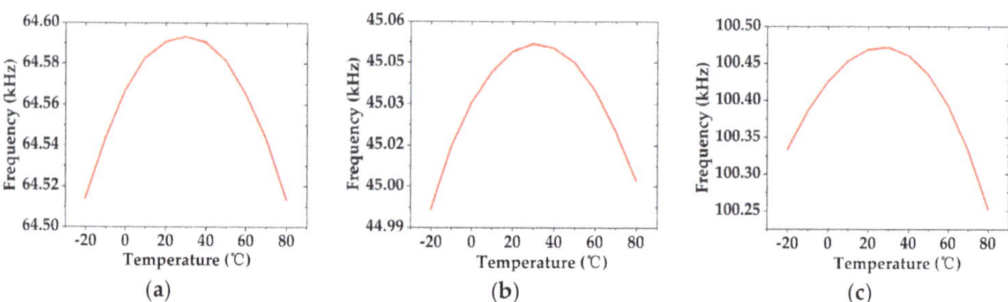

Figure 13. Temperature characteristic of output frequency at different process corners under the 1.1 V supply: (**a**) @TT corner; (**b**) @SS corner; (**c**) @FF corner.

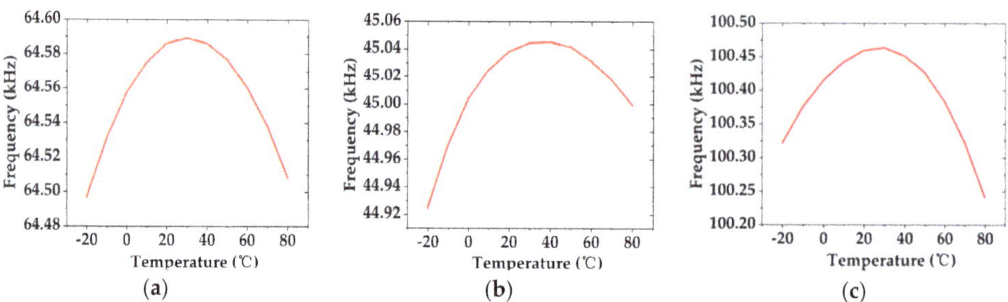

Figure 14. Temperature characteristic of output frequency at different process corners under the 1 V supply: (**a**) @TT corner; (**b**) @SS corner; (**c**) @FF corner.

Considering the parasitic effect arising from the layout issues, some model capacitors ranging from a few tens to one hundred fF are intentionally added to the critical points in each comparator, reference generator, and SR latch. The comparator displays relatively higher sensitivity due to the low bias current, while there is no significant effect from other nodes. Figure 15 shows the simulated temperature characteristic of output frequency with intentionally added parasitic capacitors in the design under the TT corner. Of particular note, the estimated capacitance from the routing for each comparator is around 18fF. Therefore, the total capacitance of several model capacitors being added in each comparator is modeled as 20 fF. The output frequency changes from 64.59 kHz to 64.04 kHz, and the T.C. is degraded from 12.4 ppm/°C to 13.7 ppm/°C. This confirms that the potential parasitic effect arising from the layout made no significant impact on the current simulation results, without incorporating layout due to the low-frequency design. Additionally, the silicon area of this design is approximated as about 4x the total active area of the components. This yields about 0.0234 mm^2, or 153 μm × 153 μm.

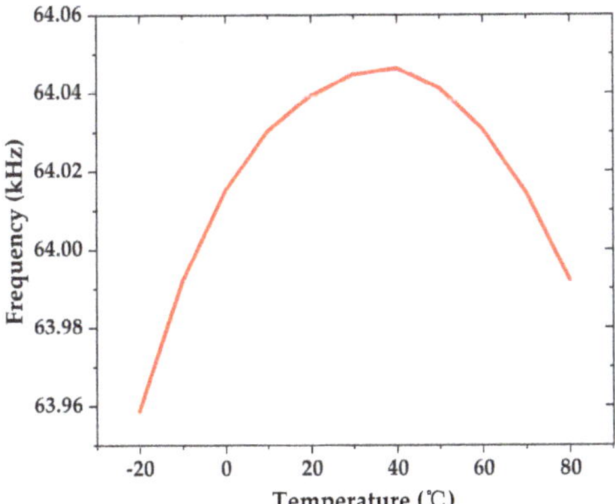

Figure 15. Temperature characteristic of output frequency with simulated parasitic capacitors at the TT corner.

The T.C. results of Monte Carlo simulation used to verify the impact of mismatch and process variations are shown in Figure 16. There are 200 samples simulated, and each sample is simulated with 11 temperature points, from −20 °C to 80 °C, resulting in 2200 points in total. The T.C. varies from 9.35 ppm/°C to 77.13 ppm/°C, with an average value of 25 ppm/°C and a standard deviation of 11.1 ppm/°C, where 75% of the samples present a T.C. smaller than 30 ppm/°C. This confirms that the output frequency of the proposed ROSC exhibits good stability under temperature change.

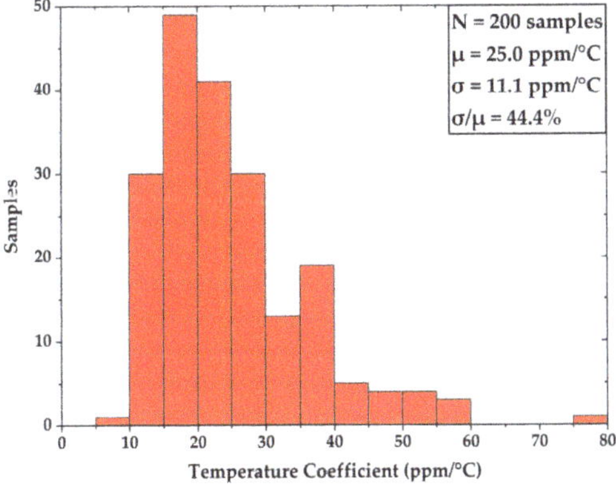

Figure 16. Monte Carlo simulation of output frequency T.C.

The supply dependence of the output frequency is depicted in Figure 17. The line sensitivity of the output frequency achieves 0.045%/V, 0.059%/V, and 0.081%/V at the TT corner, SS corner, and FF corner, respectively, from a 1.1 V to 1.6 V supply, which is attributed to V_{REF} with good PSR at low frequency and the cascode transistors shielding the variation of supply in the comparators to stabilize the response time.

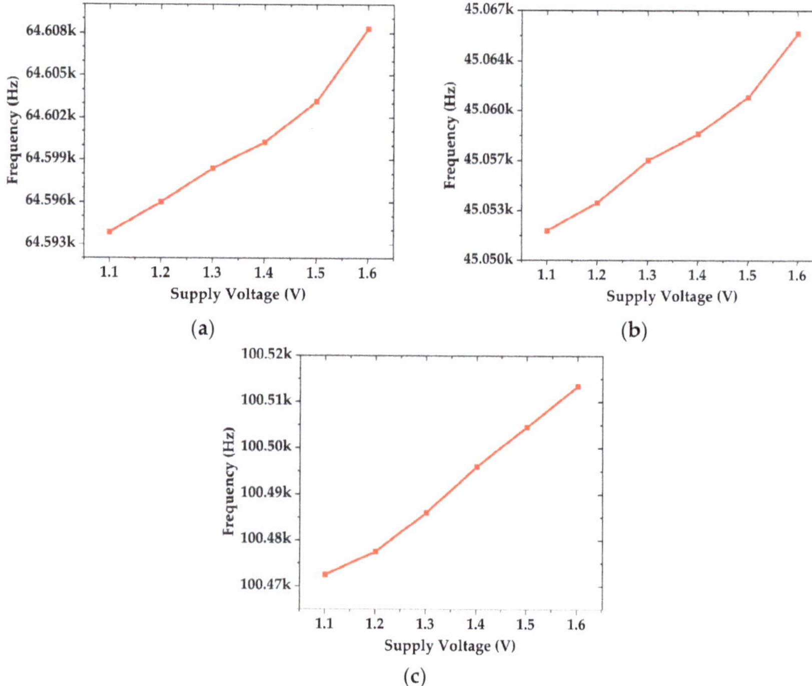

Figure 17. Supply dependence of output frequency at different process corners: (**a**) @TT corner; (**b**) @SS corner; (**c**) @FF corner.

The Monte Carlo simulation of the output frequency, with process variation and mismatch at different temperatures, is shown in Figure 18. The average values of the output frequency under different temperatures remain almost the same: 64.84 kHz, 64.95 kHz, and 64.9 kHz, with the standard deviation of 6.42 kHz, 6.42 kHz, and 6.41 kHz at $-20\,^{\circ}\mathrm{C}$, $30\,^{\circ}\mathrm{C}$, and $80\,^{\circ}\mathrm{C}$, respectively. This yields the average process sensitivity (σ/μ) of 9.88%. The output frequency is eventually dependent on the value of the composite resistor Rs in Figure 6 and the capacitors C_1 and C_2 in Figure 10. The process variation displays the moderate value, which is targeted for moderate precision applications. However, to cater to a precision design, this can be achieved by trimming the passive MoM capacitors C_1 and C_2, which are less affected by temperature.

The performance of the proposed ROSC is compared to that of the previously reported representative works using advanced process technology nodes, as shown in Table 4, and with longer channel length technology nodes, as shown in Table 5. It can be seen that the proposed relaxation oscillator, with the dynamic current consumption of 552 nA at a 1.1 V supply voltage, exhibits the best T.C. for one sample. The same goes for the Monte Carlo 200-sample result, with process variation and mismatch. Regarding line sensitivity, the proposed work a displays lower value due to the use of the cascode current mirror plus the two-transistor-based voltage reference topology, which has the feature of small line sensitivity.

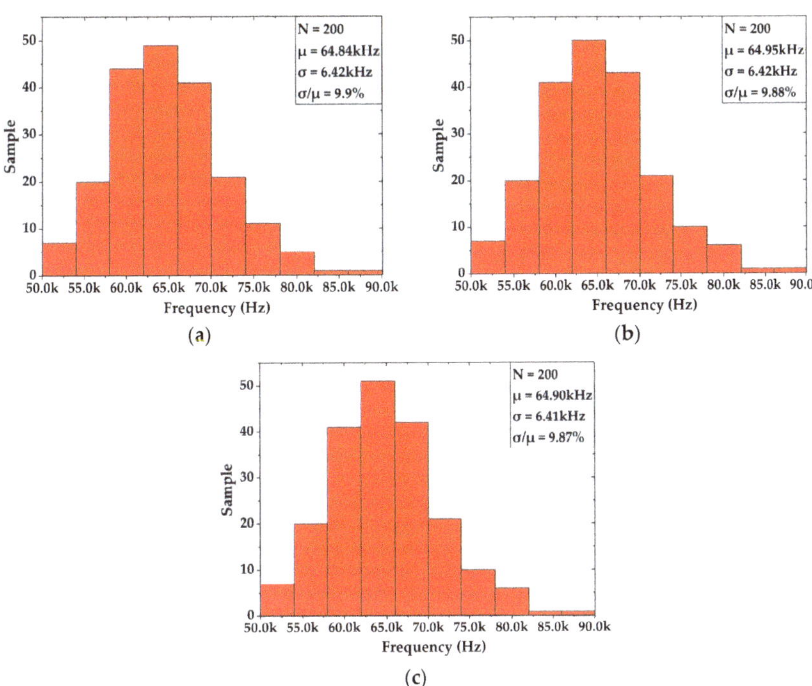

Figure 18. Monte Carlo simulation of output frequency: (**a**) @-20 °C; (**b**) @30 °C; (**c**) @80 °C.

Table 4. Performance comparison with previously reported ROSC works in advanced technology nodes.

Parameter	[7] 2016	[8] 2017	[9] 2020	[10] 2012	[19] 2017	[20] 2013	This Work
Technology	65 nm	65 nm	40 nm	60 nm	65 nm	65 nm	**40 nm**
Frequency (kHz)	64.4	64.2	32.7	32.7	32.5	18.5	**64.6**
Supply Voltage (V)	1.2	1.2	0.6	1.6	1.2	1	**1.1**
Current_TT (µA)	3.6	7.47	0.067	2.8	0.225	0.12	**0.55**
T.C._TT (ppm/°C)	144	14.7	21.7	32.4	138	22	**12.4**
T.C._MC (ppm/°C)	NA	NA	35.5	NA	NA	NA	**25**
Temp. Range (°C)	-20–100	-20–100	-40–125	-20–100	-40–80	0–90	**-20–80**
Line Sens. (%/V)	0.91	0.188	0.5	0.125	1.39	1	**0.045**
Process Sen. (σ/µ)% without trimming	3.66	NA	11.73	NA	10.4	NA	**9.86**
FoM (%)	1.549	0.169	0.247	0.344	1.54	0.32	**0.129**
Result	Simulated	Simulated	Simulated	Measured	Simulated	Measured	Simulated

The process sensitivity of the output frequency is comparable and can be improved by trimming the two capacitors. To evaluate the stability of the output frequency, with both temperature and supply variations, a figure-of-merit (FoM), including the temperature coefficient and the line sensitivity [9], is defined as

$$FoM = T.C. \times 100\ °C + Line.Sens. \times 10\% V_{DDmin} \tag{17}$$

where T.C. is the one sample value at the typical corner. The FoM of the proposed ROSC is 0.129%, which displays the best result when compared to the prior-art works shown in Table 4. Therefore, the proposed design can offer a stable frequency while providing a good tradeoff between stability and power consumption. Finally, in view of the larger leakage current, as well as the lower transistor intrinsic gain in this 40 nm technology with respect to those of technology nodes having a longer channel length, the frequency stability of the proposed work, as shown in Table 5, also shows excellent FoM. This demonstrates the usefulness of the circuit.

Table 5. Performance comparison with previously reported ROSC works employing technology nodes with longer channel length.

Parameter	[1] 2013	[23] 2010	[28] 2014	[29] 2007	This Work
Technology	180 nm	350 nm	180 nm	350 nm	**40 nm**
Frequency (kHz)	32.55	3.3	28	80	**64.6**
Supply Voltage (V)	1	1	1.2	1	**1.1**
Current_TT (μA)	0.47	0.066	0.033	1.06	**0.55**
T.C._TT (ppm/°C)	120	260	95.5	842	**12.4**
T.C._MC (ppm/°C)	NA	NA	NA	NA	**25**
Temp. Range (°C)	−40–100	−20–80	−20–80	0–80	**−20–80**
Line Sens. (%/V)	1.1	3.5	3	2.5	**0.045**
Process Sen. (σ/μ)% without trimming	1.39	NA	NA	3.95	**9.86**
FoM (%)	1.31	2.95	1.255	8.67	**0.129**
Result	Measured	Measured	Measured	Simulated	Simulated

5. Conclusions

This paper presents a 40 nm CMOS relaxation oscillator with low-current consumption. It features simple topology that comprises a two-transistor-based voltage reference generator and a simple current reference generator with a temperature-compensated composite resistor. Moreover, the comparators are designed with a CTAT delay to counteract the PTAT delay contributed by the effect of the leakage current in switches, thus improving the thermal stability of the output frequency.

The simulation results confirmed that the proposed ROSC displayed excellent output frequency stability against the changes in temperature and supply voltage. Therefore, the proposed work is suitable for low-power applications that require an oscillator with a stable frequency and moderate precision.

Author Contributions: Conceptualization, Y.L. and P.K.C.; methodology, Y.L. and P.K.C.; software, Y.L.; validation, Y.L. and P.K.C.; formal analysis, Y.L. and P.K.C.; investigation, Y.L. and P.K.C.; resources, P.K.C.; data curation, Y.L.; writing—original draft preparation, Y.L.; writing—review and editing, P.K.C.; visualization, Y.L.; supervision, P.K.C.; project administration, P.K.C. All authors have read and agreed to the published version of the manuscript.

Funding: This research received no external funding.

Data Availability Statement: The study did not report any data.

Conflicts of Interest: The authors declare no conflict of interest.

References

1. Tsubaki, K.; Hirose, T.; Kuroki, N.; Numa, M. A 32.55-kHz, 472-nW, 120ppm/°C, fully on-chip, variation tolerant CMOS relaxation oscillator for a real-time clock application. In Proceedings of the 2013 Proceedings of the ESSCIRC, Bucharest, Romania, 16–20 September 2013; pp. 315–318.
2. Wang, J.; Goh, W.L. A 13.5-MHz relaxation oscillator with ±0.5% temperature stability for RFID application. In Proceedings of the 2016 IEEE International Symposium on Circuits and Systems, Montreal, QC, Canada, 22–25 May 2016; pp. 2431–2434.
3. Cannillo, F.; Toumazou, C.; Lande, T.S. Nanopower subthreshold MCML in submicrometer CMOS technology. *IEEE Trans. Circuits Syst. I Regul. Pap.* **2009**, *56*, 1598–1611. [CrossRef]
4. Zhang, S.; Li, A.; Han, Y.; Jie, L.; Han, X.; Cheung, R.C. Temperature compensation technique for ring oscillators with tail current. *Electron. Lett.* **2016**, *52*, 1108–1110. [CrossRef]
5. Zhang, X.; Apsel, A.B. A low-power, process-and-temperature-compensated ring oscillator with addition-based current source. *IEEE Trans. Circuits Syst. I Regul. Pap.* **2010**, *58*, 868–878. [CrossRef]

6.   Ballo, A.; Pennisi, S.; Scotti, G.; Venezia, C. A 0.5 V Sub-Threshold CMOS Current-Controlled Ring Oscillator for IoT and Implantable Devices. *J. Low Power Electron.* **2022**, *12*, 16. [CrossRef]

7.   Cimbili, B.; Wang, D.; Zhang, R.C.; Tan, X.L.; Chan, P.K. A PVT-tolerant relaxation oscillator in 65nm CMOS. In Proceedings of the 2016 IEEE Region 10 Conference, Singapore, 22–25 November 2016; pp. 2315–2318.

8.   Ma, L.; Koay, K.C.; Chan, P.K. A merged window comparator based relaxation oscillator with low temperature coefficient. In Proceedings of the 2017 IEEE International Symposium on Circuits and Systems, Baltimore, MD, USA, 28–31 May 2017; pp. 1–4.

9.   Medeiros, W.T.; Klimach, H.; Bampi, S. A 40 nW 32.7 kHz CMOS Relaxation Oscillator with Comparator Offset Cancellation for Ultra-Low Power applications. In Proceedings of the 2020 IEEE 11th Latin American Symposium on Circuits & Systems, San Jose, Costa Rica, 25–28 February 2020; pp. 1–4.

10.   Hsiao, K.J. A 32.4 ppm/°C 3.2-1.6 V self-chopped relaxation oscillator with adaptive supply generation. In Proceedings of the 2012 Symposium on VLSI Circuits, Honolulu, HI, USA, 13–15 June 2012; pp. 14–15.

11.   Xu, L.; Onabajo, M. A low-power temperature-compensated relaxation oscillator for built-in test signal generation. In Proceedings of the 2015 IEEE 58th International Midwest Symposium on Circuits and Systems, Fort Collins, CO, USA, 2–5 August 2015; pp. 1–4.

12.   Chan, P.K.; Zhang, H.L. A switched-capacitor interface circuit for integrated sensor applications. In Proceedings of the 2001 IEEE International Symposium on Circuits and Systems, Sydney, NSW, Australia, 6–9 May 2001; pp. 372–375.

13.   Ye, S.Q.; Chan, P.K. A low-power switched-capacitor humidity sensor interface. In Proceedings of the 2010 IEEE Asia Pacific Conference on Circuits and Systems, Kuala Lumpur, Malaysia, 6–9 December 2010; pp. 1039–1042.

14.   Ong, G.T.; Chan, P.K. A Power-Aware Chopper-Stabilized Instrumentation Amplifier for Resistive Wheatstone Bridge Sensors. *IEEE Trans. Instrum. Meas.* **2014**, *63*, 2253–2263. [CrossRef]

15.   Chan, P.K.; Ng, K.A.; Zhang, X.L. A CMOS chopper-stabilized differential difference amplifier for biomedical integrated circuits. In Proceedings of the 2004 47th Midwest Symposium on Circuits and Systems, Hiroshima, Japan, 25–28 July 2004; pp. 3–33.

16.   Savanth, A.; Weddell, A.S.; Myers, J.; Flynn, D.; Al-Hashimi, B.M. A Sub-nW/kHz Relaxation Oscillator with Ratioed Reference and Sub-Clock Power Gated Comparator. *IEEE J. Solid-State Circuits* **2019**, *54*, 3097–3106. [CrossRef]

17.   Tsai, Y.-K.; Lu, L.-H. A 51.3-MHz 21.8-ppm/°C CMOS Relaxation Oscillator with Temperature Compensation. *IEEE Trans. Circuits Syst. II Express Briefs* **2017**, *64*, 490–494. [CrossRef]

18.   Ji, Y.; Liao, J.; Arjmandpour, S.; Novello, A.; Sim, J.-Y.; Jang, T. A Second-Order Temperature-Compensated On-Chip R-RC Oscillator Achieving 7.93 ppm/°C and 3.3pJ/Hz in −40 °C to 125 °C Temperature Range. In Proceedings of the 2022 IEEE International Solid-State Circuits Conference, San Francisco, CA, USA, 20–26 February 2022; pp. 1–3.

19.   Asano, H.; Hirose, T.; Ozaki, T.; Kuroki, N.; Numa, M. An area-efficient, 0.022-mm2, fully integrated resistor-less relaxation oscillator for ultra-low power real-time clock applications. In Proceedings of the 2017 IEEE International Symposium on Circuits and Systems, Baltimore, MD, USA, 28–31 May 2017; pp. 1–4.

20.   Paidimarri, A.; Griffith, D.; Wang, A.; Chandrakasan, A.P.; Burra, G. A 120 nW 18.5 kHz RC oscillator with comparator offset cancellation for ±0.25% temperature stability. In Proceedings of the 2013 IEEE International Solid-State Circuits Conference Digest of Technical Papers, San Francisco, CA, USA, 17–21 February 2013; pp. 184–185.

21.   Lee, J.; Cho, S. A 210 nW 29.3 ppm/°C 0.7 V voltage reference with a temperature range of −50 to 130 °C in 0.13 μm CMOS. In Proceedings of the 2011 Symposium on VLSI Circuits-Digest of Technical Papers, Kyoto, Japan, 15–17 June 2011; pp. 278–279.

22.   Allen, P.E.; Holberg, D.R. *CMOS Analog Circuit Design*; Oxford University Press: Oxford, UK, 2002.

23.   Denier, U. Analysis and design of an ultralow-power CMOS relaxation oscillator. *IEEE Trans. Circuits Syst. I Regul. Pap.* **2010**, *57*, 1973–1982. [CrossRef]

24.   Flynn, M.P.; Lidholm, S.U. A 1.2-mum CMOS current-controlled oscillator. *IEEE J. Solid-State Circuits* **1992**, *27*, 982–987. [CrossRef]

25.   Seok, M.; Kim, G.; Blaauw, D.; Sylvester, D. A portable 2-transistor picowatt temperature-compensated voltage reference operating at 0.5 V. *IEEE J. Solid-State Circuits* **2012**, *47*, 2534–2545. [CrossRef]

26.   Gregoire, B.R.; Moon, U.K. Process-independent resistor temperature-coefficients using series/parallel and parallel/series composite resistors. In Proceedings of the 2007 IEEE International Symposium on Circuits and Systems, New Orleans, LA, USA, 27–30 May 2007; pp. 2826–2829.

27.   Razavi, B. *Design of Analog CMOS Integrated Circuits*; McGraw-Hill Education: New York, NY, USA, 2005.

28.   Chiang, Y.-H.; Liu, S.-I. Nanopower CMOS Relaxation Oscillators With Sub-100 ppm/°C Temperature Coefficient. *IEEE Trans. Circuits Syst. II Express Briefs* **2014**, *61*, 661–665.

29.   De Vita, G.; Marraccini, F.; Iannaccone, G. Low-Voltage Low-Power CMOS Oscillator with Low Temperature and Process Sensitivity. In Proceedings of the 2007 IEEE International Symposium on Circuits and Systems, New Orleans, LA, USA, 27–30 May 2007; pp. 2152–2155.

MDPI
St. Alban-Anlage 66
4052 Basel
Switzerland
Tel. +41 61 683 77 34
Fax +41 61 302 89 18
www.mdpi.com

Journal of Low Power Electronics and Applications Editorial Office
E-mail: jlpea@mdpi.com
www.mdpi.com/journal/jlpea